D0777058

*Science Secrets*

# *Science Secrets*

The Truth about
Darwin's Finches, Einstein's Wife,
and Other Myths

***Alberto A. Martínez***

UNIVERSITY OF PITTSBURGH PRESS

Published by the University of Pittsburgh Press, Pittsburgh, Pa., 15260

Manufactured in the United States of America
Printed on acid-free paper
10 9 8 7 6 5 4 3 2 1

Library of Congress Cataloging-in-Publication Data

Martinez, Alberto A.
Science secrets : the truth about Darwin's finches, Einstein's wife, and other myths / Alberto A. Martinez.
p. cm.
Summary: "Accessibly written in an engaging style, this book examines classic popular stories in the history of science. Some of the myths discussed include Franklin's Kite, Newton's Apple, and Thomson's plum pudding model of the atom. Martinez successfully holds readers' attention by relying on rich documentation from primary sources to debunk speculations that have become reified over time. He argues that although scientists have disagreed with one another, the disagreements have been productive. Features includes extensive primary source documentation and detailed explanations of how to compare contradictory sources in order to determine which accounts are truly valid"—Provided by publisher.
Includes bibliographical references.
ISBN 978-0-8229-4407-2 (hardback)
1. Science—History—Miscellanea. I. Title.
Q173.M316 2011
500—dc22

2011003484

*if they list to try*
*conjecture, He his Fabric of the Heavens*
*has left to their disputes, perhaps to move*
*His laughter at their quaint Opinions wide*
*hereafter, when they come to model Heaven,*
*and calculate the Stars, how they will wield*
*the mighty frame, how build, unbuild, contrive*
*to save appearances . . .*

Milton, *Paradise Lost*

*I do not deal in conjectures.*

Isaac Newton, 1724

# *Contents*

# *List of Myths and Apparent Myths*

# *Preface*

Stories about Albert Einstein range widely: countless writers claim that he was a genius while others claim that he had a learning disability; some portray him as a saintly sufferer for humanity who wanted to read the mind of God, while others portray him as a bohemian opportunist or an atheist. In 1949, Einstein complained: "There have already been published by the bucketsful such brazen lies and utter fictions about me that I would have gone to my grave long ago if I had let myself pay attention to them."[1]

We tend to imagine scientists as gifted prodigies, lonesome heroes or martyrs, saints or sinners. Their stories acquire the shapes of myths. Giordano Bruno and Galileo appear as noble martyrs. Copernicus, Kepler, and Newton loom as marble giants in a pantheon. Some biologists invoke Darwin as a patron saint, quoting him word for word. Scientists working on eugenics appear to be a deadly sect. Many posters portray the old Einstein as an inspirational holy man.

In the newest bestselling biography of Einstein (#1 *New York Times* Best Seller), by Walter Isaacson, I find the following words in the first seven pages: tidings, halo, brilliance, genius, faith, testament, genius, miraculous, miracle, glory, reverential, faith, God, cult, genius, canonized, secular saint, halo, genius, aura, priest-like, glories, inspiring, brilliance, guardian angel, reverence, and "usher in the modern age."[2] Such language is typical, it helps to win readers, but is it necessary? Hype is not distinctive of science, it is the common ancient currency by which we tell stories of heroes and miracles. I don't want to insinuate that I dislike Isaacson's book, that's not what I mean; it's a rich biography, well worth reading. But even great writers portray the history of science with mythical tones, as if success were a matter of destiny, such as in claims that Newton was born the year Galileo died or that the young Einstein "would become" one of the greatest physicists of all time. Many writers echo traditional stories rather than dig up documentary facts, interpreting bits of evidence to match conjectures rather than to test them. Consciously or not, writers inflate tales to sell books. As rightly noted by Jürgen Neffe, another good biographer of Einstein, "speculations evolve into anecdotes that are then proliferated in book-length studies."[3] Can we write history without mythical exaggerations? Einstein did not want to be worshipped; he repudiated how people exaggerated his contributions. Likewise, Darwin felt mortified by overblown praise, and I think Pythagoras would be stunned by what writers have made of him.

A tension exists between the need to fairly describe the past and the craving to bridge gaps, to conjecture. It is not exclusive to popular books, it also appears in academic writing. Some historians portray speculations as secure findings. Other historians later deny such conjectures as unwarranted. But often they replace the old guesswork with new speculations, presented again as solid findings. Moreover, the ground that each historian tries to cover is often broad, so errors creep in wherever one relies on common knowledge and trusts other writers' words. It takes much work to authenticate some common hearsay; it takes more work to falsify, and some conjectures are too seductive to resist.[4]

Lately, some good writers of popular science have sought help from historians to check their claims. For example, while editing his *Short History of Nearly Everything*, the acclaimed author Bill Bryson, having received corrections and comments from historians, increasingly surmised the extent to which "inky embarrassments" might lurk in the pages of his book.[5] It became a best seller nonetheless, and to Bryson's credit, some critics were frustrated to not find many inaccuracies in his book. Still, many teachers and writers ignore the findings of historians. It is comparable to the degree to which most professional historians, in turn, print no comment on how some authors misrepresent history. Why do exaggerations and guesswork circulate more than any fair accounting of evidence? Is truth less interesting? No, the problem is that most specialists aim their work at a small group of peers. Yet truths are so much more attractive than myths that specialists spend *thousands of hours* absorbed in efforts to illuminate the past. But some of us should challenge misconceptions in order to delineate legends from substantive findings that are also fascinating.

This book analyzes several famous topics in the history of science: the lore of Pythagoras, the Copernican revolution, the alchemical quest for the Philosophers' Stone, Darwin's path to evolution, the mysteries of electricity, Einstein's relativity, and the rise of eugenics. Each topic involves great stories that we tell and tell again, as we should. And then we change the stories, they evolve. But they sometimes lose genuine aspects and gain an opaque gloss. Again we feel compelled to analyze them, so I too have now tried to untangle hearsay. Of course, I'm not immune to mistakes either; for each error addressed in this book, I too have made many others, but I've struggled to keep mine out, to not clutter these pages. Mistakes are more interesting when issued by prominent sources, so I discuss few of my own. But again and again I rewrote; what I had somewhere first read crumbled later on close inspection. Some of what I had taught for years turned out to be baseless, plausible fictions.

Layers of conjecture have caked stories about dead scientists like plaster, paint, and acrylic gloss. Yet there are some sparkling facts underneath. Precious, neglected gems of history compensate for the loss of the surface sheen. So we will do plenty of nitpicking. Much of it will be tacit, hidden in the struggle to fairly represent evidence; but in other instances, we must explicitly pinpoint the nits to be picked—because nits, the tiny eggs of parasitic insects such as lice, are not as harmless as they seem. They hatch and they feed and they grow and they breed. How much illusory history would we have been spared if writers centuries ago had nitpicked seemingly harmless infidelities before they spread?

Besides, nitpicking can be fun. There is a long tradition of writers who have written corrections to popular myths in science and math. Most recently, I have enjoyed books by Tony Rothman and John Waller.[6] As those authors did, I now cover an idiosyncratic selection of topics. In this regard, Rothman frankly noted: "The contents of this book are entirely arbitrary, having been determined by accident."[7] In my case, the present topics are connected by the theme of how stories evolve, and to various degrees, by Pythagoras and Einstein: legendary heroes from ancient and more recent times. Pythagoras is particularly noteworthy because his name strangely recurs in many fields. We still learn about him in early school courses on math, music, and astronomy. He led a religion in the sixth century BCE and allegedly he made many contributions to philosophy, society, the sciences, and supposedly he performed miracles, did not eat beans, and he never laughed. But actually, all or nearly all of the mathematical discoveries that are often attributed to Pythagoras are fiction, the result of writers' guesswork. In the present book, I trace his mythical influence in the sciences. As in math, it is again striking to see how often scientists attributed discoveries to Pythagoras, so he will again and again show up in these pages. Pythagoras emerges as the patron saint of the urge to conjecture, to pretend to know the past.

So we pit stories against evidence. Stories powerfully influence the public conception of science; true or false, they leave a deep impression on the imagination. Therefore, those issues and others will arise in the pages that follow, some will be briefly addressed by pointing to the appropriate evidence while others will be disentangled in detail. Every chapter relies on primary sources and new translations.

I'm impressed by an excellent recent book where the authors frankly state that although the word *myth* has a sophisticated academic meaning, they yet use it "as done in everyday conversation—to designate a claim that is false."[8] I too will tend to follow this common usage, though I will also refer to myths as valuable fiction. Myths disguise our ignorance, but they also link powerful notions.

Yet the present book is not essentially about debunking myths; that's a secondary goal. It is more an effort to retell history, to squint at it through plausible myths. Instead of just collecting errors, we must try again to tell the past fairly. And to do so, I will follow the structure of successful stories, true or false. Myths become pervasive because they work, they function or satisfy. Despite being historically untrue, they convey deep truths about human psychology. We enjoy some stories more than others partly because they are tales of individuals: touching vignettes of often solitary explorers struggling to overcome difficulties to win unlikely success. So we can reconstruct remarkable happenings in the moving frames of individuals and their struggles. This book studies how stories evolve, and it contributes to that process.

Above all, this book shows why opinions that were once secret and seemingly impossible became scientifically compelling. Ink that might go to the usual hero-worship or to academic spin will instead be used to clearly *explain*. For it is unfair for historians to demand that scientists include more proper history in their teaching if by contrast historians do not include science when writing history. Specialists have carved deep divisions among their fields, disconnecting disciplines from their past. Even famous scientists leave a wake of confusions. Darwin acknowledged that the plan of his argument in *Origin of Species* could be improved. Einstein was plagued by a barrage of curious inquirers who did not understand relativity, and he blamed himself for initiating many of the confusions that arose.

Incidentally, I once heard the following phrase on television: "it's magic—and like science you don't have to actually understand it, you just have to believe it."[9] There it was, right there, the problem. Some people give up on understanding. Despite the efforts of many writers, there remain many readers who, having fairly tried to understand the reasonings of dead scientists, remain quite unconvinced that reason was there at all. They conclude that science constitutes just another blind cult.

My concern is not that they remain unconvinced. People may well study something such as Einstein's relativity and rightly remain skeptical, as did Einstein, that it constitutes a satisfactory solution. Instead, I'm concerned that often writers don't explain the order of the arguments and the historical conditions that give meaning to scientific knowledge. Contrary to common impressions, science books do not all say the same things. For example, in some physics books the constancy of the speed of light shows up as a fundamental *assumption*, while in others it is an *experimental fact*. Which is it? In some books, Darwin was crucially inspired to evolution thanks to finches, while in others he was most influenced by giant tortoises, and in others by pigeons or by mockingbirds. Which is it?

In his lively book on historical myths, physicist Tony Rothman repeatedly admitted: "I have few illusions that this little book can make a dent in just-so history. . . . I have few illusions that this book will make the slightest puncture in the status quo."[10] But in the end, to the contrary, I find it inspiring that despite our failings, writers do jointly struggle to capture and convey truth, to advance our understanding. Stories change, and it is wonderful to contribute to the process.

Contrary to the common phobia against science and math, many people are immensely attracted to pseudoscience. We should therefore decipher how the secrets and specifics of history can meet the cravings that pull readers to the realms of conjectures. As brilliantly suggested by Karl Popper: "science must begin with myths, and with the criticism of myths; neither with the collection of observations, nor with the invention of experiments, but with the critical discussion of myths."[11] Alongside the precious findings of inquiry, there are also neglected dimensions of stories—trinkets and skeletons in the backroom closets of science and math, not entirely the sort of thing that some teachers would like you to see, but nevertheless just as seductive as good myths.

# Acknowledgments

I warmly thank the following professors, friends, relatives, and students for kindly sharing thoughtful comments and helpful suggestions on different aspects of this project: Eric Almaraz, Ronald Anderson, Theodore Arabatzis, Casey Baker, Nicole Banacka, Miriam Bodian, Susan Boettcher, Christine Bruton, Frank Benn, H.W. Brands, Dolph Briscoe IV, Jed Z. Buchwald, Heidi Buckner, Jochen Büttner, Lori Carlson, Richard Chantlos, Etienne Colón Ríos, Alexis R. Conn, Robert Crease, Randy Diehl, Jason Edwards, Allen Esterson, Paul Forman, Allan Franklin, Jennifer Rose Fredson, Jeremy Gray, Roger P. Hart, Rodrigo Fernós, Connie Folse, Jess Haugh, Judy Hogan, Rob W. Holmes, Bruce J. Hunt, Daphne Ireland, Robert Iliffe, Joshua Johnson, Fumihide Kanaya, Daniel Kennefick, Brian Levack, Mary Long, Abigail J. Lustig, Michael Marder, Ronald Martínez Cuevas, Rubén Martínez, Lillian Montalvo Conde, Roberto Ortíz, John Overholt, Jorge Luis Pardo, Lawrence M. Principe, Shawn Piasecki, Lynn Robitaille, César Rodríguez, Illary Quinteros, Jenna Saldaña, Paige Sartin, Neven Sesardic, Hannah Siemens-Luthy, Claudia Salzberg, Alan E. Shapiro, Angela Smith, Pamela Smith, John Stachel, Sarah Steinbock-Pratt, Dee Stonberg, Ariel Taylor, Eve Tulbert, Heather Turner, Amy Umberger, James Vaughn, John Volk, Scott Walter, Sarah D. Wilson, Diane Wu, and James A. Wilson Jr. I also thank Roger H. Stuewer for having taught me how to do history of science.

I also thank the helpful librarians of the following universities: the University of Texas at Austin, Harvard University, the California Institute of Technology, the Einstein Archives, the Dibner Library of the National Museum of American History, Smithsonian, the Burndy Library (when it was located at the Dibner Institute for the History of Science and Technology at MIT), Boston University, the Landesbibliothek Bern, the Babson College Horn Library, the Université Nancy 2, and the Archives and Special Collections of the Bibliothek of the Eidgenössische

Technische Hochschule in Zürich. I thank the Hebrew University of Jerusalem for granting permission to quote translations from documents in the Albert Einstein Archives; the content of such original documents are copyrighted by Princeton University Press and the Hebrew University of Jerusalem. Next, I thank the following editors for their kind advice, support, and suggestions: Ann Downer-Hazell, Elizabeth Knoll, Michael Fisher. Moreover, I also thank editors Phyllis Cohen, Vickie Kearn, and Maria Guarnaschelli. I likewise appreciate the early help of two editorial assistants: Vanessa Hayes and Matthew J. Hills. And, I thank Heather Rothman for kindly and attentively helping to copyedit an early version of the manuscript. I also thank ten anonymous reviewers for their various comments and suggestions.

I am especially thankful to the University of Pittsburgh Press for kindly adopting my evolving manuscript. Their acquisitions editor, Beth Davis, enthusiastically supported the project and steadily contributed to its fruition and shape. The success of most books also depends greatly on the support provided by the marketing crew, and hence I warmly thank Lowell Britson, David Baumann, and the staff at the University of Pittsburgh Press for their gracious support. Next, I am very pleased with the cover for the book, created by Gary Gore, I thank him for the elegant and lively artwork, and I also thank Ann Walston for overseeing the book's production and design. I also greatly appreciate the generous support in the form of a University Co-operative Society Subvention Grant, awarded by the University of Texas at Austin, which has helped to make this book more affordable to all readers. Finally, I thank Deborah Meade for her painstaking care in copyediting the manuscript, for her rigorous demands wherever source information was lacking, and for her appropriate and helpful suggestions.

*Science Secrets*

# 1

## *Galileo and the Leaning Tower of Pisa*

ALTHOUGH many writers dismiss this tall tale as apocryphal, some writers, teachers, and physicists still claim that Galileo Galilei carried out experiments on gravity by dropping objects from the Leaning Tower of Pisa. Here is an old and dramatic version of the story:

> Members of the University of Pisa, and other onlookers, are assembled in the space at the foot of the wonderful leaning tower of white marble in that city one morning in the year 1591. A young professor [Galileo] climbs the spiral staircase until he reaches the gallery surmounting the seventh tier of arches. The people below watch him as he balances two balls on the edge of the gallery, one weighing a hundred times more than the other. The balls are released at the same instant, and are seen to keep together as they fall through the air until they are heard to strike the ground at the same moment. Nature has spoken with no uncertain sound, and has given an immediate answer to a question debated for two thousand years.
>
> "This meddlesome man Galileo must be suppressed," murmured the University fathers as they left the square. "Does he think that by showing us that a heavy and a light ball fall to the ground together he can shake our belief in the philosophy which teaches that a ball weighing one hundred pounds would fall one hundred times faster than one weighing a single pound? Such disregard of authority is dangerous and we will see that

> it goes no further." So they returned to their books to explain away the evidence of their senses; and they hated the man who had disturbed their philosophic serenity. For putting belief to the test of experiment, and founding conclusions upon observation, Galileo's reward in his old age was imprisonment by the Inquisition, and a broken heart. That is how a new scientific method is regarded by guardians of traditional doctrine.[1]

A different writer claimed, "Galileo's older colleagues knew nothing of experiments. The very idea implied to them a sort of hideous witchcraft—a profanation of the sanctity of the Aristotelian doctrine."[2] Likewise, another account stated, "The Aristotelians ridiculed such 'blasphemy,' but Galileo determined to make his adversaries see the fact with their own eyes."[3]

Most recent historians do not believe this elaborate, dramatic story because it is not based on any evidence; it is mainly the product of writers' lively imaginations. Still, plenty of books and teachers still echo its basic elements.[4] Therefore, fair questions remain: What fragments of the story are true? How did this story grow?

Figure 1.1. The Leaning Tower of Pisa.

In a book of 1935, *Aristotle, Galileo, and the Tower of Pisa,* Professor Lane Cooper boldly denounced the story as false: he argued that Galileo did not drop objects from the tower. Nevertheless, some later historians have disagreed. For example, an expert scholar on Galileo, Stillman Drake, argued that the core of the story might be true, although Drake had no contemporary accounts stating that Galileo ever dropped anything from the tower of the cathedral at Pisa.[5]

The story about Galileo and the Leaning Tower first appeared in a biography of Galileo written by Vincenzio Viviani, the young secretary who served him in his final years of blindness and home imprisonment, from 1639 until Galileo's death in 1642. Viviani, who apparently drafted his account sometime between 1654 and 1657 (although it was not published until 1717), described events that allegedly happened six decades earlier, but which he did not witness:

> At this time [ca. 1590] it seemed to him [Galileo], that the investigation of natural effects necessarily require a true cognition of the nature of motion, there being a philosophical and popular axiom: *ignorance of motion, ignorance of nature*, he gave himself entirely to contemplate it: and then with great discord from all the philosophers, by means of experiences and good demonstrations and discourses he convinced them of the falsehood of many conclusions of Aristotle himself on the topic of motion, which up to that time were held as most clear, and indubitable, as among others, that the speeds of unequal weights of the same material, moving through the same medium, did not at all keep the proportion of their heaviness, assigned by Aristotle, but instead, these all moved with the same speed, he demonstrated this by repeated experiments made from the height of the Bell-Tower of Pisa, in the presence of all the Lecturers and Philosophers, and of all the Students.[6]

Stillman Drake argued that in this account: "Viviani was repeating his recollection of what Galileo himself had told him." But this kind of wording makes me doubt Drake. Really, we don't know what Galileo told Viviani, nor whether Viviani faithfully repeated it fifteen years later. Regardless, to make his case, Drake used more speculative words that make me part company with him: "would have been," "probably meant," "presumably argued," "it would then be natural," "more probably," and so forth.[7] I think that there are no probabilities in the past; events either happened or they did not. In history, we either have enough evidence to claim that an event transpired, or we do not. Faced with uncertainty, one might choose to imagine whatever sounds most plausible, but it seems preferable to build our accounts of the past on the basis of evidence over conjecture.

**'*They were seen to fall evenly.*'**

Figure 1.2. Apparently, one of the balls was very large.

By neglecting chronology, and presenting selected pieces of evidence in a contrived order, writers often convey whatever impressions they wish. As with other evolving stories in the history of science, this story benefits from establishing a chronology of events.

In 1544, the Florentine historian Benedetto Varchi alluded to experimental tests showing that traditional views on falling objects were wrong, in that heavier objects do not actually fall faster in proportion to their weight:

> the custom of modern philosophers is to believe always, and never to test all that which is found written by the good authors, and above all in Aristotle, but that does not mean that it would not be more certain, and more delightful to do otherwise, and to sometimes descend to experience in some things, as for example, in the motion of heavy objects, about which Aristotle, and all other Philosophers, without ever doubting it, have believed and affirmed that as a body is more heavy, so much more quickly it descends, which evidence demonstrates is not true.[8]

Varchi opposed the claim that the heavier a body is, the faster it falls. So, *long before Galileo was even born,* Aristotle's claims on motion had already been challenged by experiments on falling objects.

Later, Giuseppe Moletti, a professor of mathematics at the University of Padua, carried out experiments on falling objects, and in 1576 he reported that falling objects composed of the same material but having different weights reach the ground together.

> Aristotle . . . [seems to argue that] if from the top of a tower we release two balls, one of lead of twenty pounds and the other equally of lead but of one pound, the motion of the larger will be twenty times greater than that of the smaller. . . . My dear sir, you would err; they both arrive at the same time and I have made the test of it not once but many times. And there is more: a ball of wood more or less the same size as a lead one, and released from the same height, descends and reaches the ground soil in the same moment of time.[9]

Moletti thus claimed that objects of roughly the same volume but of different materials and weights reach the ground simultaneously. However, more recent experimenters have shown that this is not necessarily the case. If they are released by hands [palms down], the lighter ball is usually released slightly prior to the heavier ball, and thus the two do not fall quite side by side, the lighter ball moves ahead initially.[10]

During his time at the University of Pisa (1589–1592), Galileo began to write *On Motion*, a book that was published long after his death. In that work, he included a question about hurling objects from a tower:

> How ridiculous is this opinion [Aristotle's] is clearer than daylight: Because who would ever believe, for example, that if two lead balls were dropped from the orb of the moon, one a hundred times larger than the other, if the larger reached Earth in one hour, the smaller would take a hundred hours in its motion? Or also, if from a high tower two stones, one stone twice the size of the other, were flung simultaneously, that when the smaller was halfway down the tower, the larger would have already reached the ground?[11]

In this manuscript, Galileo repeatedly referred to objects falling from a tower, but he did not specify the Leaning Tower of Pisa, and he did not describe any experiments in detail. Moreover, Galileo there argued explicitly that objects of different weights fall at different speeds! He claimed that he dropped two objects, one of wood and one of lead from the top of a high tower, and that "the lead one moves far out in front. This is something I have often tested."[12] Galileo then believed that the speed of falling objects is proportional to their density (instead of their weight, as Aristotle seemed to argue).

Sometime in the late 1500s, Simon Stevin, a Flemish mathematician and engineer, apparently became convinced that falling objects of different weights hit the ground at the same time. He claimed to let two balls of lead, one ten times heavier than the other, "from a point about 30 feet high" to a plank below, and that "the lighter one falls not ten times later than the heavier, but that both seem to hit the plank in a single thump."[13] In 1605, Stevin published a book in which he claimed that he had carried out this experiment "long ago" with his friend John Grotius, with the aim of showing that Aristotle was mistaken in his *Physics* and *On the Heavens*.

In 1597, Jacopo Mazzoni, a friend of Galileo, published a book in which he advocated Galileo's early ideas on motion, against Aristotle's; but he did not mention any experiments at any tower.[14]

In 1604, Galileo sent a letter to Paolo Sarpi, in which he stated that different objects fall at the same rate.[15] At the time, Galileo mistakenly thought that speed is proportional to distance of fall (instead of time).

In 1612, Giorgio Coresio, a professor of Greek, attacked Mazzoni's claims by complaining that Mazzoni had experimented with falling objects from an insufficient height. Coresio then briefly noted that *by dropping objects from the top of the Tower of Pisa,* Coresio himself had shown Aristotle to be correct: a whole body falling faster than a separate piece of it:

> Mazzoni commits anew two other errors of no slight importance. First, he denies a matter of experiment, that, with one and the same material, the whole moves more swiftly than the part. Herein his mistake arose because, perhaps, he made his experiment from his window, and because the window was low all his heavy substances went down evenly. But we did it from the top of the cathedral tower of Pisa, actually testing the statement of Aristotle that the whole of the same material in a figure proportional to the part descends more quickly than the part. The place, in truth, was very suitable, since, if there were wind, it could by its impulse alter the result; but in that place there could be no danger. And thus was confirmed the statement of Aristotle, in the first book of *On the Heavens*, that the larger body of the same material moves more swiftly than the smaller, and in proportion as the weight increases so does the velocity.[16]

Coresio did not attribute any experiments at the Tower of Pisa to Galileo.

Decades later, in March 1641, Vincenzo Renieri, professor of mathematics at Pisa, sent a letter to Galileo stating that he (Renieri) had carried out experiments by dropping objects from the Leaning Tower of Pisa, and he asked Galileo to interpret them. Renieri wrote:

> We have had occasion here to make an experiment of two weights falling from a height, of diverse material, namely one of wood and one of lead, but of the same size; because a certain

> Jesuit [Niccolò Cabeo] writes that they descend at the same time, and with equal velocity reach the earth; and a certain Englishman affirms that Liceti here set a problem, and gave the explanation of it. But finally we have found the fact in the contrary, because from the summit of the Campanile of the Cathedral [at Pisa], between the ball of lead and the ball of wood there occur at least three cubits of difference. Experiments also were made with two balls of lead, one of a bigness equal to a cannon-ball and the other to a musket-ball, and there was observed the biggest and the smallest, from the height of the same Campanile, to be a good palm's difference by which the biggest preceded the smallest.[17]

Not all the letters in the correspondence between Renieri and Galileo still exist, but those that exist show no evidence that Galileo ever conducted any such experiments himself.

At that very time, when Renieri wrote to the old, blind Galileo about experiments from the Leaning Tower, Viviani was Galileo's secretary. A year later, Galileo died. Fifteen years later, Viviani claimed that Galileo dropped objects from the Tower of Pisa. In two drafts, Viviani followed his claim about Galileo's experiments from the Leaning Tower by noting that Galileo discussed the subject in his work on the *Two New Sciences*. But really, Galileo there did not refer to any specific tower and he discussed only *thought*-experiments.[18]

None of the many letters and manuscripts that Galileo wrote over decades claim that he dropped anything from the Leaning Tower. And none of his contemporaries who would have witnessed that event reported any such thing. None of Viviani's alleged lecturers, philosophers, or the many students at Pisa seem to have left any clue that they were aware of such an experiment carried out by anyone in the 1590s.

Historians have found that Viviani was not entirely credible. For example, in accord with the Pythagorean idea of the transmigration of souls, Viviani misrepresented the date of Galileo's birth, as being not on

February 15, 1564, but four days later—so that it would seem to follow the death of the great artist Michelangelo, on February 18, 1564.[19]

Regardless, some prominent historians chose to believe Viviani's account about the tower. In particular, Antonio Favaro, the prestigious editor of the collected works of Galileo, claimed that Viviani "must have heard it from the lips of Galileo himself, affirmed in a manner so certain and explicit that it cannot be called into question."[20]

*Must have?* And, it *cannot* be questioned?

Lane Cooper traced how various versions of the tale added details that were not even present in Viviani's story. Some writers claimed that the demonstration happened "in the morning."[21] Why in the morning? Was it the dawn of a new era? Others claimed that Galileo dropped one ball that weighed a pound and another that weighed a hundred pounds. Another writer claimed that Galileo placed different materials in "exactly similar boxes," before dropping them. Fiction. And yet another writer claimed that the results were "epoch-making."[22] Cooper rightly complained: "Why 'epoch-making'? I have yet to learn what communal scientific advance arose out of Galileo's alleged experimentation from the tower of Pisa; there was no mention of it that can be traced before 1654; and if indeed the thing took place, it seems to have been overlooked by the world at large for sixty years and more."[23] Even at Pisa, Galileo's own admirer, Renieri, made claims quite contrary to Galileo's.

Lane Cooper was not a historian, nor a physicist. He was a professor of English. Hence, when he published his book showing that the story about the tower is not supported by evidence, many historians just dismissed it. For decades they belittled him and ignored his book. But Cooper used evidence to make his argument. It is encouraging that outsiders can contribute to a professional field such as history, especially by focusing on documentary facts rather than speculations. Legends about Galileo have propagated partly because people were willing to parrot the claims of specialists, believing authority, rather than evidence. Thus, the irony of this myth is that whereas it purports to *criticize* philosophers who blindly believed in the authority of Aristotle, in actuality it instead

**Table 1.1** Experiments and explanations, actual or imagined, about falling objects and the Leaning Tower of Pisa

| | | |
|---|---|---|
| 1589–1592 | Galileo Galilei | Taught mathematics at the University of Pisa. |
| 1612 | Giorgio Coresio | By dropping objects from the Tower of Pisa, I have shown that Aristotle's theory of motion is *correct:* a body falls faster than a piece of it. |
| 1638 | Galileo's fictional dialogue *Two New Sciences* | Salviati: "Aristotle says that an iron ball of one hundred cubits, reaches the ground before a one-pound ball has fallen a single cubit." [Actually, Aristotle did not specify this.] . . ."I [Salviati] say that they arrive at the same time." |
| 1638 | Galileo's fictional dialogue *Two New Sciences* | Aristotle is refuted by a logical contradiction [*a thought-experiment*]: the sum of a smaller stone and a larger stone should slow down the larger body, but they should also make a larger, faster body. |
| 1646–1647 | Niccolò Cabeo and Giovanni Baliani | Prior to Galileo, the experiments of Giovanni Battista Baliani (1630s) showed that two objects of different weights take the same amount of time to fall to the ground, irrespective of the medium. |
| 1641 | Vincenzo Renieri | From the top of the Tower of Pisa, I have dropped wood objects and lead objects, of different weights but the same size, and the heavier ball hit the ground first. I have also dropped a large lead ball, and a small lead ball, and the larger one hit the ground first. |
| 1651 | Giovanni Riccioli | Cabeo is wrong, because in 1634, I noticed that the heavier stone dropped from a tower reached the ground slightly faster. |
| 1650s | Vincenzio Viviani | When Galileo taught at Pisa, he repeatedly demonstrated by experiments, in front of all the lecturers, philosophers, and students, that objects dropped from the bell tower of Pisa landed on the ground simultaneously, proving that Aristotle was wrong. All this is treated by Galileo in his *Two New Sciences* [Not true]. |
| 1717 | Vincenzio Viviani | Viviani's claim about the Tower (above) is finally printed, but without the footnote mistakenly claiming that Galileo himself reported it in *Two New Sciences*. |
| 1890s | Rafaello Caverni | Galileo lied to Viviani about the Leaning Tower. |
| 1916 | Richard Gregory | In 1591, the young Galileo dropped balls of different weights from the Tower of Pisa, proving that they land together, and thus Nature clearly and instantly answered a debate that had lasted two thousand years. |
| 1934 | Lane Cooper | There is no evidence that Viviani's account was true. |
| 1978 | Stillman Drake | "Viviani was repeating his recollection of what Galileo himself had told him." |

exhibits the gullibility of persons who repeat the tale, their readiness to believe on the basis of authority.

The story of the Leaning Tower of Pisa captured people's imagination, I think, because it conveyed the dramatic illusion that in a single moment, one young man publicly dared to challenge an arbitrary tradition that had dominated for over a thousand years, and supposedly, his successful experiment abruptly and clearly revealed the truth. Allegedly, this simple, dramatic event constituted a turning point in the history of science; nature had spoken, delivering the death-blow to an arbitrary traditional belief. This epic mythical image was conveyed well in the early 1900s:

> As Galileo's statement was flouted by the body of professors, he determined to put it to a public test. So he invited the whole University to witness the experiment which he was about to perform from the leaning tower. On the morning of the day fixed, Galileo, in the presence of the assembled University and townsfolk, mounted to the top of the tower, carrying with him two balls, one weighing one hundred pounds and the other weighing one pound. Balancing the balls carefully on the edge of the parapet, he rolled them over together; they were seen to fall evenly, and the next instant, with a loud clang, they struck the ground together. The old tradition was false, and modern science, in the person of the young discoverer, had vindicated her position.[24]

In the 1990s, experimental physicist Leon Lederman wrote a section of a book titled "The Truth of the Tower." Lederman embraced the story, and described it as a media happening, the first great scientific publicity stunt: "Galileo knew in advance how it was going to come out. I can see him climbing the tower in total darkness at three in the morning and tossing a couple of lead balls down at his postdoc assistants."[25]

Sixty years prior to Lederman's narrative, Lane Cooper, perplexed by the widespread currency of the Leaning Tower story, commented on people's willingness to believe across the centuries:

Again, there are many who as aforesaid believe the story about Galileo and the tower of Pisa, and have no better ground for accepting the story than hearsay. They have read or vaguely heard that it was so just as they believe that Aristotle said a certain thing about falling bodies of different weights, and that every one down to Galileo believed the same thing on the authority of Aristotle. As casual readers accept some modern authority for an opinion about the speed of bodies heavy and light in falling, so they accept upon authority, however vague, the tale about Galileo at Pisa.[26]

# 2

# *Galileo's Pythagorean Heresy*

EVEN without evidence, most people believe in Earth's motion. When asked why Earth moves, some college students reply: "because of the seasons" or "because astronauts see it from spaceships." But seen from Earth, the spaceships seem to move instead, so who's right? And for thousands of years, it seemed that the seasons change just because the sun rises at different places on the horizon, taking different paths; so why not say that seasons are caused by the sun's motions? Centuries ago, teachers taught that the sun circles Earth, and it seemed obvious; now they teach the contrary, and it seems obvious. It's easy to accept basic scientific knowledge without knowing why scientists believe it, but that would make science seem like just another doctrine.

Many myths are associated with the Copernican revolution. I'll now recount the dramatic struggles by which people changed their minds about Earth's motion, while also pointing out several myths. But mainly, I will trace the many connections to legends about Pythagoras and his followers. For centuries, writers have often associated the idea of Earth's motion with Pythagoras of Samos, a philosopher who lived in southern Italy in the sixth century BCE. But is there truth to such claims? Strangely, I find not a single historical account that systematically connects the diverse legends about Pythagoras and his followers to the Copernican revolution. Making that connection reveals a surprising dimension of Galileo's famous heresy.

According to Aristotle, the philosopher with the aquiline nose, the so-called Pythagoreans believed that Earth moves in circles around the center of space: "they say that fire is at the center and that Earth is one of the stars, and that moving in a circle about the center it produces night and day."[1] But their arguments did not convince Aristotle, who had many good arguments to the contrary. For example, he proposed, if Earth moves across the heavens, then we should observe certain changes in the background of stars. Specifically, we should see apparent changes in the gaps between the stars. Consider a few objects in a row, as seen by someone running along. There's a line of sight to each object. The objects closest to the person will seem to be farther apart from one another, the two objects farthest away will seem to be closest together. But once the person has reached those farthest objects, the effect will be reversed.[2]

Likewise, if Earth moves across the heavens, we would expect that the relative distances between the stars should change. Ancient astronomers could expect to see such effects, and more, if Earth were moving. But no such effects were observed. So Aristotle fairly argued that Earth does not move at all.

Still, decades after Aristotle's death (in 322 BCE), the astronomer Aristarchus of Samos advocated that Earth travels in circles around a center, the sun. No treatise by Aristarchus (who died around 230 BCE) on this matter seems to have survived, but there are brief, indirect accounts. In particular, Archimedes mentioned the theories of Aristarchus in this address:

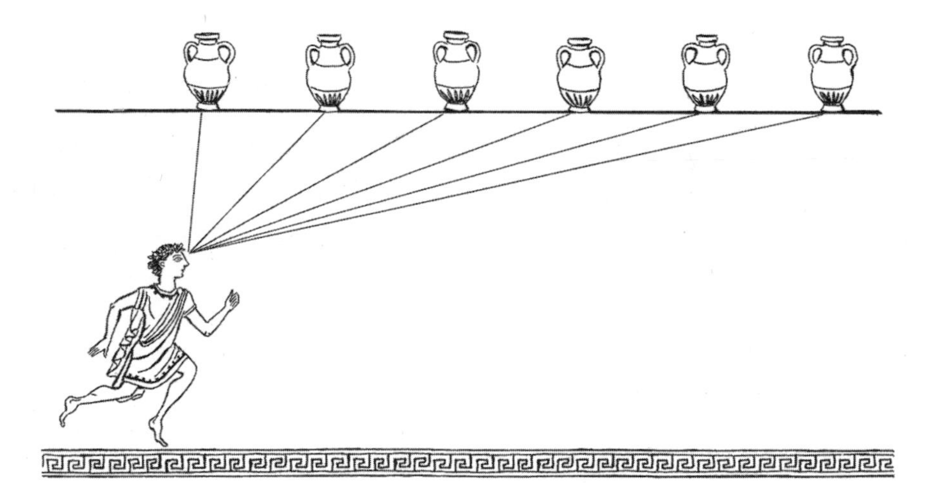

Figure 2.1. Lines of sight to a group of objects. As the man moves, the gaps between the objects seem to change

> You, King Gelon, know that 'universe' is the name given by most astronomers to the sphere the center of which is the center of the Earth, while its radius is equal to the straight line between the center of the Sun and the center of the Earth. This is the common account as you have heard from astronomers. But Aristarchus brought out a book consisting of certain hypotheses, wherein it appears, as a consequence of the assumptions made, that the universe is many times greater than the 'universe' just mentioned. His hypotheses are that the fixed stars and the Sun remain unmoved, that the Earth revolves about the Sun in the circumference of a circle, the Sun lying in the middle of the orbit, and that the sphere of fixed stars, situated about the same center as the Sun, is so great that the circle in which he supposes the Earth to revolve bears such a proportion to the distance of the fixed stars as the center of the sphere bears to its surface.[3]

Archimedes rejected Aristarchus's theory, as did most mathematicians and astronomers. One problem was that it seemed to require a ridiculously immense universe. In order for there to be no visible shifting of the distances between the stars, such distances had to be almost unimaginably huge compared to the orbit of Earth around the sun. The last line of the quotation states that Aristarchus apparently claimed that the stars were virtually infinitely far away from Earth.

By contrast, Pliny the Elder claimed by 77 CE that "the penetrating genius of Pythagoras" had inferred that the distance from the sun to the zodiac stars is merely three times the distance between Earth and the moon.[4] If Pythagoras ever made such a claim, it was grossly incompatible with the later, allegedly Pythagorean, claim that Earth moves, because its motion would then cause apparent changes in the distances between the stars. In any case, Pliny ridiculed Pythagoras for "occasionally" drawing on music theory to describe the distances between Earth and the heavenly bodies as tones and half-tones, producing "a universal harmony" of seven tones, "a refinement more entertaining than convincing."[5] I do not know what credibility Pliny had, but to undermine our reflex to respect and believe ancient authorities, it is useful to bear in mind any of the charming bits of nonsense that some readily proclaimed. For example, Pliny claimed that the saliva of a fasting human was the best safeguard against serpents and he praised the various magical and medicinal properties of spitting: on epileptics, on leprous sores, on insects in an ear, and upon your right shoe.[6] Nonetheless, Pliny was a careful compiler of knowledge that he found in many books. The main problem is that he wrote more than five hundred years after Pythagoras died. The accuracy of Pliny's unspecified source on Pythagoras as an astronomer is called into question by the fact that the many prior scholarly commentators on astronomy, such as Aristotle, seem to have made no such claims about Pythagoras.

Aristotle formulated the conception of the heavens that became widespread for centuries, and it was revised systematically around 150 CE by the Roman citizen Ptolemy, in Alexandria, Egypt. Earth was said to remain stationary at the center of the universe while the moon, the sun, the planets, and the stars all moved in circles around Earth.[7] Everything below the moon was said to be in the terrestrial sphere, where things change and decay and violent motions, death, and rot take place. By contrast, the moon and everything above it was said to be perfect, unchanging, eternal.[8] All heavenly motions seemed unforced and circular. Like rain and clouds, comets were assumed to happen not above the moon, but in Earth's atmosphere, since they too were fleeting, temporary phenomena.

Every night the stars and planets cross the sky in a seemingly circular, westward path. But the planets lag slightly eastward, week after week, and when we track the positions of a planet such as Mars against the background of stars, we find that sometimes, instead of moving uniformly eastward in a circular path, Mars seems to slow down and pause, then go back. It loops backward and then again it continues eastward. This is called retrograde motion, and the ancient astronomers wanted to explain it.

Apparently it was disturbing to think that heavenly things start and stop. In the first century BCE, the Greek writer Geminos noted:

> The Pythagoreans, who first approached such investigations, hypothesized that the movements of the Sun, Moon, and the

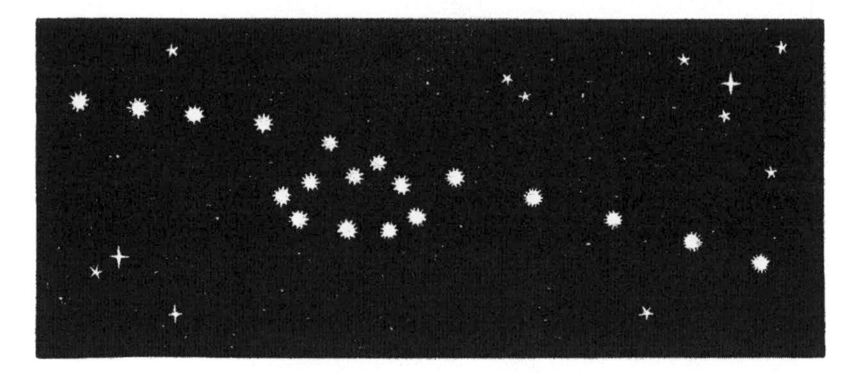

Figure 2.2. Retrograde motion: Mars is pictured repeatedly against a background of stars over six months, as it traces a loop in the sky. Every 780 days, Mars begins a retrograde loop that lasts almost twelve weeks.

> 5 wandering stars are circular and uniform. For they did not accept, in things divine and eternal, such disorder as moving sometimes more quickly, sometimes more slowly, and sometimes standing still. . . . One would not accept such an anomaly of movement in the goings of an orderly and well-mannered man. This business of life is often the cause of slowness or of swiftness for men. But in the case of the incorruptible nature of the stars, it is not possible to adduce any cause of swiftness or slowness. For this reason, they put forward the question: how would the phenomena be accounted for by means of uniform and circular motions?[9]

Later, Ptolemy also tried to explain retrograde motion. He argued that a planet such as Mars is carried along not in one circular motion alone but by two. Imagine that there is a great sphere that rotates around Earth and it traces a path as large as the orbit of Mars. And on this sphere there is a wheel, centered on a point on the sphere's surface, and Mars is attached to this wheel, as the wheel spins slowly around its center. Thus Mars traces a small loop on the sky as it moves carried by the sphere and wheel (epicycle).

Additional geometric devices were also used to account for variations in the paths and speeds of the heavenly bodies. Centuries later, a legend

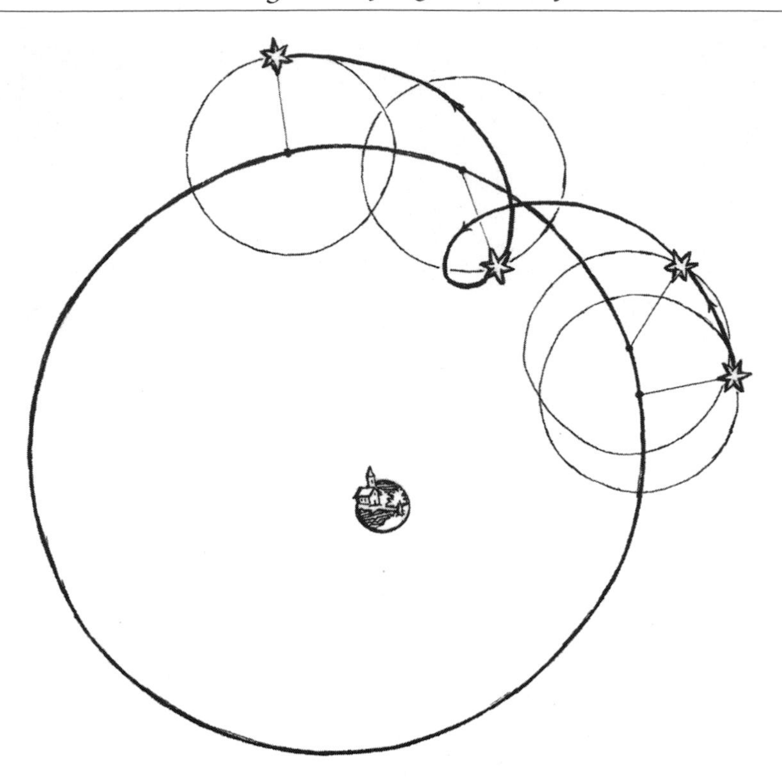

Figure 2.3. Cycle and epicycle carry Mars in loops around Earth. To the ancients, Mars was a kind of star.

developed: that whenever astronomers found that two circles did not suffice to account for the motion of a particular planet, they added more epicycles. By 1969 the *Encyclopedia Britannica* claimed that in Ptolemy's system, each planet required forty to sixty epicycles. Yet historian Owen Gingerich has shown that there's no evidence that *anyone* used more than one epicycle per planet.[10] The story was a myth that grew, making it seem as though Ptolemy's scheme was ridiculously complicated and had collapsed under its encumbrances.[11]

We tend to imagine that obsolete scientific theories were ridiculously complicated whereas contemporary theories are elegant and reasonable. Yet Ptolemy's geocentric account was extremely useful and successful. It was used for over 1,400 years. But there were growing problems. Based on Ptolemy's geometric system, astronomers devised numerical tables that helped to predict astronomical events, such as eclipses, alignment of planets, and equinoxes.[12] By the 1500s, these tables didn't agree with the calendar that had been established by Julius Caesar (in 45 BCE). In the year 325 CE, for example, the spring equinox happened on 21 March, its official date according to the calendar—but over the centuries the spring equinox arrived progressively earlier, and by 1500 it fell instead on 11 March. This was a problem because the astronomical observations no longer matched the calendar. Yet people set the date of some important events, such as Easter, by reference to both. The Catholic Church was disturbed by the mismatch between the seasons and the calendar. And so, in 1514, Pope Leo X requested the help of astronomers.

The Polish astronomer, Nicolaus Copernik, was one of the experts who provided advice on the problem. "Copernicus" was unsatisfied with the structure of Ptolemy's account and its limited precision. He disliked that some of Ptolemy's circular motions were not uniform relative to their centers. Therefore, Copernicus had studied ancient writings to find an alternative. In the works of Cicero and those attributed to Plutarch, he found reference to various Pythagoreans, including Philolaus, who claimed that Earth spins or moves.[13] Thus Copernicus came to privately formulate a theory similar to that of Aristarchus, reasoning that it would be preferable to assume that Earth was not at the center of the universe, but that it moved like a planet, and that the planets orbited the sun. Remarkably, by centering the sun, it turns out that the orbital speeds of the planets, including Earth, fall in an ordered sequence: the speed of each planet varies according to its distance from the sun, the closer the planet is to the sun, the faster it moves. Copernicus realized that if indeed Earth circles the sun, then we would observe the *apparent* retrograde motions of the planets, since Earth sometimes overtakes them as it moves around the sun.

The heavenly bodies circled the sun, the moon circled Earth. All moved in circular paths, as in Ptolemy's account. Earlier, Plato had claimed that God "made the universe a circle moving in a circle, one and solitary," with its parts moving in circles.[14] Copernicus argued that the planets move because "it is in the nature of perfect circles to rotate forever." Also, he admitted epicycles into his system to account for deviations

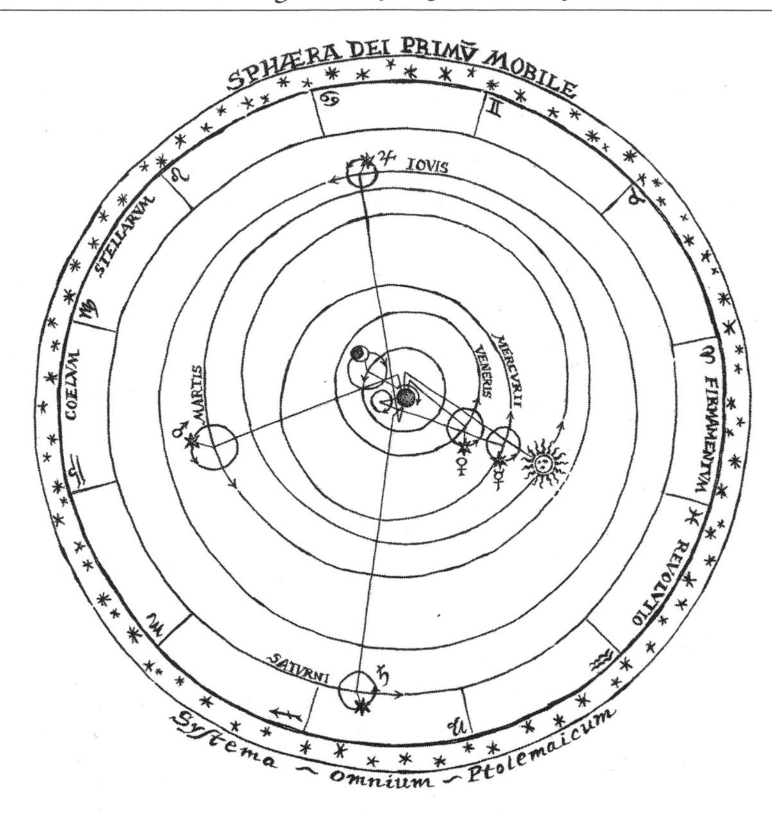


Figure 2.4. Ptolemy's scheme, ca. 150 CE. Earth is immobile at the center of the universe. Everything in the heavens moves in circles.

in the paths of the planets.[15] Yet, his theory departed from tradition, in part, by breaking the barrier between the terrestrial and the celestial, by placing Earth in the heavens.

Present methods of education lead us to see Copernicus's scheme as natural, simple, and reasonable. However, it had features that seemed utterly repulsive to Copernicus's contemporaries. For example, if Earth is a heavenly body, then why does it spin? None of the others seemed to spin. And if Earth is spinning, why aren't things flung off? Furthermore, if Earth is moving, why don't we feel it? Shouldn't physics and astronomy be based on our experiences? Also, if Earth were truly a planet, then would planets also be worlds like Earth? Would there be humans

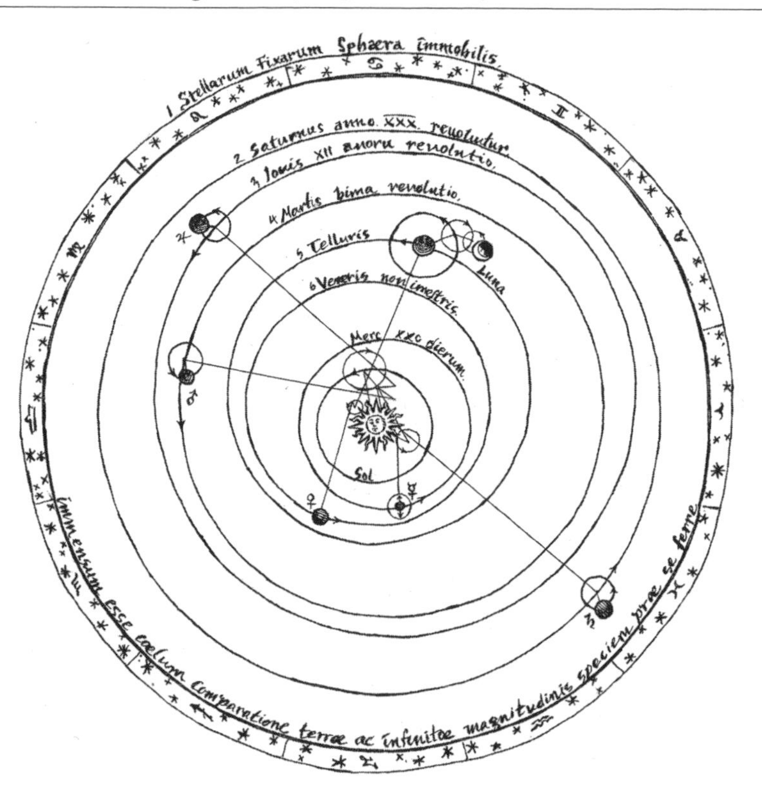


Figure 2.5. Copernicus's circles, 1543. The sun lies near the center, immobile. Earth and the planets orbit the sun. The stars are immobile.

in other planets? Why would God create many worlds? Did Jesus Christ visit them too? Any such questions would be unsettling.

Also, Copernicus's system was hardly geometrically simpler than Ptolemy's, partly because it too had epicycles. The small epicycles of Copernicus did not entail retrograde motions; instead, they served to account for irregularities in the changing velocities of each planet.

Moreover, the motion of Earth entailed that there should be apparent changes in the separations between the stars throughout the year, but since no such changes were observed, Copernicus argued, like Aristarchus, that the distance between the planets and the stars was spectacularly immense. That means that the gap between Saturn and the stars should *not* be similar to the gap between two planets' orbit but much, much greater. That gap between Saturn and the stars should be at least around thirty thousand times greater than the distance between the sun and Saturn. Try to draw that. Why would there be such a huge and horrifyingly empty void between the planets and the stars? Why would God create such extraordinarily large emptiness? As later remarked by Blaise Pascal: "The eternal silence of these infinite spaces frightens me."[16]

Copernicus anticipated that his theory would be severely criticized, partly because the Catholic Church embraced the Aristotelian account. He also followed the reputed practice of the Pythagoreans to keep honest opinions private and reveal them only to friends.[17] Therefore, he kept his work relatively secret for decades. Still, for Copernicus, the sun-centered system was elegant proof of God's plan and the divine order of things. In his manuscript *On the Revolutions*, he noted, "what better place could be found for this lamp in this exquisite temple than where it can simultaneously illuminate everything?"[18] Yet his astronomical calculations led him also to acknowledge that the center of the sun is not located exactly at the center of the planetary orbits.

Copernicus delayed publication of his work for many years, though he circulated a few manuscripts. Some of his friends, including the cardinal of Capua and the bishop of Culm, encouraged him to publish, but he resisted. By 1539, the "evangelical" leader Martin Luther had heard of Copernicus's theory. There is a story that Luther complained, "The fool will overturn the whole art of astronomy."[19] But that claim was made by someone who did not hear Luther's words. Someone who *did* hear Luther firsthand reported instead that Luther said "the new astrologer" wished to turn astronomy upside down, while "I believe the Holy Scriptures, for Joshua commanded the Sun to stand still, and not the Earth."[20]

In 1541, the sixty-eight-year-old Copernicus finally agreed to publish his book, the labor of three decades. He dedicated it to Pope Paul III, expressing his prior fears of scorn and slander from "idle talkers wholly ignorant of mathematics," those whose "inherent stupidity among philosophers holds as drones among bees."[21] Late in 1542, Copernicus suffered a stroke that paralyzed his right side. While he was convalescing, a Lutheran theologian who helped to oversee the printing of his book added an anonymous disclaimer, a foreword arguing that there is no certainty in astronomy and that hence the work consisted essentially of "hypotheses" convenient for mathematical calculations.[22] Ill and confined to his bed, Copernicus finally received a copy of his book on 24 May 1543. He died a few hours later.

The idea of Earth's motion was generally unwelcome. The "Reformed" Christian leader, John Calvin, denounced those who "pervert the order of nature. We shall see some who are so deranged, not only in religion but who show in all that they have a monstrous nature, that they say the sun does not move, and that it is the earth that moves and turns. When we see such persons, it must be said that the devil has possessed them."[23] Calvin was not necessarily referring to Copernicus, whom he might not have read, but to anyone who made such claims.

The bestselling writer Arthur Koestler famously described *On the Revolutions* as "the book that nobody read," but historian Owen Gingerich spent decades tracking copies of the book and concluded that Copernicus's work found plenty of readers. One was the prominent astronomer Erasmus Reinhold, who used it to calculate tables giving the positions of the heavenly bodies in time, which later served to revise the calendar. In his heavily annotated copy of Copernicus's book, Reinhold exulted, "Celestial motions are circular and uniform or composed of circular and uniform parts."[24]

Another who came to believe in the theory was the English mathematician Thomas Digges. In 1572, a bright new light was seen in the heavens. Digges inferred that it was a star, and he concluded that for ages astronomers had been wrong: the heavens were not unchanging. The region beneath the moon remained "the Empire of Death," in his words, but there was change in the heavens, too. He thus thought it prudent to consider alternative accounts of the heavens, and so he adopted the "perfect description of the celestial orbs" which he attributed to the ancient Pythagoreans and to Copernicus.[25]

Also in 1572, a Danish astronomer, Tycho Brahe, was likewise impressed by the bright new light in the heavens. Brahe was a wealthy nobleman who wore a lump of silver and gold in a hole in the center of his face, since a chunk of his nose, the bridge, had been cut off in a duel.[26] This man with a golden nose devised instruments for measuring the positions of the stars and planets with great accuracy. And he found that the new bright light did not shift relative to the other stars, so it wasn't an atmospheric effect beneath the moon. It was indeed another star. Again, this meant that Aristotle had been wrong about the nature of the heavens.

Since there was change in the heavens, as in Earth, then Earth might conceivably be in the heavens. Thomas Digges published a summary of the Copernican scheme in English, and he added the notion that the stars are not embedded in a sphere, but instead are spread throughout an infinite space beyond the planets. For Digges, the spherical and infinite orb of the stars was "the very court of coelestiall angels devoid of greefe and replenished with perfite endlesse joye the habitacle of the elect."[27]

If Earth were truly moving, astronomers expected to see some observable proof of this. So, throughout the years, Digges expected that the new star would become periodically brighter or dimmer as Earth presumably moved toward it or away from it. But no such effect was observed.

Meanwhile, Catholic astronomers reformulated the calendar to fix the dates of religious events, such as Easter. In 1582, Pope Gregory XIII decreed that 4 October be followed immediately by 15 October, to bring the calendar into accord with the astronomical phenomena. Ten days were zapped out of existence. What happened on 5 October 1582? Nothing happened: that day did not exist (that is, in the few Catholic countries that immediately adopted the calendar, such as Italy and Spain). Tycho Brahe, a Lutheran, appreciated the value of the new Gregorian calendar.

Like Copernicus, Brahe disliked some of Ptolemy's geometrical devices. Brahe appreciated the work of Copernicus, but only as a mathematical contrivance, he did not believe that Earth moves. Astronomers expected that if Earth moves, we should be able to observe some relative shifting of the stars: parallax. Copernicus knew that his hypothesis would entail parallax, but since no such effect was known, he had argued, like Aristarchus, that the stars were extraordinarily far away and far apart from one another. Apparently Brahe did not look for stellar parallax as a way to test Copernicus, but in 1583 he did try to detect parallax in the orbit of Mars, to test whether Mars is ever closer to Earth than Ptolemy expected.[28] He found no such effect, so he rejected Earth's motion, which he also ridiculed as contrary to Scripture.

But a few clergymen appreciated the idea of Earth's motion. In 1584, the Spanish theologian Diego de Zuñiga published a commentary on the book of Job, in which he noted that the characterization of God in Job 9:6 as "He who moves the Earth from its place, and its pillars are shaken," seemed to say that Earth moves.[29] Zuñiga claimed that this passage could be explained by "the opinion of the Pythagoreans" and of Copernicus. Seemingly contrary passages, such as Ecclesiastes 1:4, "Generations will come, and generations will pass away, but the Earth remains forever," argued Zuñiga, referred not to any motion of Earth but to its durability.

Meanwhile, Tycho Brahe also observed comets. According to Aristotle and Ptolemy, comets were not celestial phenomena, because comets exhibited change: they appeared and disappeared, and they did not travel in circles. So, astronomers and philosophers said that comets transpire beneath the orbit of the moon, as just another effect of Earth's atmosphere. But in 1577, Brahe observed a comet and carefully measured its parallax relative to the stars. He found that it exhibited too little parallax to lie beneath the orbit of the moon. He concluded that the comet was far above. This meant that Aristotle was mistaken: comets were not atmospheric phenomena. The ancient accounts turned out to be wrong—there could be change in the heavens along with noncircular motions.

Some writers—who were not astronomers—had misconstrued Aristotle and Ptolemy, as if the ancient accounts claimed that the planets are embedded in crystalline material orbs, harder than diamond, "absolutely solid."[30] Brahe believed such misinterpretations. But by analyzing a comet in 1585, Christopher Rothman found that it traveled across the orbits of the planets, proving that no solid orbs separate them. In letters to Rothman, Brahe later appropriated this discovery to himself. He claimed to have refuted the common opinion of Pythagoras, Aristotle, and even Copernicus. Actually, there is no evidence that *any* of them ever held such an opinion about celestial orbs. Unfortunately, Brahe had little historical acumen, he "would not spare the time to scrutinize the writings of his predecessors to inquire who said what or when."[31] Accordingly, later historians wrongly echoed that the crystalline spheres were an ancient belief, propagating grandiose generalizations, reiterating mistakes.

Since Brahe found inaccuracies in Ptolemy's theory, he formulated his own account of the heavens. The man with the golden nose argued that experience compels us to accept that Earth stays at the center of the universe while the moon, the sun, and the stars circle around it. And, he claimed, the planets circle the sun. Brahe's system made sense: it matched physical perceptions and it echoed some of the mathematical convenience of Copernicus's system. Brahe's system required the same phenomena entailed by the Copernican scheme, including a slight parallax for Mars. In 1587, Brahe detected that parallax and interpreted it as evidence in favor of his own system.

In 1599, Brahe was appointed imperial mathematician to Emperor Rudolph II, in Prague, then-capital of the Holy Roman Empire. But Brahe died two years later, partly due to courtesy and gluttony. His assistant, Johannes Kepler, recounted Brahe's last days:

> On 13 October, Tycho Brahe dined with the illustrious Master Mincowitz, in the company of Mater Rosenberch. Holding his urine longer than usual, Brahe remained seated. He drank too much and felt pressure on his bladder but he valued civility. When he returned home, he could still barely urinate with the greatest pain, and nevertheless obstructed. The insomnia continued, with intense fever and gradually leading to delirium, and the food he ate, which could not be kept from him, exacerbated the evil. . . . On his last night, in his delirium, which was completely gentle, he repeated these words, like someone composing a verse: *Not to seem to have lived in vain.*[32]

The man with the golden nose died on 24 October 1601. Thus ended the painstaking observations that he carried out and recorded for about thirty-eight years. Four centuries later, forensic analyses of hairs from

Brahe's corpse showed that he had ingested toxic amounts of mercury, which caused or accelerated his illness and death.[33]

Then, Brahe's collaborator, Kepler, became imperial mathematician, inheriting the planetary data. He was a German mathematician who was deeply religious. In the work of Copernicus, Kepler had found what he regarded as God's true design of the heavens.

Years before Brahe's death, seeking to find the mathematical ordering

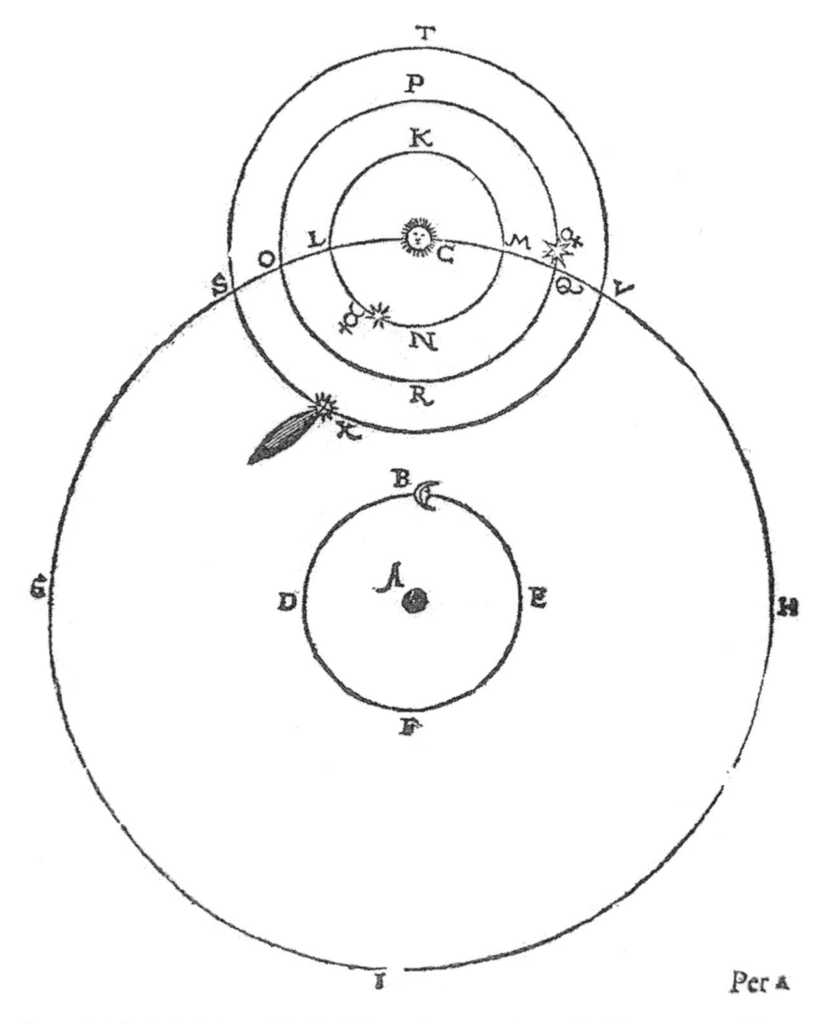


Figure 2.6. Brahe's circles, 1588. Earth lies at the center, immobile. The moon and the sun circle Earth, while the planets circle the sun.

of the universe, Kepler had investigated why there were only six planets, including Earth. He also wondered why the distances between the six planets had various definite sizes. Kepler had studied old writings that showed how Plato, and supposedly the early Pythagoreans, ascribed great importance to the five regular solids in the order of the universe. The regular solids are figures that are each made of only identical sides and each side has identical side-lengths: pyramid (four sides), cube (six sides), octahedron (eight sides), dodecahedron (twelve sides), and icosahedron (twenty sides). According to Proclus, Pythagoras had discovered "the structure of the cosmic figures," though we have no evidence to support this.[34] Impressed by such ideas, Kepler believed that he could explain the quantity of planets and their relative separations by inferring that their orbits were interspaced by the five regular solids.

Consider a sphere, standing for the orbit of Saturn, and inside of it posit a cube, inside of which there would be another sphere, as large as can fit in the cube. That second sphere would be the orbit of Jupiter, and in turn, if a tetrahedron (a four-sided pyramid) were inscribed in that sphere, and another sphere were inscribed in the tetrahedron, the latter would give the orbit of Mars. Inside, there would be a dodecahedron, then the orbit of Earth, followed by an icosahedron, then the orbit of Venus, and finally an octahedron and in it the orbit of Mercury.

Interspaced between the orbits of the six planets, the five regular solids in the sequence of six, four, twelve, twenty, eight, actually gave the relative separations between the planets. Kepler thought he had found evidence of God's divine plan of the universe.

This geometrical scheme may seem ridiculous now. But it is remarkable in hindsight because it does replicate pretty closely the distances between the planets. Kepler published his *Cosmographical Mystery* in 1596. He had wanted to include an introductory chapter explaining the consistency of Copernicus's theory with the Bible, but the senate of the University of Tübingen requested that he omit that part. So Kepler quietly acquiesced: "We shall imitate the Pythagoreans also in their customs. If someone asks us for our opinion in private, then we wish to analyze our theory clearly for him. In public, though, we wish to be silent."[35]

Echoing some ancient writers, Kepler also alluded to the claim that

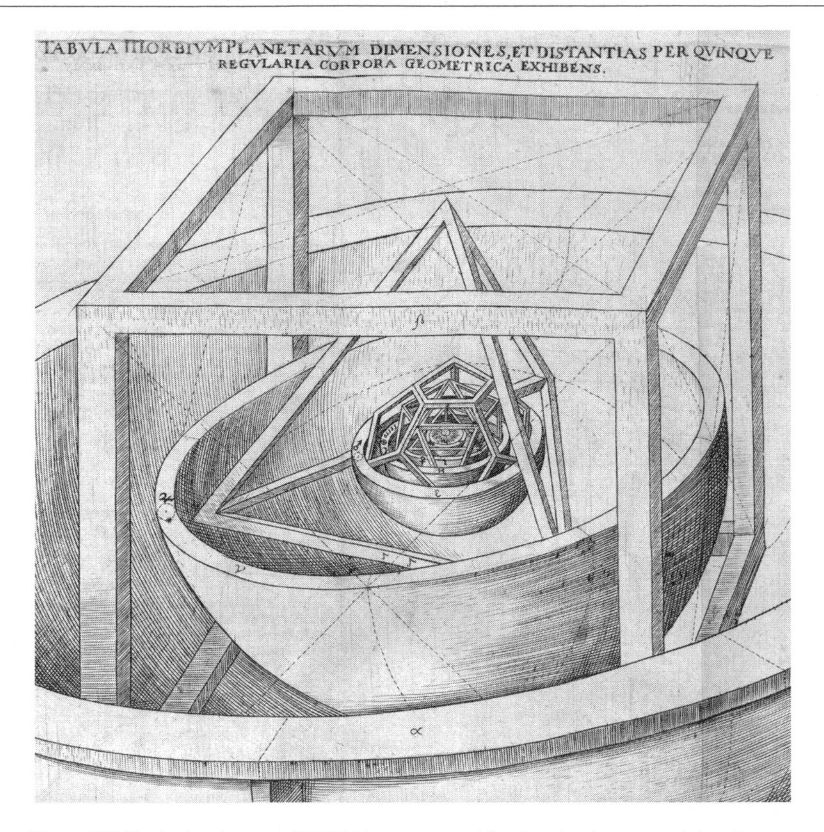


Figure 2.7. Kepler's scheme of 1596. He accounted for the six planets and the distances between their heliocentric, circular orbits by interspacing the five regular solids.

Pythagoras could hear a universal harmony, a "music of the spheres" emitted by the motions of the planets.[36] And if anything did not quite work in the supposed numerical and geometrical harmony of the universe, Kepler hoped that Pythagoras might perhaps rise from the dead to help him—which did not happen, he wrote, "unless perhaps his soul has transmigrated into me."[37]

In the meantime, Copernicus's conception had drawn the attention of a few other eccentrics. In particular, an Italian philosopher and priest, Giordano Bruno, shared similar views; he adopted and expanded the theory. Bruno despised the theory of epicycles, considering them mere crutches for Aristotle's erroneous theory.[38] He claimed that no body is entirely spherical and that all natural motions deviate considerably from uniform circular motion around a center.[39] Bruno denied the perfect circularity of celestial motions, claiming instead that planetary paths were more like spirals. He further reasoned that the stars are suns, that the universe is infinite, and that there exist other worlds like Earth. Centuries earlier, some Pythagoreans had allegedly claimed that the stars were worlds in the infinite space.[40] Bruno studied texts that had been prohibited by the Catholic Church and attacked Aristotle's physics and his finite universe.

Like the Pythagoreans, Bruno claimed that souls are repeatedly reborn, even into animals. Moreover, he entertained unorthodox ideas about Jesus Christ, whom he viewed as a clever magician: he doubted that Jesus was born from a virgin; he denied that Jesus was actually God. Bruno also behaved as if knowledge found by reason is superior to knowledge attained by faith. Because of such opinions, the Catholic Church prepared to put him on trial and excommunicate him. Bruno fled the Church and was excommunicated in absentia. He then befriended the Calvinists, but they too felt antagonized by him, and proceeded to arrest and excommunicate him. Bruno then approached the Lutheran Church, but since he argued that the various churches should coexist peacefully, the Lutherans excommunicated him in 1589. In 1592, he was taken out of his bed in the night by the Catholic Inquisition, who imprisoned him and put him on trial for years. Ultimately he refused to recant his theological and philosophical views, so they declared him guilty of heresy and sentenced him to death. On February 17, 1601, the officers of the Inquisition tied Bruno and conveyed him to the Campo dei Fiori, or Field of Flowers, where they silenced him with a gag on his tongue. Then they tied Bruno to a wooden stake, before onlookers, and burned him alive.

Writers often claim that Bruno was executed because of his belief in Copernicus and in the infinity of worlds.[41] Was Bruno a martyr for science? Actually, historians do not know for certain, because much of Bruno's file is missing from the records of the Inquisition. At least he was a martyr for the freedom of expression.

Galileo Galilei was another Italian who adopted the Copernican scheme. In 1592 he became a professor of mathematics at the university of Padua, a position that Bruno had sought unsuccessfully. In 1609, Galileo learned that a Dutchman devised a spyglass to see distant things. So Galileo figured out how to construct and improve a telescope for seeing even farther, by grinding lenses designed according to the principles of optics. At the end of 1609, Galileo used his innovative telescope to observe several astonishing astronomical phenomena. In the words of one historian, "In about two months, December and January, he made more discoveries that changed the world than anyone has ever made before or since."[42]

In May 1610, Galileo published his observations in a book titled *Message from the Stars*. He described mountains on the moon and stated that the Milky Way consists of stars and that four shiny bodies orbit Jupiter. These findings were spectacular and lent support to the Copernican theory, which Galileo also believed had been the theory of Pythagoras.[43] Copernicus had credited only the Pythagoreans; Galileo arbitrarily took an extra step.

Galileo saw mountains and valleys on the moon, which showed that, contrary to the beliefs of the Aristotelians, Ptolemy, and so many others, the moon was not a sphere. Therefore, the objects in the heavens were not all perfect. The moon resembled Earth—apparently it was *another world*. Since at least some heavenly bodies were like Earth, Earth too might be a heavenly body, as Copernicus seemed to require. Likewise, if the Milky Way indeed consisted of distant stars, then the universe would be much larger than most astronomers imagined.

Galileo also saw that there were four lights wandering around Jupiter; hence, Jupiter was a center of heavenly motions. The bodies around it could be like the moon. Therefore Earth and Jupiter were alike, so Earth might be a planet. Moreover, Ptolemy's system was now wrong or incomplete: not all motion centered on Earth. If there were multiple centers of motion, then Earth might imaginably orbit the Sun. And further, Jupiter might well be another world.

Earlier tradition alleged that some Pythagoreans believed that the moon was a world, like Earth, inhabited by large animals and beautiful plants.[44] For years, Kepler had been drafting a story in which he dreamed of a journey to the moon. In the story, an old mother revealed to her son that she could converse with the moon and she showed him the secret art of conjuring spirits. Then an alien daemon quickly conveyed the boy and his mother on a dangerous voyage up to the moon to meet its inhabitants (who hid in caves), discuss their astronomy, and witness monstrous snakelike creatures that shun the sun and others who daily die by daylight and return to life at night.[45]

Now Kepler heard of Galileo's findings. He immediately saw that they supported the Copernican outlook. Having learned that the moon really had mountains, Kepler inferred that, indeed, there were beings living on the moon! He speculated that they were large and spent their time building barriers with clay to protect themselves from the insufferable heat from the sun.[46] But fascination was mixed with dread. At first,

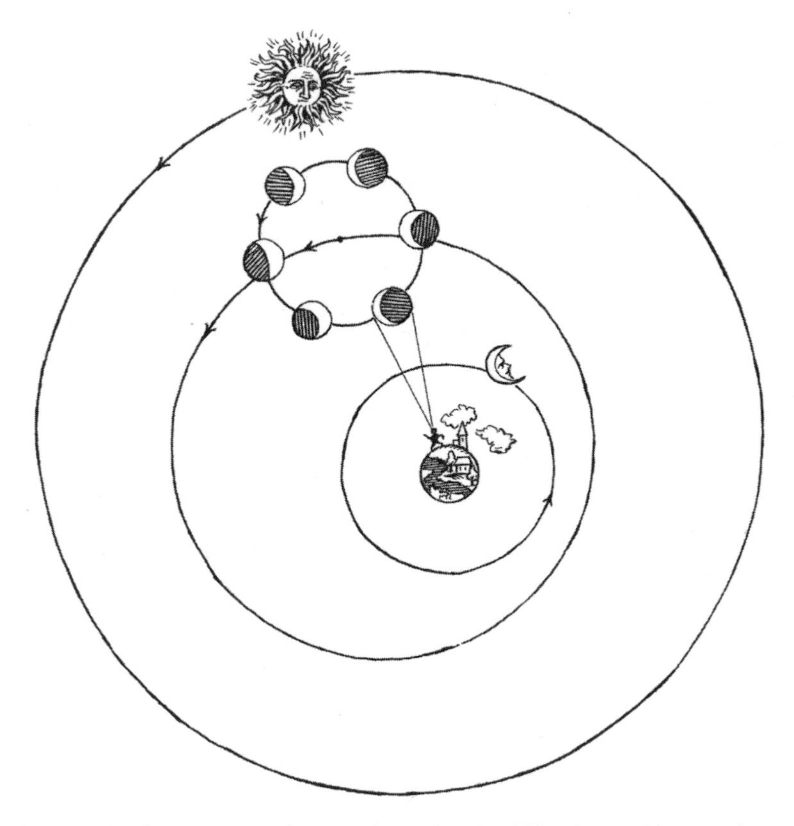

Figure 2.8. Ptolemy's system, showing phases that would be observed from Earth as Venus travels on its epicycle.

Kepler feared that the lights near Jupiter might actually be planets moving not around Jupiter but around a distant star. Kepler imagined life on Jupiter, but it seemed horrifying that the sun might not be the center of creation, that the countless stars might be other suns with many other worlds around them. He dreaded the views of Bruno.

In December 1610, Galileo announced a startling new discovery. Venus exhibited phases, like the moon. He presented this finding as strong evidence in favor of Copernicus.

Astronomers since antiquity knew that Venus always stayed close to the sun. Supposedly Pythagoras had discovered that what appeared to be two separate stars—the morning and evening stars, each appearing near the sun—were actually one planet: Venus.[47] Even if Venus were travel-

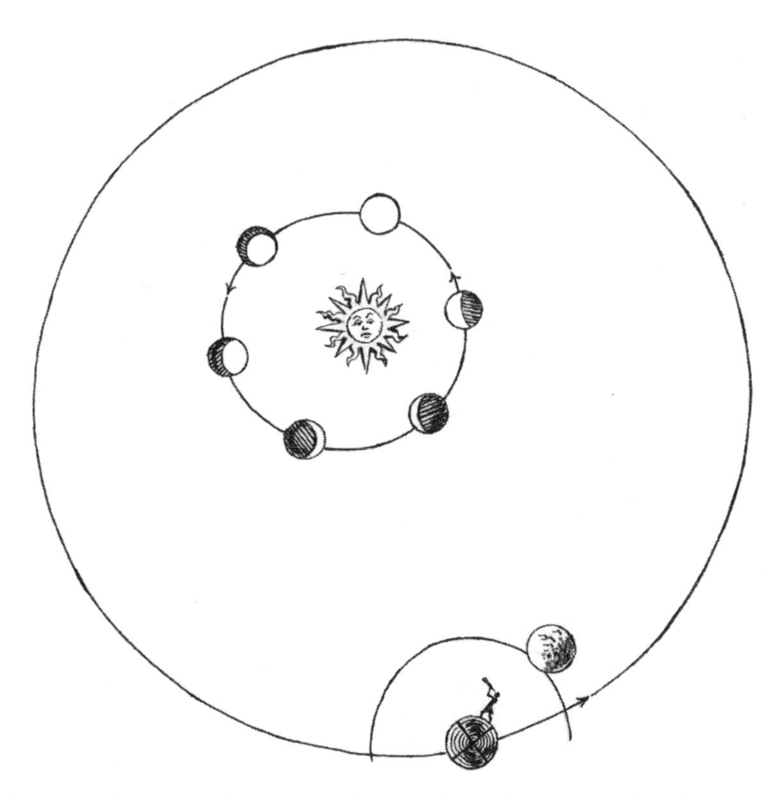

Figure 2.9. The system of Copernicus, showing phases of Venus as it orbits the sun as seen from Earth.

ing on an epicycle, it would never be on the side of the heavens opposite to the sun. Therefore, in Ptolemy's system, Venus should never seem to reflect a fully sunny face onto Earth. By the 1600s, Ptolemy's system had been construed in various versions: in some, the orbit of Venus was beneath the orbit of the sun; in other versions, its orbit was above the orbit of the sun, and in others, Venus did not travel on any epicycle. In none of these geocentric accounts could Venus exhibit the sequence of phases that Galileo observed through his telescope. Instead, he saw phases corresponding to the motion of Venus around the sun. He concluded that the phases of Venus showed the truth of the Copernican system.

Galileo's findings were startling and quickly gained public attention, yet there was plenty of skepticism, and most people remained unconvinced. Some professors did not believe in the images produced by Galileo's telescope. Some even refused to look into his instrument. Thus he complained to Kepler, in a letter: "My dear Kepler, I wish we could laugh at the extraordinary stupidity of the mob. What say you about the foremost philosophers of this University, who with the obstinacy of a stuffed snake, and despite my attempts and invitations a thousand times they have refused to look at the planets, or the Moon, or my telescope?"[48] Still Galileo continued to refine his instrument and to make new observations. He used it to project images of the sun and found, surprisingly, that the sun had spots. It was not a perfectly uniform body as people thought. Plus, the spots were moving! He found that the spots moved all around the sun in about twenty-seven days and concluded that the sun must be spinning on its axis. Therefore, he inferred, this proved that at least some heavenly bodies do spin; thus, that Earth also spins, just as Copernicus had claimed. Again Galileo's claims were criticized. For example, Christoph Scheiner, a Jesuit, hypothesized that the spots were not on the sun itself, but consisted instead of swarms of moons.

Despite such disparagement, Galileo's refined image of the universe took shape. It was basically the account advanced by Copernicus, but Galileo added four moons onto Jupiter and also advanced the notion that the sun spins at the center of the universe. To some people, the Copernican account seemed to clash with the Bible, which includes passages that refer to the motion of the sun. Kepler and Galileo did not consider this a problem, because they argued that Holy Scripture was written in the language of human perceptions, just as today we still say that the sun rises in the morning. According to Kepler and Galileo, the Bible was not meant to teach about astronomy, and astronomy could be pursued in the service of Christianity, to better understand Scripture.

A problem, however, was that the various churches were fighting over how to properly interpret Scripture. Since 1546, in the Council of Trent, the Catholic Church had attributed to itself alone the right to interpret Scripture, contrary to the church reformers' claim that the Bible could be comprehended by individuals who read it literally. The "heretics" claimed the right to read God's book; Galileo, in 1613, claimed the right to read the book of nature, "this great book of the world," against the interpretations of Aristotle.[49] And Galileo's opinions were disrespectful to the Jesuits. Galileo was a devout Catholic, but he implied that astronomers and mathematicians could understand God's creation independently of the decrees of theologians.

Meanwhile, in Naples, a Father of the Carmelite Order wrote a defense of "the opinion of Pythagoras and of Copernicus." Father Paolo Antonio Foscarini admitted that their opinion that the Earth moves seemed to be "sheer madness" and "one of the oddest and most monstrous paradoxes"—*but*, that it was denied mainly by habit rather than reason and that new astronomical evidence might well show, again, that "the majestic white beards of the ancients were wrong; they have been believed too easily and their false imaginations solemnized."[50] Foscarini argued at length that the Pythagorean opinion could be reconciled with the various Biblical passages that seem to say that Earth does not move and so forth. He denounced the patchwork of spheres and circles: "epicycles, equants, deferents, eccentrics, and a thousand other fantasies and chimeras, which are more like beings of reason than real things."

Despite explicitly submitting his argument reverently to the judgment of the Holy Church, Foscarini quoted a line from the Epistles of Horace: "I am not bound to swear as any master dictates," because he argued that whereas the Church could not possibly err in matters of faith and salvation, it could err in practical and philosophical judgments.[51] Foscarini explained that the Bible includes various metaphors and common ways of speaking, such as when it refers to God as walking, having a face, eyes, anger, and similar images, or to death as eating, moving, having a voice, and having a shadow. He defended the "monstrous and extravagant" notion that hell, at Earth's center, revolves through heaven around the sun by insisting that the heavens of the planets are distinct from the spiritual heaven, which is above everything. In light of the findings of Copernicus, Kepler, and Galileo, Foscarini concluded that the opinion of Pythagoras was quite probable and not necessarily in conflict with Scripture.

Responding to accusations against Foscarini and Galileo, the Catholic Church intervened. The Inquisition met in 1616 to discuss the Copernican theory. Cardinal Roberto Bellarmino, who had been the inquisitor in the deadly trial against Giordano Bruno, participated. There is a recurring myth to the effect that the Catholic theologians were mainly annoyed by the notion that humanity was not literally the center of God's creation—but historical research shows no evidence of that.[52] The center of the universe, inside Earth, was not a particularly privileged place to Catholics such as Bellarmino, who believed it to be the actual location of hell.[53] The Copernican scheme seemed offensive because Galileo advocated it as a source of authority for interpreting Scripture, against the interpretations of the Church leaders. Cardinal Bellarmino noted that, although the subject matter of Earth's motion was indeed not a subject of faith, it was yet a matter of *faith on the speakers*: the apostles, prophets, and commentators, who had all agreed that the sun is in the heavens and moves around Earth. Bellarmino argued that, there being yet no proof, no demonstration of Earth's motion, it was inappropriate to abandon the traditional interpretation of Scripture or to depart from the Council of Trent.[54]

In 1616, the Holy Congregation denounced "the false Pythagorean doctrine, altogether contrary to Holy Scripture."[55] The cardinals concluded that the proposition of a stationary sun is "foolish and absurd in philosophy, and formally heretical since it explicitly contradicts in many places the sense of Holy Scripture, according to the literal interpretation of the words."[56] Therefore, the sun-centered theory could not be defended or held. The Holy Congregation promptly banned the books of Copernicus and Zuñiga until each could be corrected. More forcefully, to defend Catholic dogma, the Holy Congregation "completely prohibited and condemned" Foscarini's Pythagorean booklet and all future works like it.[57] Similarly, inquisitors banned Galileo from believing in Copernican ideas, and Cardinal Bellarmino gave Galileo a certificate forbidding him from believing or defending the idea that Earth moves. But Galileo surmised that he could at least entertain and teach the theory as a speculative hypothesis. After all, the clergymen acknowledged that the scheme of Copernicus had certain mathematical advantages: it simplified calculations of the positions of the planets.

Meanwhile, Kepler faced other troubles. His wife suffered from Hungarian spotted fever and epileptic seizures. Three of his children had died, and later his wife died too. He remarried, but then three more newborns died. He also had religious problems. Kepler was a Lutheran but he wanted a reconciliation among the divisions of Christianity: "It makes me heartsick that the three big factions have so miserably torn up the truth among themselves."[58] Kepler refused to sign the Formula of Concord, the orthodox beliefs that were allegedly required of all Lutherans. (In fact, some Lutherans did not accept it, almost half of the ministers who prepared the document did not sign it.) He mainly disagreed with their claim that the physical body and blood of Jesus Christ actually combines with the bread and wine in the ceremony of the Eucharist. The Lutherans had denied the Catholics' idea of transubstantiation: that bread and wine literally become the body and blood of Christ. The Lutherans asserted instead the so-called "sacramental union" (or consubstantiation): that four substances combine during the ceremony: bread with the body of Christ, and wine with His blood. Kepler disagreed and sympathized instead with the Calvinist idea that the Eucharist was essentially a symbolic commemoration infused by the spiritual presence of Christ. Kepler still wanted to take communion; he argued that this was a minor disagreement. But the Lutheran ministers disagreed. In 1619, they excommunicated him from their Church, to his chagrin.[59]

That same year, the Catholics banned Kepler's Copernican writings. Not only had Kepler affirmed the reality of the Copernican system, he openly pondered numerology and voiced his theological opinions. He argued that mathematics contains hidden divine knowledge: "Plato teaches us many remarkable things about the nature of the gods through the appearance of mathematical things; and the Pythagorean philosophy disguises its teaching on divine matters with these, so to speak, veils."[60] Kepler conjectured that the ancient Pythagoreans had known and hidden the connection between five regular solids and the order of the planets.[61] He also argued against Aristotle and Cicero for having denied that the heavens generate inaudible harmonies: "These preconceived opinions are a considerable obstacle to readers who are striving towards the inner secrets of Nature, and could frighten off many who have great powers of judgment and are seekers after truth, to such an extent that they would disdain those Pythagorean pipedreams, scarcely recognized at arm's length, and throw away the book unread."[62]

At the same time, Kepler had to deal with legal attacks upon his mother, who was accused of being a witch. His great-aunt had earlier been burned at the stake for witchcraft. And for seven years, superstitious townspeople and neighbors spread gossip and suspicions against Kepler's wrinkled old mother. "The lies had been repeated so many times by so many people that they had begun to be accepted as truth."[63] His mother—allegedly—had poisoned people with potions, she rode a calf to death, she terrorized livestock, and she had injured and killed children by touching them. Gossip became admitted into prosecution records. Actually, she had at least done a few odd acts, such as asking a gravedigger for her father's skull, which was illegal, to make a goblet for her son. For years, Kepler defended his mother against forty-nine articles that could have led to her execution. Still, she was sentenced to be verbally terrorized by the executioner showing his instruments of torture so that she would confess to witchcraft. But she did not. Finally they exonerated her of the charges. But then promptly she died in 1622. And in 1623, another newborn son of Kepler died.

Meanwhile, Galileo continued to pursue the Copernican theory as if it were true, despite having said that he would accept the decrees of the Inquisition. He also continued to argue as if there were paths to truth independent of the authority of the Catholic Church. In 1623, he wrote:

> Philosophy is written in this grand book, the universe, which stands continually open to our gaze, but the book cannot be

> understood unless one first learns to comprehend the language and read the letters in which it is composed. It is written in the language of mathematics, and its characters are triangles, circles, and other geometric figures without which it is humanly impossible to understand a single word of it; without these, one wanders about in a dark labyrinth.[64]

Again, he seemed to be saying that mathematicians such as himself had a special skill for understanding nature, the work of God.

At the same time, a friend of Galileo, Maffeo Barberini, was elected as the new leader of the Catholic Church, becoming Pope Urban VIII in 1623. Galileo subsequently was kindly invited to papal audiences on six occasions. Thus, he seems to have grown bolder in his confidence that the Copernican account might eventually be favored. Meanwhile, Kepler, though ill, planned to finally publish his belated but much expanded "dream" about the moon. But he died in 1630, survived by his second wife and six children.

In 1632, the sixty-eight-year-old Galileo published his *Dialogue on the Two Chief Systems of the World—Ptolemaic and Copernican*. In it, Galileo portrayed a discussion among three characters, one of whom argued persuasively in favor of the Copernican theory. The character who defended the Aristotelian views appeared as an idiot. Promptly thereafter, Galileo was summoned by the Inquisition to a trial in Rome to determine whether he had violated the conditions of its 1616 decrees.

Pope Urban VIII became bitterly enraged that Galileo had deceived him, that Galileo "dared entering where he should not have, into the most serious and dangerous subjects which could be stirred up at this time. . . . matters, involving great harm to religion (indeed the worst ever conceived), . . . . the most perverse subject that one could ever come across."[65] Likewise, representatives of the Society of Jesus said that Galileo's vile book was "more harmful to the Holy Church than the writings of Luther and Calvin."[66] *But why?* How could anyone say that? Were they just exaggerating? The Catholic Church had lost *half of Europe* because of the influence of Luther and Calvin. How could a scientific question such as Earth's motion seem in any way more dangerous than the transgressions of the heretics? Having studied this issue for years, I still could not understand why. Finally I realized that one particular word has some neglected significance. *Pythagorean.*

Consider again the Decree of the Inquisition of 1616, which banned "the false Pythagorean doctrine, altogether contrary to Holy Scripture." Historians and scientists have read this as an allusion to an ancient tradition of mathematicians, scientists and astronomers.[67] As usual, writers give no evidence that Pythagoras really ever claimed that the Earth moves; as there is none. But what about the religious connotations? If instead the sentence read, "the false *Lutheran* doctrine, altogether contrary to Holy Scripture," it would be evident that there was some religious issue here. But since the Pythagorean profile has been reinterpreted over centuries, this religious dimension seems to have become invisible.

Nowadays, Pythagoras is mostly portrayed as an ancient Greek mathematician. But that is a fame that he did not clearly deserve. Instead, in ancient times, and for many centuries, he was known primarily as the leader of a secretive religious cult that believed in multiple lives after death. They praised Pythagoras as a demigod.

Perhaps the Catholic theologians who disparaged Galileo's heliocentric ideas as "Pythagorean" surmised a repulsive threat. Pythagoras was an alien pagan influence: he worshipped Greek gods, especially the sun god, Apollo.[68] One ancient poem claimed that Pythagoras was the son of Apollo, who visited his mother: "Pythagoras, whom Pythias bore for Apollo, dear to Zeus, she who was the loveliest of the Samians."[69] Iamblichus claimed that Pythagoras was "the most handsome and godlike of those ever recorded in history."[70] Above all, Pythagoras was a super-human sent from the domain of Apollo to enlighten mortals to live properly. Early Christian Church fathers vigorously criticized Pythagorean philosophy as heresy and lies, in which "the Creator of all alleged existence is the Great Geometrician and Calculator a Sun; and that this one has been fixed in the whole world"—and they denounced various heretics as "disciples not of Christ but of Pythagoras."[71] Whereas Christians believed that Jesus and *only* Jesus had reincarnated, supposedly Pythagoras had been reborn several times. And Pythagoras taught that other human souls are also reborn repeatedly, even in animals. Allegedly he taught that human souls come from the Milky Way, where the infernal regions begin, and that animal souls come from the stars.[72] And he apparently taught that we should not eat flesh, but the Bible told believers to eat meat, and at Mass Christians ate the body of Christ. Al-

**Table 2.1** The Pythagoreans were portrayed as mistaken on astronomy, or, as ancient authorities who knew the true structure of the universe

| | | |
|---|---|---|
| ca. 350 BCE | Aristotle | The Pythagoreans wrongly say that Earth is one of the stars and that, moving in a circle around a central fire, it produces night and day. |
| ca. 220 BCE | Archimedes | Aristarchus wrongly hypothesized that the fixed stars and the sun remain unmoved while Earth revolves about the sun. |
| ca. 150 CE | Ptolemy | Earth is at the center of the universe and the planets move around it in eccentric circles and epicycles. There exists a discernible celestial harmony. |
| ca. 150 CE | "Plutarch" | Some Pythagoreans claimed that stars are worlds in infinite space. |
| ca. 300 CE | Iamblichus | Through Pythagoras there came to be a true understanding of everything in the universe, including movements of the spheres and stars, eclipses, eccentrics and epicycles. Pythagoras could hear the universal harmony and music of the spheres and of the stars. |
| *Twelve centuries later . . .* | | |
| ca. 1540 | Copernicus | The secretive Pythagoreans rightly argued that Earth moves around the sun. |
| 1572 | Thomas Digges | The Pythagoreans had a perfect description of the celestial orbs. |
| ca. 1570 | Giordano Bruno | The Pythagoreans argued that souls are repeatedly reborn, even into animals. Stars are worlds in infinite space. |
| ca. 1590 | Tycho Brahe | The Pythagoreans wrongly believed that solid impenetrable orbs separate the planets' orbits. |
| 1590s | Johannes Kepler | The Pythagoreans appreciated the five regular solids in the order of the universe. There is harmony in the motions of the planets. And "perhaps his soul [Pythagoras] has transmigrated into me [Kepler]." |
| 1611 | Galileo Galilei | Pythagoras rightly believed that Earth and the planets orbit the sun. |
| 1616, 1632 | Catholic theologians | Pythagorean ideas are vile and dangerous and should be condemned. |

legedly Pythagoras had performed many miracles too.[73] His supposed teachings and divine revelations were featured, for instance, in Ovid's pagan poem *Metamorphoses*.[74] Diogenes Laertius noted that Pythagoras theorized that "the Sun, and the Moon, and the stars were all Gods" and echoed that Pythagoras "was actually the god Apollo, and that he had spent two hundred and seven years in the underworld, had there seen all the men who ever died ("he saw the soul of Hesiod bound to a brazen pillar, and gnashing its teeth; and that of Homer suspended from a tree, and snakes around it, as a punishment"), that he remembered his sufferings in hell—and that Pluto, god of the dead, ate only with the Pythagoreans.[75]

Porphyry too, a staunch advocate of polytheism, celebrated Pythagoras as a moral and divine superhuman, more eminent than anyone.[76] In one of his earliest works, Porphyry claimed that the god "Apollo exposed the incurable corruption of the Christians, saying that the Jews, rather than the Christians, recognized God."[77] Porphyry's notorious fifteen-volume work *Against the Christians* was banned by the first Christian Roman emperor; nearly all copies were destroyed by the Christians, burned, only fragments remain.[78] Because of his extensive, historical-critical attacks, Porphyry became known as one of the greatest early enemies of Christianity; his name became synonymous with blasphemy. When Pope Leo X condemned the writings of Martin Luther as "blinded in mind by the father of lies," the Devil, he denounced Luther as "a new Porphyry."[79] Saint Augustine, in his widely read *City of God Against the Pagans*, had criticized Pythagoras for being a necromancer, one who tried to divine the future by communicating with inhabitants of the netherworld: the dead, or demons pretending to be gods.[80]

Allegedly Pythagoras used numbers to divine the future.[81] A Christian satirist ridiculed Pythagoras for claiming that souls are made of "number in motion."[82] One of the most famous professed followers of Pythagoras, Apollonius of Tyana, a reputed prognosticator and exorcist, had been repeatedly denounced as practicing demonic magic to imitate Christ's miracles.[83] In the late 1500s, the Vatican launched an attack on the occult arts. At the time, various books on magic discussed the powers of Pythagoras.[84] Fortune-tellers used the so-called "Wheel of Pythagoras" for divination.[85] They also used "Pythagorean" numerology that related the letters of the alphabet to the planets, the days of the week, and the signs of the zodiac.[86]

These extensive, esoteric connotations could scarcely be avoided by simplistically referring to Pythagoras as a geometer. Even the word *mathematician*, so proudly used by Galileo and Kepler, still involved implications of astrology, occult numerology, and divination. Saint Augustine had cautioned: "The good Christian should beware of mathematicians and anyone who practices unholy divination, most especially if they speak truly, lest the soul be ensnared by consorting with demons."[87]

Hence, for centuries, some Catholic theologians construed the Pythagorean outlook as overtly anti-Christian. In 1616, the Inquisition "completely prohibited and condemned" Father Foscarini's scriptural defense of the "New Pythagorean System of the World," a rejection more forceful and damning than the injunctions against Copernicus and Galileo. Likewise, in 1622, a book on Pythagorean symbols was banned by the Index of the Inquisition for religious reasons.[88] Thus it is understandable that some Jesuits denounced Galileo's book as more vile and harmful than the writings of the heretical reformers. At least the Protestants did not defend a pagan system of the world.[89]

Contrary to the certificate that Galileo had received from Cardinal Bellarmino, the Inquisition now claimed that Galileo had been granted no implicit permission to teach and discuss the Copernican theory. They professed a document to that effect, though it lacked the signature of Bellarmino, who had died in the meantime. Then, in the trial depositions, Galileo lied: he pretended that in his *Dialogue* he was "refuting" the theory of Copernicus, showing that it was "invalid and inconclusive," with only the "purest intention."[90] Officials of the Inquisition then showed, to the contrary, that his book vigorously defended and taught the theory that he actually seemed to believe. The Inquisition ruled that Galileo indeed had trespassed against the Church.

The Holy Roman Inquisition then declared Galileo guilty of heresy—punishable with torture, prison, or death. They forced him to kneel and recant his claims. He then spoke: "with a sincere heart and unfeigned faith I abjure, curse and detest the above-mentioned errors and heresies [the Copernican account], and in general each and every other error, heresy, and sect contrary to the Holy Church."[91]

Yet according to legend, as he finally stood up Galileo also muttered "Still it moves." But there is no historical evidence supporting that, which would have been an utterly brash and dangerous act. Apparently, the story of Galileo's defiant words first appeared in print in 1757, in English, by a native of Turin, Giuseppe Baretti, who wrote: "The moment he was set at liberty, he looked up at the sky and down to the ground, and, stamping with his foot, in a contemplative mood, said *Eppur si move*."[92] There is no evidence of this, but a painting from the 1640s portrays an old Galileo in a dark dungeon holding a nail with which he has scratched a few figures on the wall, Earth circling the sun, and the words: *E pur si muove*.[93]

The cardinals condemned Galileo to lifelong imprisonment. They also banned his book. He was not imprisoned in a dungeon, although that myth, like the painting above, has prevailed.[94] He was allowed to serve his sentence by staying confined to his house in Florence, watched by guards. In the summer of 1633, Melchior Inchofer, a Jesuit theologian who had testified against Galileo published a book justifying the Catholic opposition to the heliocentric theory, "to rally everyone as soldiers of religion." Among various objections, Inchofer complained that "since the Pythagoreans have gradually come to oppose the faith, it must be shown that the truth is found in the Scriptures, and as our major authors knew, is opposed to them." He required that "the Copernican theory and its related Pythagorean philosophy should not be taught at all." The Commissary General of the Roman Curia promptly approved the book for publication, noting, "This theologian has given a Christian refutation of these Pythagoreans. And he shows rightly that mathematics and the human sciences should be subordinated to the rule of Sacred Scripture."[95]

The traditional story about Galileo says that in ancient times, Pythagoras argued that Earth and the planets orbit the sun; his theory was later adopted and refined by Copernicus, and it led Galileo to clash with the Catholic Church. This story is defective because half of it is fictitious. Instead, we can replace it with the following: Galileo attributed the heliocentric theory to Pythagoras, but this association entailed pagan connotations that could hardly help its acceptance among Catholics. This sentence does not summarize the main aspects of the Galileo affair, but it is the sort of thing we can fairly say if we wish to remark on its connection to a Pythagorean context.

Three and a half centuries later, the Catholic Church admitted that theologians had made some errors in the trial of Galileo.[96]

In any case, we have seen that in the phases of Venus, for example, Galileo found convincing evidence against the accounts of Aristotle and Ptolemy. Given that, plus all his other findings and arguments, had he actually proven the truth of the Copernican scheme? Not at all. Tycho Brahe's account explained the phases of Venus equally well. Galileo's strategy in his *Dialogue* had been essentially to ignore Brahe's account. Yet Jesuit astronomers were well aware of it, and they knew that it served well enough to accommodate Galileo's findings. By the 1620s, the Jesuit astronomers had adopted Brahe's system. While despising the "Pythagorean" scheme, these Catholics had adopted a worldview contrived by—a Lutheran.

As for the moons of Jupiter, Ptolemy had not predicted them, but neither had Copernicus. They could be added onto any account, though ruining the Aristotelian claim that there is only one center of motion in the universe. Moreover, some of Galileo's favorite mathematical and physical arguments that purportedly showed Earth's motion were wrong. In particular, his *Dialogue* argued as though the most compelling proof of Earth's motion were the oceans' tides. Galileo reasoned that if Earth were perfectly still, the waters would also be still, and that, like any vessel carrying water, its motion affects the water. He knew that Kepler argued that the changing tides were correlated to the motion of the moon. But Galileo denied it. Kepler was right, yet Galileo dismissed those arguments as if Kepler were childishly speculating about occult astrological influences of the heavenly bodies on terrestrial phenomena. Thus Galileo's main "evidence," supposedly showing Earth's motion, was wrong.

Hence, astronomers and the Catholic Church were not irrational in criticizing Galileo, for he advocated more than was certain at the time. Again, if Earth spins constantly to the east, then why is it that when things are thrown up into the sky they do not deviate from a straight downward path? Galileo had no good answer. We now know that, actually, things do deviate as they fall, only very slightly so. This effect was demonstrated beautifully in Paris in 1851 by Jean Bernard Léon Foucault. He showed that as a pendulum swings repeatedly, its direction changes gradually throughout the day, just as if Earth indeed spins to the east. An example is that if you were in a moving car, and you tossed something straight forward toward the windshield, just as the driver turned the steering wheel to the left, you would see that the object you tossed would not hit the spot where you aimed it, but would tend instead to the right. Likewise, as a pendulum swings, its direction seems to change very slightly, which we attribute to Earth's spin. Lacking such results, Galileo had no clear proof of Earth's motion, just many interesting findings and analogies, along with some bad ones. Just as Aristotle and Ptolemy were wrong about many things, so were Copernicus and Galileo. The sun is not immobile, it is not at the center of the universe. Contrary to their expectations, the sun *is* a star. The stars are not immobile, either; they are not embedded in a sphere. And importantly, the orbits of the planets are not circular.

Returning to the start, Aristotle had explained that if Earth moves we should be able to detect some shifts in the relative positions of the stars. Like Brahe, Galileo saw no such shift. Such effects do exist, but Galileo's telescopes were far too weak to detect them. Thanks to the use of improved instruments, Friedrich Bessel, a school-dropout largely self-trained in astronomy and math, successfully detected and measured stellar parallax in 1838.[97] Aristotle was right! If Earth really moves, the stars *should* seem to shift—and they do.

# 3

## *Newton's Apple and the Tree of Knowledge*

LET'S dispel another widespread myth: that Newton was born in the same year Galileo died. This mistake continues to appear in recent literature, such as in this example: "He died in 1642, within a few days of Isaac Newton's birth."[1] Writers often claim that both events happened in 1642, which, by the way, would fit nicely with the Pythagorean idea of the transmigration of souls. I have even found some writers who claimed that Newton "was born on the very day on which Galileo died."[2] As other writers know, the mistake stems from using the Gregorian calendar to date the death of Galileo while using the old Julian calendar to date the birth of Newton. Actually, there transpired almost a year's difference. In the Gregorian calendar, Galileo died on 8 January 1642 and Newton was born on 4 January 1643. In the old Julian calendar, Galileo died on 29 December 1641 whereas Newton was born on Christmas Day, 25 December 1642.

It's actually necessary to write the exact dates, as otherwise, we meet additional confusions. For example, bestselling physicist Stephen W. Hawking wrote the following: "Galileo died on 8 January 1642, exactly three hundred years before the day I was born. Isaac Newton was born on Christmas Day of that year in the English industrial town of Woolsthorpe, Lincolnshire. He would later become Lucasian Professor of Mathematics at Cambridge University, the chair I now hold."[3] This wonderful passage has several mythic dimensions. The coincidences of space and time seem to imply a connection between Galileo, Newton, and Hawking. But Hawking was confused about Newton's birth; using the Gregorian calendar it was not on Christmas Day and it was not the same year as when Galileo died. Woolsthorpe was not an "industrial town," it's a small village. Note also the use of the phrase "He would later become," as if Newton was destined at birth to become great. Other writers echo the mistake about dates, and some even do it on purpose. One textbook, using the same dates given by Hawking, adds this footnote: "Because England had not yet reformed its calendar, 25 December 1642, in England was 4 January 1643, in Europe. It is only a small deception to use the English date."[4] Instead of just fixing the factual mistake, the editors preferred to keep it.

Anyhow, let's talk about the apple. For more than two centuries, hundreds of commentators have written thousands of words about the story of Newton's apple. But most of these are based on very little evidence. Even good biographers who cite evidence tend to just allude to documents, rather than quote them, and they ignore or omit various sources. Since there exists no comprehensive account of the historical evidence and its early interpretations, I will now give one.[5]

By *comprehensive*, I do not mean *exhaustive*, I merely mean that I've assembled more evidence on the matter than is available in any other source. To do so, I have abstained from any of the many psychological, speculative, and literary themes that many writers develop when referring to this topic. While I find many of those commentaries to be engaging and insightful, I will focus on a plain accounting of documentary evidence. I hope that the following material helps to facilitate the systematic study of myths in science and how they grow.

In 1662, the nineteen-year-old Isaac was experiencing intense religious concerns. To confess his sins in a private, hidden way, he wrote them in a brief, cryptic code. He listed sins that, across the years, he had occasionally committed against God. First on Newton's list was "Using the word 'God' openly"; he also included, "Not loving Thee for Thy self," "Not desiring Thy ordinances," "Fearing man above Thee," and "Caring for worldly things more than God."[6] Newton also listed various acts that he should not have done on God's day: twisting a cord, making a mousetrap, making pies, idle chatting, squirting water, swimming, and, second on his list: "Eating an apple at Thy house." He also included graver sins: lying, stealing, robbing his mother's box of plums and sugar, putting a pin in someone's hat to prick him, punching his sister, "Striking many" people, wishing death to some, and threatening to set his stepfather and mother on fire: "Threatening my father and mother Smith to burne them and the house over them."

The fiery threat dates from more than ten years earlier. His actual father had died shortly before his birth, and the boy despised his mother's second husband: the Reverend Barnabas Smith, who, anyhow, died before the boy's eleventh birthday. The fiery threat was the thirteenth sin on his list; eating an apple in Church was, again, second.

In the summer of 1665, at twenty-two years of age, Isaac kept away from his college, Trinity at Cambridge, to escape from the spreading crisis: the bubonic plague that was killing thousands of people throughout England. He subsequently spent about two years in relative isolation. While living at his family farm in Woolsthorpe at Lincolnshire, he spent time thinking and working on problems in physics and mathematics.

One of the most famous stories in the history of science is that Newton was inspired to think about universal gravitation by seeing an apple fall in his garden at Woolsthorpe in 1666. Some writers, for various reasons, accept this story. Other writers dismiss it as mere legend. Yet there is evidence that Newton himself told it. In a manuscript, Newton's friend William Stukeley reported that on 15 April 1726, he dined with the then-very-old Newton and that:

> after dinner, the weather being warm, we went into the garden, & drank tea under the shade of some appletrees; only he & my self. amidst other discourse, he told me, he was just in the same situation, as when formerly, the notion of gravitation came into his mind. "why should that apple always descend perpendicularly to the ground," thought he to himself; occasioned by the fall of an apple, as he sat in a contemplative mood. "why should it not go sideways, or upwards? but constantly to the earth's center? assuredly, the reason is, that the earth draws it. there must be a drawing power in matter. & the sum of the drawing power in the matter of the earth must be in the earth's center, not in any side of the earth. therefore dos this apple fall perpen-

dicularly or toward the center. if matter thus draws mater; it must be in proportion of its quantity. therefore the apple draws the earth, as well as the earth draws the apple." that there is a power like that we here call gravity, wch extends its self thro' the universe.[7]

Figure 3.1. The young Isaac Newton in the garden.

Newton died in 1727, aged eighty-four. That same year, Robert Greene reported in print, in Latin, that his friend Martin Folkes (vice president of the Royal Society when Newton was president) had told him that Newton's idea of universal gravity was inspired by an apple: "which famous proposition, all things considered, originates, as disclosed to our knowledge, from an apple; that which I learned from a most ingenious & most learned man, and also the finest, and friendliest to me, *Martin Folkes* Esquire, truly meritorious Fellow of the Royal Society."[8] Similarly, a memo drafted by Newton's friend John Conduitt in 1727 or 1728 reported, "& in the year 1665 when he retired to his own estate on account of the Plague he first thought of his system of gravity which he hit upon by observing an apple fall from a tree—."[9] In another draft of the same, Conduitt wrote "he discovered his system of gravity / he took the first hint of it from seeing an apple fall from a tree."[10] And in yet another manuscript, Conduitt wrote (crossing out some words, as noted, and inserting others, noted here between slashes):

> In the year ~~1666~~ he retired /again/ from Cambridge on acc$^{t}$ of the plague to his mother ~~at Boothby in~~ Lincolnshire & whilst he was musing in a garden it came into his thought that the /same/ power of gravity (w$^{ch}$ ~~brought~~ /made/ an apple /fall/ from the tree to the ground) was not limited to a certain distance from the earth but ~~that this power~~ must extend much farther than was usually thought—Why not as high as the Moon said he to himself & if so /that must influence/ her motion ~~must be influenced by it~~ /&/ perhaps retain her in her orbit, . . .[11]

One more writer recorded the story at this early time. And in a way, it is the most epic, dramatic version. When Newton died, Voltaire was visiting England. He spent time with a group of individuals who were more or less close to the famous old physicist, including his niece. And Voltaire promptly reported what he heard about Newton. Like Greene, he first published it 1727, writing in English. Voltaire was praising the great skills of the poet John Milton when he inserted one sentence about Newton. Voltaire wrote that Milton, in his youth, was traveling in Italy and saw a dreadful play. It was a comedy about "the *Fall of Man*; the Actors, God, the Devils, the Angels, *Adam*, *Eve*, the Serpent, Death, and the seven mortal Sins."[12] Voltaire recounted that the play began with an extravagant chorus of angels discoursing about the rainbow, planets, time, and winds, all making music, rising in a "Profusion of Impertinence." He wrote that Milton then saw through the absurdity of that show, and sensed the great, hidden Majesty of the subject, which years later, led to his epic poem, *Paradise Lost*. Right then, Voltaire mentioned Newton and the apple, as he discussed Milton's response to the extravagant play:

> He [Milton] took from that ridiculous Trifle the first Hint of the noblest Work, which human Imagination hath ever attempted, and which he executed more than twenty Years after.
>
> In the like manner, *Pythagoras* ow'd the Invention of Musick to the Noise of the Hammer of a Blacksmith. And thus in our Days, Sir *Isaac Newton* walking in his Gardens, had the first Thought of his System of Gravitation, upon seeing an Apple falling from a Tree.
>
> If the Difference of Genius between Nation and Nation, ever appear'd in its full Light, 'tis in *Milton's Paradise Lost*.
>
> The *French* answer with a scornful Smile, when they are told there is in *England* an *Epick* Poem, the Subject whereof is the Devil fighting against God, and *Adam* and *Eve* eating an Apple at the Persuasion of a Snake.[13]

Thus the story about Newton and the apple appeared in print, layered between Milton, Pythagoras, the Devil, God, Adam, Eve, and the snake. One could hardly wish for a more wonderful public birth for this scientific tale! A few years later, without writing about Adam or Eve, Voltaire elaborated his account of Newton's creative moment:

> He retreated in 1666, because of the plague, to the countryside near Cambridge, one day he was walking in his garden, & he saw fruits fall from a tree, he let himself go into a deep meditation about such Gravity about which all the philosophers have

> sought the cause for so long a time in vain, & in which the common people do not even sense the mystery; he said to himself, from any height in our hemisphere that these bodies fall, their descent will certainly be at the rate discovered by Galilei; & the spaces covered by them will be as the squares of the time. This power which makes heavy bodies descend, is it the same without any sensible diminution to any depth that one be in the Earth, & atop the highest mountain; why might not that power extend right up to the Moon? And if it is true that it reaches that far, is there no great likelihood that this power holds her in orbit & determines her motion?[14]

Here, Voltaire just wrote about "fruits," rather than specifying an apple. He noted that he had learned of this event from Newton's niece, Catherine Barton, wife of John Conduitt (the couple lived with Newton in his house in London, until he died).[15]

One more report is worth mentioning. Henry Pemberton had interviewed Newton about the thoughts that led him to his theory of gravity. And Pemberton too reported:

> As he sat alone in the garden, he fell into a speculation on the power of gravity: that, as this power is not found sensibly diminished at the remotest distance from the center of the earth, to which we can rise, neither at the tops of the loftiest buildings, nor even on the summits of the highest mountains; it appeared to him reasonable to conclude, that this power must extend much farther then was usually thought; why not as high as the moon, said he to himself? And if so, her motion must be influenced by it; perhaps she is retained in her orbit thereby.[16]

Pemberton did not mention an apple, but he reported that Newton had inspirational thoughts while alone in the garden.

An important aspect of the apple story is that, over time, in contrast to other questionable stories, evidence for it has *increased*. In 1831, David Brewster published a biography of Newton, and he noted, "The anecdote of the falling apple is mentioned neither by Dr Stukely nor by Mr Conduit, and, as I have not been able to find any authority for it whatever, I did not feel myself at liberty to use it."[17] Later, Brewster did find one account by Conduitt; and others later found Stukeley's detailed, handwritten account.

Henry Pemberton, William Stukeley, John Conduitt, Catherine Barton, and Martin Folkes all apparently claimed that Newton said that he was inspired at the garden in 1665 or 1666, and apparently four of them referred to a fruit or apple. If we had a document by Newton himself, we would regard it as a firsthand account; hence, two of our sources are at least secondhand (Stukeley and Conduitt), and two others (Voltaire and Greene) are at best thirdhand accounts. Still, the nine manuscripts and published accounts converge on one conclusion: the story came from Newton.

It might seem as if I just asserted that really Newton was inspired by a falling apple in the year of the plague, 1666. But note that I did not actually claim that. We have no good reason to assume that Voltaire and others accurately described the sequence of thoughts, the chain of reasoning, that apparently was triggered by an apple, sixty years earlier. Instead, there are plenty of manuscripts that historians have used to fairly trace Newton's gradual progress. And in any case, it took Newton many years of work to formulate his mathematical theory of universal gravitation.

Another problem is that in order to gauge Newton's honesty, we need to question his originality. It is inspiring to imagine that the young lad, sitting alone in his garden, was the first person to imagine that the force of gravity extends all the way up to the moon. But actually, such arguments were relatively common in astronomy at the time. For example, in 1609, Kepler published a work in which he explained gravity and the tides as follows:

> If the Earth should cease to attract its waters to itself, all the sea waters would be raised, and would gush up to the body of the Moon. . . . The sphere of influence of the attractive virtue in the Moon extends out onto the Earth, and it lifts up the waters in the torrid zone. . . . Therefore, if the Moon's attractive

> virtue extends as far as the Earth, it follows with greater reason that the Earth's attractive virtue extends to the Moon, and far beyond; and thus, nothing that consists of earthly material whichever, though lifted up to any height, can ever escape the powerful action of this attractive virtue.[18]

There it was, already—the idea that gravity extends beyond the height of mountains, up to the moon, and that there is a mutual attraction between Earth and the moon. Following William Gilbert, Kepler had tried to explain gravity as caused by a magnetic power within bodies. Kepler's work was one of the most widely circulated texts in astronomy. Kepler did not claim that the planets attract one another, only kindred bodies such as Earth and the Moon. Kepler did not have a notion of universal gravity, but Newton's reportedly initial reflections about the apple were quite consonant with previous astronomers' works.[19]

Another problem is that Newton was secretive, paranoid, and quite capable of making up stories to backdate his discoveries and to claim priority over someone else. For example, his law of gravity was so impressive that there were debates and speculations about its origins.

Independently of Newton's work, Robert Hooke guessed that the force of gravity decreases inversely as the square of the distance.[20] He inferred and complained, wrongly, that Newton had plagiarized his account.[21] Deeply annoyed, Newton proceeded to delete his previous acknowledgments to Hooke from his *Principia* and look instead for an older, more prestigious pedigree for his discovery. To Newton, the inverse square law seemed reminiscent of the brilliant accomplishments of the ancient philosophers. He surmised that Thales and Pythagoras knew well that some bodies exhibit magnetism, electricity, and gravity—acting at a distance upon one another—because they are all infused by the animating soul of God.[22]

Newton began to associate his notion of universal gravity with Pythagorean legends. Tradition claimed that the Pythagoreans had found that the same tension on a string half as long acts four times as powerfully. Newton conjectured that the Pythagoreans had hidden their knowledge about gravity and the solar system in such statements.[23] Newton knew of the old claim that Pythagoras experimented with intestines of sheep and sinews of oxen stretched by hanging weights, and conjectured that Pythagoras had discovered a musical proportion which Pythagoras: "applied to the heavens and consequently by comparing those weights with the weights of the Planets and the lengths of the strings with the distances of the Planets, he [Pythagoras] understood by means of the harmony of the heavens that the weights of the Planets towards Sun were reciprocally as the squares of their distances from the Sun."[24] Newton based his conjectures on claims by Macrobius: that "Pythagoras was the first of all the Greeks" to grasp that the rotations of the heavenly spheres emit harmonious sounds, and, that Pythagoras discovered "the great secret" of the numerical ratios of the fundamental musical concords by experimenting with hammers, stringed instruments, and intestines and sinews stretched by weights.[25]

It is merely fiction that Pythagoras knew the law of gravity. The accounts of how he allegedly discovered the musical ratios are also false because, contrary to the claims by ancient writers, replications of the described experiments do not reveal the effects sought. For example, the vibrations of strings are not proportional to the number of units of weight on the strings, but to the square roots of the units of weight. (Accordingly, Newton improved his version of what Pythagoras allegedly found in music.) Likewise, hammers of different weights do not necessarily produce a different tone or pitch. For example, it is difficult to distinguish the sound of a one-pound hammer striking iron on an anvil in contrast to a two-pound hammer.

Nevertheless, Newton credited Pythagoras. Accordingly, his friend Fatio de Duillier noted that Newton believed that Pythagoras, Plato, and others were well aware of the inverse square law of gravity, and had anticipated *all* of Newton's demonstrations.[26] John Conduitt noted in a manuscript that "Sir I. admired Pythagoras [and] thought his Musick was gravity."[27] Another confidant, David Gregory, went even further by attributing the law of gravity to Pythagoras by the same speculative conjectures, without evidence—in a textbook.[28]

This example illustrates how the authority of a successful scientist, Newton, enabled his arbitrary historical conjecture infiltrate a science

**Table 3.1** After Newton, the inverse square law seemed to branch to the future and to the past

| | | |
|---|---|---|
| ca. 430 CE | Macrobius | Pythagoras was the first of the Greeks to discover that the rotations of the heavenly spheres produce harmonies. |
| ca. 450 CE | Proclus | The Pythagoreans called the center "the prison of Jupiter." |
| 1609 | Johannes Kepler | The sun has a soul and it moves the planets, like a magnet, with a power that decreases inversely with the distance. |
| 1645 | Ismaël Boulliau | If a power from the sun holds the planets it is weaker inversely as the square of the distance, but no such power exists. |
| 1666–71? | Isaac Newton | The planets' "tendencies to recede from the Sun will be reciprocally as the squares of their distances from the Sun." |
| 1674 | Robert Hooke | "All Coelestial Bodies whatsoever, have an attraction or gravitating power towards their own centers . . . they do also attract all other Coelestial Bodies." |
| 1680 | Robert Hooke | "My supposition is that the Attraction always is in a duplicate proportion to the Distance from the Center Reciprocall." |
| 1666–70s | Isaac Newton | Universal gravity follows an inverse square law, from the centers of spherical bodies. |
| 1686–90s | Robert Hooke | Newton plagiarized the inverse square law. |
| early 1690s | Isaac Newton | "Pythagoras, on account of its immense force of attraction, said that the Sun was the prison of Zeus . . ." And Pythagoras realized "that the weights of the Planets towards Sun were reciprocally as the squares of their distances from the Sun." |
| 1692 | Fatio de Duillier | Newton believes that Pythagoras possessed all his demonstrations from the system of the world, based on gravity diminishing inversely as the squares of increasing distances. |
| 1715 | David Gregory | "Pythagoras . . . understood, as it were by the Harmony of the Heavens, that the Gravity of the planets towards the Sun (according to whose measures the Planets move) were reciprocally as the Squares of their Distances from the Sun." |
| 1767 | Joseph Priestley | "Electricity is subject to the same laws with that of gravitation, and is therefore according to the squares of the distances." |
| 1785 | Charles Coulomb | Experimentally, electricity follows an inverse square law. |

textbook, a common happening in the history of physics. Fortunately in this particular case, other authors did not copy Gregory's account, so Pythagoras did not become the widely acknowledged author of the inverse square law.

Now, Voltaire mentioned Newton while discussing *Paradise Lost.*

Milton's poem includes science: it refers to Galileo, it ponders whether the sun is the center, whether Earth has three motions, as Copernicus argued. It refers to the sun's magnetic "attractive virtue," like Kepler. In Milton's Garden of Eden, the serpent claimed to have eaten apples from the tree of knowledge of good and evil.[29] It gained speech and reasoned about "things visible in Heav'n or Earth." Leading curious Eve to the prohibited "root of all woe," the serpent praised this tree, "Mother of Science," which gave knowledge of causes. Eating, Eve became inebriated by the "sciential sap."[30] Let's summarize similarities between stories:

> Risking death, young Eve wandered in her Father's garden. She conversed with the serpent by the apple tree. Eve ate an apple and began to think differently about heaven and Earth. She shared this new knowledge; she proceeded to sin. Affected by a natural trifle, Eve ruined the harmony of the world.
>
> Avoiding the plague, young Isaac retreated to his mother's garden. He sat alone in contemplation, near an apple tree. An apple fell, Isaac began to think differently about heaven and Earth. Keeping this new knowledge secret, he proceeded to science. Affected by a natural trifle, Isaac found the harmony of the world.

In both stories, a God-fearing youth discovers extraordinary truth about the world thanks to an ordinary thing.

Shortly before dying, Newton apparently said: "I don't know what I may seem to the world; but as to myself, I seem to have been only like a boy playing on the sea-shore, and diverting myself in now and then finding a smoother pebble or a prettier shell than ordinary, whilst the great ocean of truth lay all undiscovered before me."[31] He seemed to echo Milton:

> Deep versed in books, and shallow in himself,
> Crude or intoxicate, collecting toys
> And trifles for choice matters, with a spunge,
> As children gathering pebbles on the shore.[32]

By 1827, an amusing claim arose: "When Newton read 'Paradise Lost,' he calmly remarked, 'It is a fine poem, but what does it prove?'"[33]

Figure 3.2. Eve, fruit, and the serpent.

It is very plausible that Newton saw an apple fall; for there were apple trees in his farm. Regardless of whether the event happened, it effectively has the shape of a myth: the simple start of a great development, a major transformation. It therefore resonated and gained currency. Consider now some traces of how the original account evolved.

In 1760, the mathematician Leonhard Euler wrote many letters explaining science to the Princess of Anhalt Dessau, niece of the King of Prussia. In one such letter, Euler conveyed a version of this story that later became common:

> This great English philosopher & mathematician, one day was laid down in a garden, under an apple-tree, an apple fell on his head, & gave him occasion to have several thoughts. He knew well that heaviness had made the apple fall, after it detached from the branch, maybe by the wind or by another cause. This idea seemed very natural, & maybe any peasant would have had the same thought, but the English philosopher went much further. . . . If Newton had not laid down in a garden under an apple-tree, & if by chance an apple had not fallen on his head, maybe we would still find ourselves in the same ignorance about the motion of heavenly bodies, & about an infinity of other phenomena that depend on that.[34]

Thus by 1760, it seemed that the apple had actually hit Newton on the head! Euler's letters were published and translated into several languages.

In 1791, a work was published anonymously, titled *Curiosities of Liter-*

*ature*. The author argued, "Accident has frequently occasioned the most eminent geniuses to display their powers."[35] He claimed that great poets, philosophers, and artists alike are made by sudden incidents:

> It is also well known, that we owe the labours of the immortal Newton to a very trivial accident. "When, in his younger days, he was a student at Cambridge, he had retired during the time of the plague into the country. As he was reading under an apple-tree, one of the fruit fell, and struck him a smart blow on the head. When he observed the smallness of the apple, he was surprised at the force of the stroke. This led him to consider the accelerating motion of falling bodies; from whence he deduced the principles of gravity, and laid the foundation of his philosophy."[36]

The anonymous author was Isaac D'Israeli, a Doctor in Civil Law at the University of Oxford. As an essayist, he became very popular, and was celebrated by George Byron (later Lord Byron), Sir Walter Scott, and other eminent men. Regarding his comments on Newton, apparently D'Israeli was quoting a prior account; it was not his own invention. Still, his *Curiosities* became widely reprinted and reissued.

Afterward, the poet Baron Byron composed several lines about Newton and the apple. The lines are part of his poem *Don Juan*, published in 1823. A beautiful thing about these lines, in hindsight, is that they foretell of journeys in spaceships to the moon:

> When Newton saw an apple fall, he found
> In that slight startle from his contemplation—
> 'Tis *said* (for I'll not answer above ground
> For any sage's creed or calculation—)
> A mode of proving that the earth turned round
> In a most natural whirl, called "Gravitation;"
> And this is the sole mortal who could grapple,
> Since Adam, with a fall, or with an apple.
>
> Man fell with apples, and with apples rose,
> If this be true; for we must deem the mode

In which Sir Isaac Newton could disclose
Through the then unpaved stars the turnpike road,
A thing to counterbalance human woes:
For ever since immortal man hath glowed
With all kinds of mechanics, and full soon
Steam-engines will conduct him to the Moon.[37]

Meanwhile, a few writers became annoyed by D'Israeli's account of Newton and of other historical figures. One writer suggested that D'Israeli should write about "the *Inconsistencies* of authors" and include himself as an example.[38] Another, the literary critic Bolton Corney, sought to refute D'Israeli's distortions of many topics. Corney had first read *Curiosities* at a young age and had greatly enjoyed it, but he was perplexed by its many errors when he again read it in its ninth edition of 1834. To rebut D'Israeli's account of Newton, Corney quoted the writings of Conduitt, Voltaire, Pemberton, and a few later writers. Corney complained against men who accept, value, and echo hearsay of hearsay. He said that it's important to correct the errors of popular works and that "it may also be expedient to unveil the deception and conceit of its author." He estimated the propagation of errors in this particular book's many editions: "we must conclude that D'Israeli, in this instance alone, *had misled more than twenty thousand of his readers!*"[39]

Instead of simply admitting his mistakes, the famous D'Israeli defended himself in a pamphlet.[40] He belittled Corney as incompetent without giving any evidence to ground his old claims. Corney replied: "I am not apt to be *reckless* when the truth is at stake; nor was I *reckless* on this occasion. . . . If I may now be permitted to hint to the *man of ideas,* that ideas unsupported by facts are mere day-dreams. . . . I have built my sixth-rate bark of the best materials within reach; have equipped her to the best of my humble skill for the service required; and she shall float on the stream of time in spite of the broad-sides of the *write-with-ease* school."[41] In this seemingly trivial and irrelevant corner of knowledge, someone fought seriously for a little bit of truth. In short, the famous man had argued by authority, and just shrugged off the critique. But Corney was right, his humble vessel has remained afloat, even in relative darkness, while D'Israeli's account is now known as mere fiction.[42]

The mathematician Augustus De Morgan also commented fairly about the apple story, and he too criticized D'Israeli: "I cannot imagine whence D'Israeli got the rap on the head, I mean got it for Newton: this is very unlike his usual accounts of things. The story is pleasant and possible: its only defect is that various writings, well known to Newton, a very *learned* mathematician, had given more suggestion than a whole sack of apples could have done, if they had tumbled on that mighty head all at once."[43] Some writers viewed Newton's insight as caused by the accident; others denied that great discoveries are triggered by accidents, and instead insisted, like De Morgan, that Newton's discoveries stemmed from his extraordinary studies; and still other writers attributed such breakthroughs to the innate quality of men. One anonymous writer argued: "these discoveries are not due to the accident of this or that occurrence taking place at this or that time, but almost entirely to their taking place before the eyes of some fit observer—some one of those spirits which are ever looking from the watch-towers of science into the dark fields of the unknown before them—that a Newton was in the garden when the apple fell—applies with great force."[44] At the time, many people were interested in the would-be science of phrenology, the study of how the shape of a person's head is correlated to intelligence and behavior. Phrenologists generally disagreed with D'Israeli's account: it was not that the apple had hit Newton's head and *caused* his mental development; it was that his innate mental features enabled him to extract knowledge from the falling apple.

One phrenologist argued that a specific part of the brain handles weight, and that the corresponding facial bump, along the eyebrows, mapped the individual's ability to analyze mechanical forces. He commented: "It is particularly large in the mask of Sir Isaac Newton: the falling of the apple attracted the attention of this organ, and led the philosopher into a train of thought, which resulted in developing the true theory of gravitation. If the organ had been small in his head, the falling of the apple would never have excited in his mind such a train of thought."[45] Allegedly, the apple could not inspire anyone who did not already have an acute physical propensity for the analysis of weights.

In 1870, an article in the *Phrenological Journal* remarked on the "happy moment" when Newton observed an apple fall, and conceived the law of gravitation; and, the "happy moment" when Ben Franklin sent up his kite to bring down electricity; comparing these to the "happy moment" when Franz Joseph Gall, as a schoolboy, noticed that boys with big eyes were better at memorization than him, and that actors too had the same trait, and thus he realized that features of heads correspond to abilities—founding later the science of phrenology.[46] In 1897, the *Phrenological Journal* duly quoted D'Israeli's account about Newton, perhaps with humor.[47] Soon, another writer joked that a phrenologist might well say that the apple, guided by the goddess Minerva, had struck Newton exactly on the right bump.[48]

Was Newton really inspired by a falling apple? Because the evidence is inconclusive, various writers choose whichever stance on the story seems preferable to them: true, partly true, or false. For example, the mathematician Carl Friedrich Gauss dismissed it: "The story of the apple is too simple," he said, "one can believe whether the apple fell or remained, that thereby such a discovery was delayed or accelerated, but the business is really as follows: a stupid, pushy man once came to Newton, to ask him, how he had arrived at his great discoveries. But then Newton realized that he faced a childish mind, and wanting to get rid of the man, he answered: that an apple fell on his nose, whatever, the man left satisfied by that, completely enlightened."[49] Notice that Gauss, too, freely proceeded to invent additional details and to presume some certain knowledge. Such variations of the story of Newton's apple are worthwhile because although they do not portray the past, they enable us to glimpse aspects of the person who tells the tale.

The story has evolved in various directions. In a 1997 book, *Isaac Newton: The Last Sorcerer*, Michael White argued that Newton, working in physics in general and gravity in particular, was greatly influenced by alchemy. Accordingly, White said that the story of the apple "was almost certainly fabricated by Newton to disguise the truth," he argued that the apple story is "at least an exaggeration designed for a specific purpose—almost certainly to suppress the fact that much of the inspiration for the theory of gravity came from his subsequent alchemical work."[50]

**Table 3.2** Some variations in the evolution of the story of Newton's apple

| | | |
|---|---|---|
| 1665–1666 | Isaac Newton worked on physics and mathematics at his family farm, alone. | |
| *Sixty years later . . .* | | |
| 1720s | William Stukeley | In 1726, Newton told me that he began to think about gravity when he was sitting in a contemplative mood, and an apple fell. |
| 1727 | Robert Greene | Martin Folkes told me that Newton's idea of universal gravity was inspired by an apple. |
| 1727 | Voltaire | "Newton walking in his Gardens, had the first Thought of his System of Gravitation, upon seeing an Apple falling from a Tree." |
| 1727–28 | John Conduitt | Musing in a garden, it occurred to Newton that gravity (which brought an apple from the tree to the ground) was not limited to a certain distance from the earth… |
| 1728 | Henry Pemberton | "As he sat alone in the garden, he fell into a speculation on the power of gravity…" |
| 1734 | Voltaire | "In 1666 . . . one day he was walking in his garden, & he saw fruits fall from a tree, he let himself go into a deep meditation about such Gravity." |
| 1760 | Leonhard Euler | Newton was in a garden, laying under a tree, then an apple fell on his head, and set him thinking about gravity. |
| 1797 | Isaac D'Israeli | When the young Newton was reading under a tree, an apple fell and hit him hard on the head. |
| 1838 | Bolton Corney | There is no evidence that an apple hit Newton on the head. |
| 1838 | Thomas Chalmers | "…sitting one day in his garden, he saw an apple fall from a tree at his feet." |
| 1800s | Carl Gauss | In order to drive away a stupid, pushy man, Newton said that an apple fell and hit him on the nose. |
| 1980 | Richard Westfall | The apple story "is too well attested to be thrown out of court." |
| 1997 | Michael White | Newton's apple story was almost certainly a lie to hide the fact that his theory of gravity came from alchemy. |
| 1999 | A. Rupert Hall | Conduitt learned the story of the apple from his wife Catherine. |

It suits White's narrative to interpret Newton's story as a cover-up, a disguise for the alleged narrative that White seeks to support. Notice the repeated expression "almost certainly," an exaggeration to camouflage what there would be much more appropriate: *with no certainty*. Another writer, renowned historian A. Rupert Hall, claimed that Conduitt had learned the story about the apple from his wife, Catherine, Newton's niece.[51] But there is no evidence to that either. And there are other similar examples.

I do not know whether Newton's original apple story is true. But at least we have it from the old man himself. Plus, there is much more evidence in support of the apple story than other stories that are widely repeated, such as that Pythagoras discovered the arithmetic of music, that Galileo dropped objects from the tower of Pisa, or that Darwin was inspired by finches. Regardless of its truth, the story became an attractive hook that moved many people to meditate about scientific creativity; not a bad thing at all.

Whatever happened to Newton's apple tree? Legends produce relics. After Newton died, his estate and possessions were divided among his four half-nephews and four half-nieces, the grandchildren of his mother and of Reverend Smith. Before dying, Newton gave some lands to members of the Conduitt family, and he bequeathed his estates at Woolsthorpe and Sustern to John Newton, whose great-grandfather was Isaac's uncle. Just five years later, in 1732, John Newton sold the manor house and grounds at Woolsthorpe to Thomas Alcock who next sold it to Edmund Turnor the following year. Subsequently, generations of the Turnor family kept the property. Almost a century later, in a biography of Isaac Newton published in 1831, David Brewster reported what happened to a tree in the Woolsthorpe property: "The celebrated apple tree, the fall of one of the apples of which is said to have turned the attention of Newton to the subject of gravity, was destroyed by wind about four years ago; but Mr Turnor has preserved it in the form of a chair."[52]

The square chair is now in a private collection. Later, in Brewster's more extensive biography of Newton, published in 1855, he omitted mentioning the chair and instead added a footnote that relayed a different account: "We saw the apple tree in 1814, and brought away a portion of one of its roots. The tree was so much decayed that it was taken down in 1820, and the wood of it carefully preserved by Mr. Turnor."[53]

Two accounts by Brewster: in one the tree was knocked down by wind in 1827, and turned into a chair; and in the other, the old tree decayed and was taken down in 1820, the wood carefully preserved.

In any case, Augustus De Morgan wryly commented: "One particular tree at Woolsthorpe has been selected as the gallows of the apple-shaped goddess: it died in 1820, and Mr. Turnor kept the wood; but Sir D. Brewster bought away a bit of root in 1814, and must have had it in his conscience for 43 years that he may have killed the tree."[54]

Additional details in the history of Newton's apple tree have been painstakingly traced for decades by Richard Keesing.[55] Before Brewster wrote about the tree, another Edmund Turnor published a single footnote line about it, in 1806. He merely noted that "The apple tree is now remaining and is showed to strangers."[56] That was eighty years after Newton's death. Some years later, a boy who went to school at Lincolnshire reportedly saw the apple tree after it was knocked down by a storm. This account was written by William Walker, who wrote about his father, the boy Richard Walker (born in 1807):

> My father told me that while he was at school there, there was a very severe storm of wind one night, and that in the morning news came that Sir Isaac Newton's apple tree had blown down at Woolsthorpe. The school master, Mr. Pearson and several of the boys at once set off for Woolsthorpe, where Sir Isaac's house was, and which is not far from Stoke, and just on the Lincolnshire side of Belvoir Castle. When they arrived there they saw the old apple tree lying on the ground. It had been propped up all round for many years, and every effort had been made to preserve it. My father said it lay there, having by the force of the wind, blown over its props. He said that he did not know by what authority Mr. Pearson acted, but that he obtained a saw from somewhere and sawed a good many logs of wood from the branches. My father got one of these pieces,

> which he always kept as being a most interesting relic. Various friends and other people often tried to induce my father to part with this, but he always refused, as he prized it very much indeed.[57]

In 1912, William Walker gifted this account, along with his father's small log, to the Royal Astronomical Society in London.

Researching the history of the tree, Keesing notes that he has found some drawings of the broken-down tree, one apparently from 1816, another dated 1820, and a third with no date. Keesing conjectures that there was only one apple tree in Newton's garden (whereas there were many in the orchard), and that therefore that was *the* famous tree. However, a drawing of Newton's manor house, by William Stukeley, sketched in 1721, shows no tree where a later drawing shows an apple tree.

Meanwhile, Brewster's pieces of wood were apparently lost. Reportedly, he gave a few pieces of the tree to J. D. Forbes, professor of natural philosophy at the University of Edinburgh. Forbes passed the relics to his son, George Forbes, who later recounted:

> The writer inherited from his father (Professor J. D. Forbes) a small box containing a bit of wood and a slip of paper, which had been presented to him by Sir David Brewster. On the paper Sir David had written these words: "If there be any truth in the story that Newton was led to the theory of gravitation by the fall of an apple, this bit of wood is probably a piece of the apple tree from which Newton saw the apple fall. When I was on a pilgrimage to the house in which Newton was born, I cut it off an ancient apple tree growing in his garden." When lecturing in Glasgow, about 1875, the writer showed it to his audience. The next morning, when removing his property from the lecture table, he found that his precious relic had been stolen. It would be interesting to know who has it now![58]

Keesing found that in 1840, a Charles Turnor claimed that he had taken some grafts from Newton's apple tree (but had it not died almost a century earlier?) and had cultivated "two thriving apple trees" from it. It is not known what happened to those trees, but by 1937, a Christopher Turnor (then owner of Woolsthorpe Manor) claimed that there was a scion of Newton's apple tree at Belton Park. Keesing writes: "Subsequently a scion of this tree was grafted at the Fruit Research Station at East Malling and it is from this material that most of Newton apple trees planted worldwide come."

In 1977, in his quest to find what happened to the tree, Richard Keesing visited Woolsthorpe, hoping to take a photograph of the manor house that would resemble an old drawing that showed the apple tree: "I was walking backwards composing the scene from the Turnor drawing through the viewfinder of a camera when I found myself lying upon my back. Regaining my feet I looking round and was amazed to find that I had fallen over what appeared to be the tree illustrated in the drawing. Even now the memory of the event is disorienting for I recall the confusion of not knowing whether I was in the year 1820 or 1977."[59]

Comparing his photograph with Charles Turnor's drawing of 1820, Keesing concluded that the images were so similar that the present tree is none other than an outgrowth from a piece of the 1820 trunk that had fallen after the storm. In 1997, Keesing conjectured: "I would like to suggest that the prone hollow trunk which is rooted at each end and is today still growing at Woolsthorpe Manor is the prone branch of the tree drawn by Charles Turnor in 1820, and is one and the same tree which was identified from which Newton saw an apple fall in the year 1665/6. If this is the case, the apple tree must now be about 350 years old."

While this account contains uncertainty, Keesing's doubts seemed to have diminished by 2010. Without citing any new evidence, he claims in a website of the Department of Physics of the University of York: "Despite all their efforts to prop the aged tree up, it blew down in a storm in 1816. Some branches were removed but the major portion of the tree was left and re-rooted. The surprising fact is that this tree is still growing at Woolsthorpe Manor today and now must be over 350 years old."[60]

*It must be?* Some apple trees become very decrepit after 60 years. Some others have been reported to live 100 or even 200 years. But 360-plus? Despite Keesing's sympathy for his favorite tree at Woolsthorpe, there is no evidence that *that* was the tree. Like many other people, he sought to find something, for decades, and to his satisfaction at least, he found it. The tree that is absent from a drawing of 1721, the same old tree that had badly decayed and was torn down in 1820, the same old tree that was "destroyed" by storm winds in 1827 (or perhaps eariler), miraculously is now *alive*, and is more than 350 years old.

First it ended up as a chair, then pieces of its roots were allegedly preserved, then pieces of its branches, then living grafts, and today, there are descendants of Newton's tree in many places: There's one on the lawn to the right of Great Gate of Trinity College Cambridge, there's another at the University of York, and another by the Babson College Library in Massachusetts, and so forth. The apples produced by these trees are of a rare variety known as "Flower of Kent," they are somewhat pear shaped, flavorless, and colored red with streaks of yellow and green.

The latest destination for the tree is beyond planet Earth, *outer space*. The Royal Society archives granted permission to British-born astronaut Piers Sellers to take a valuable piece of their collections on a space mission. A news article by the Royal Society reported that "The section of wood, taken from the original tree that inspired Newton to formulate his theory of gravitation" would be taken on a NASA mission to space; and Sellers commented: "We're delighted to take this piece of Sir Isaac Newton's apple tree to orbit. While it's up there, it will be experiencing no gravity, so if it had an apple on it, the apple wouldn't fall."[61] Another astronaut, described the space shuttle *Atlantis* as "the single most incredible machine humanity has ever built."[62] And in May 2010, it took a piece of old wood into space.

# 4

## The Stone of the Ancients

In 1657, Oswald Crollie referred to alchemy as follows: "This visible and invisible fellowship of Nature is that golden chaine so much commended, this is the marriage of heaven and riches, these are Plato's rings, this is that dark and close Phylosophy so hard to be known in the most inward and secret parts of Nature, for the gaining whereof Democritus, Pythagoras, Apollonius, &c. have travelled to the Brachmans and Gymnosophists in the Indies, and to Hermes his Pillars in Æegypt."[1] By contrast, chemists later viewed alchemy in the same way that astronomers viewed astrology—as an irrational, ancestral fraud. Galileo and Kepler used astrology to gain money from wealthy patrons, though they both had secret reservations against it. It had more validity than Galileo imagined, as Kepler rightly reasoned that some things in the heavens do affect life on Earth: the moon does cause the tides. Alchemists believed that metals, too, were linked to the heavenly bodies. Some believed that dull metals were diseased but could be cured, perfected. They thought that metals live and grow in the depths of Earth, and they sought the elusive occult "seed," the so-called Philosophers' Stone, that would allegedly transmute "diseased" dull metals into perfect gold.[2] Alchemists sought this mysterious red powder in order to concoct a drinkable gold that would give health, extend life, and allow the alchemist a closer communion with God. They worked in basements and caves, with animal skeletons hanging overhead, stirring molten metals over fires and cooking smelly substances that would supposedly sweat and salivate and grow. They veiled their findings in secrecy, cloaked in mythical imagery.

Some alchemists honestly worked to unravel the secrets of nature, but they too wrote in cryptic, allegorical terms. For example, consider the following lines: "The wind is the bath of the Sun and the Moon, and Mercurius, and the Dragon, and the Fire that succeeds in the third place as the governor of the work: and the earth is the nurse, Latona, washed and cleansed, whom the Egyptians assuredly had for the nurse of Diana and Apollo, that is, the white and red tinctures. This is the source of all the perfection of the whole world."[3] Such fancies seem to lie quite some distance away from science.

The main works of alchemy were said to be ancient, written by a mysterious author known as Hermes Trismegistus ("three times great"), a legendary mixture of the Greek god Hermes (Mercury) and the Egyptian god of wisdom. And alchemists were secretive, like the cult of Pythagoras. Hence, throughout the centuries, some seekers of alchemical knowledge increasingly suspected that Pythagoras, who allegedly visited Egypt, had been privy to such secrets. Since historians already have often written about alchemy by focusing on the figure of Hermes, I will now analyze the history of transmutation partly in regard instead to the legendary Pythagoras. For a long time, the transmutation of the elements seemed to be a secret or a myth.

In the poem *Metamorphoses* (ca. 8 CE), Ovid told the story of Midas, King of Lydia, who received a drunken old satyr captured by peasants. Midas treated the satyr hospitably and entertained him for ten days and nights. On the eleventh day, King Midas returned the satyr to the young god Bacchus, the satyr's foster son, who in turn offered to grant Midas one wish, "which was pleasing, but futile, since he was doomed to make poor use of his reward. 'Make it so that whatever I touch with my body, turns to yellow gold,' he said. Bacchus accepted his choice, and gave him the harmful gift, sad that he had not asked for anything better."[4] King Midas was overjoyed by the gift, at first, turning twigs and stones to gold. But then he was horrified to find that his food also turned to gold, wine and water too. He fled in misery, hating his gift, thirsting and starving in despair. And "justly, he is tortured by the hateful gold. Lifting his shining hands and arms to heaven, he cries out: 'Father, Bacchus, forgive me! I have sinned. But have pity on me, I beg you, and save me from this costly evil!' The will of the gods is kindly. Bacchus, when he confessed his fault restored him," by sending him to a foaming river to wash away his sin. Hence, Midas came to hate wealth, and he stayed in the woods and caves. Yet he remained dull-witted and foolish, destined to hurt himself again.[5]

Allegedly not all men were as miserable with transmutation. Also in the *Metamorphoses*, Ovid portrayed Pythagoras as a master of change, and ascribed these words to him: "nothing in all the world remains unchanged. All things are in a state of flux, all shapes receive a changing nature." Ovid's Pythagoras spoke of "the change of times from gold to iron," and of various transmutations: a stream of water that turned the drinker's entrails to stone, other streams that "will turn the hair to something like clear amber or bright gold," rotting horses generating hornets, dead humans' spine marrow mutating into snakes, and mud giving birth to green frogs.[6] He claimed that even the elements can change. Yet Ovid's poem was grossly ahistorical: the character of Pythagoras, for example, described events that did not happen until centuries after his death. Pythagoras, proponent of the idea that souls are repeatedly reborn, had a unique connection to gold. In Lucian's stories (ca. 170 CE) Pythagoras appeared as having a golden thigh, which proved his relation to the sun god Apollo. In one place Lucian described Pythagoras, after seven transmigrations of the soul, this way: "the entire right half of him was gold."[7]

The association of Pythagoras with alchemy was also plausible because some writers claimed that he had studied magic and mastered the medicinal properties of plants. Many alchemists claimed that minerals had vegetative qualities. Around 78 CE, Pliny the Elder claimed that Pythagoras had written an entire book on the medicinal and magical properties of bulbs and herbs, assigning his discoveries to Apollo.[8] According to Pliny, Pythagoras celebrated the virtues of cabbage and mustard, declared that epileptic seizures cannot happen to one who holds anise in hand, and he claimed that squills hung on a doorway would keep out evil spirits.[9] Apparently Pythagoras also claimed that when cabbage, wheat, hemlock, and violets are in bloom, certain diseases strangely attack the human body. Another ancient writer noted: "Pythagoras and the group of those who said of matter, that it is susceptible to change, asserted a becoming and dissolution in reality. For they believed, that *the becoming* happens because of the transmutation and motion of elements."[10]

Moreover, according to Heraclides Ponticus, Pythagoras was reputed to be the son of Mercury. This god had granted him the gifts of "the perpetual transmigration of his soul, so that it was constantly transmigrating and passing into whatever plants or animals it pleased; and he had also received the gift of knowing and recollecting all that his soul had suffered in hell, and what sufferings too are endured by the rest of the souls." Pythagoras seemed to be a master of the occult, because he had "requested that whether living or dead, he might preserve the memory of what had happened to him."[11]

Hence, one of the earliest Latin texts on alchemy featured Pythagoras. It appeared in the thirteenth century, translated from Arabic, and tells of a gathering of nine philosophers convened by Pythagoras to clarify obscurities in ancient alchemical books. Pythagoras spoke of "the stone that is not a stone," common but hidden, and known by many names: Spume of the Moon and Heart of the Sun. Then a fellow philosopher described the process of transmutation: "when first cooked it becomes water; then the longer it is cooked, the more it thickens until it becomes a stone, as the envious call it, but really it is an egg that tends to become a metal. Afterward it becomes saturated and breaks, when you must roast it in a fire even more intense, until it gains the color of blood, when it is placed on coins and changes them into gold, according to Divine desire."[12] Pythagoras argued that the philosophers used marvelously varied expressions to convey the same art and to keep hiding the precious art from the vulgar and foolish. The characters were historical, but the dialogue was fictitious—though it echoed viewpoints of pre-Socratic philosophers.

The growing legend of Pythagoras crept into the history of alchemy and chemistry.[13] The association of Pythagoras with the secrets of the Egyptians illustrates the kind of syncretism of Greek and Egyptian myths that is often embodied by the legendary figure of Hermes Trismegistus. In the words of Johannes Kepler, "either Pythagoras hermeticizes, or Hermes pythagorizes."[14]

For centuries, writers have claimed that Pythagoras traveled to Egypt and learned various secrets from the Egyptian priests.[15] But which disciplines he supposedly studied there vary: geometry or astronomy, architecture or alchemy, religion or others. Actually, we do not even know whether he really ever traveled to Egypt. The earliest extant account of such travels, written more than a century after his death (Isocrates, ca. 375 BCE), only states, with derision, that in Egypt Pythagoras studied religious rituals, not necessarily to impress the gods but to enhance his reputation.[16] Earlier, Herodotus made several claims (ca. 430 BCE) that from Egypt some of the Greeks copied practices, ceremonies, and names of gods.[17] And Herodotus, who traveled to Egypt, made a comparison between burial practices that were "Egyptian and Pythagorean: for it is impious, too, for one partaking of these rites to be buried in woolen wrappings. There is a sacred legend about this."[18]

Herodotus did not mention whether Pythagoras traveled to Egypt or whether he studied alchemy—which does not mean that he did not do

**Table 4.1** How the legend of Pythagoras spread to alchemy and chemistry

| | | |
|---|---|---|
| ca. 430 BCE | Herodotus | The Greeks borrowed many practices and ceremonies from Egypt. |
| ca. 375 BCE | Isocrates | Pythagoras visited Egypt and studied religious ceremonies to at least enhance his reputation among men. |
| ca. 8 CE | Ovid's poem, in the voice of Pythagoras | "Nothing in all the world remains unchanged. All things are in a state of flux, all shapes receive a changing nature." Even the elements can change. |
| ca. 150 CE | anonymous, attributed to "Plutarch" | Pythagoras asserted that matter is susceptible to change, that becoming happens by transmutation of the elements. |
| ca. 900 CE | *Turba Philosophorum* | Pythagoras knew the secret of the Stone that can make gold. |
| 1575 | Paracelsus | Arabs and Greeks deciphered occult mysteries of Persians and Egyptians, and called it "the arcanum of the Philosophers' Stone according to the counsel and judgment of Pythagoras." |
| 1620 | *Gloria Mundi* | Pythagoras quietly possessed the medicine of the Blessed Stone and used it to extend his life to more than nine hundred years. |
| 1657 | Oswald Crollie | Pythagoras traveled to India and Egypt to learn the alchemical secrets of nature. |
| 1776 | Edward Gibbon | "These ancient books, so liberally ascribed to Pythagoras, to Solomon, or to Hermes, were the pious frauds of more recent adepts." |
| 1980 | Carl Sagan | "Chemistry is simply numbers, an idea Pythagoras would have liked." |

those things. Isocrates later claimed that Pythagoras traveled to Egypt—which does not necessarily mean that he truly did. What matters to me is that in this connection a recent commentator on Isocrates, Niall Livingstone, neatly pinpointed a pattern by which stories grow, how associations give rise to legends. Livingstone suggests that Herodotus indirectly originated the story that Pythagoras visited Egypt "by taking the first two steps of the reasoning which leads to the idea of a visit (similar feature, therefore same feature; same feature, therefore borrowed feature)." Herodotus noted that the burial practices of the Pythagoreans and the Egyptians were similar, and hence he seems to imply that they were identical. Then our interpretive drift leads us to imagine a causal connection, that the Pythagoreans copied the Egyptians: "Someone, perhaps Isocrates, then followed Herodotus' hint and took the third step of asserting that Pythagoras visited Egypt, an idea which was eagerly accepted by later Pythagoreans and biographers of Pythagoras."[19]

I do not mean to argue that Pythagoras did not visit Egypt, I simply do not know. What I do want to highlight is that this pattern—similar therefore same, same therefore borrowed—seems to recur in the growth of the legends about alchemy. Moreover, such tales serve to apparently give substance to the early roots of a discipline, as such stories about Pythagoras's imports from Egypt have propagated in fields as varied as mathematics, architecture, astronomy, religion, and alchemy. Regardless of whether Pythagoras actually traveled anywhere, writers added colorful and significant details to such tales over time. The most moving and inspirational version I have read is the following—in a book dedicated "To the Sacred Majesty of Truth." Note how it connects to common notions of a hero's journey, a lost paradise, the noble struggle against deaf mediocrity:

> Led by a desire of this, as by some guiding star, Pythagoras travelled into Egypt, and cheerfully encountered the greatest difficulties, and maintained the most obstinate perseverance, until at length he happily penetrated the depths of Egyptian wisdom, and brought into Greece a treasury of truth for future speculation. But these were happy days; this was the period

> destined to the reign of true philosophy, and to the advancement of the human soul to the greatest perfection its union with the terrene body can admit. For in our times, the voice of wisdom is no longer heard in the silence of sacred solitude; but folly usurping her place, has filled every quarter with the barbarous and deafening clamours of despicable sectaries; while the brutal hand of commerce has blinded the liberal eye of divine contemplation. For unfortunately, the circle of time, as it produces continual variations, at length reverses the objects of pursuit; and hence, that which was once deservedly first, becomes at length, by a degraded revolution, the last in the general esteem.[20]

One of the legends that grew up around Pythagoras was that "Pythagoras and many others possessed in silence the Medicine of the Blessed Stone, and neither used it for evil purposes, nor revealed it to the wicked; just as God himself always has concealed this knowledge from the proud, the impure, and the presumptuous." Allegedly some used it to extend their lives, including the Biblical figure of Adam (said to live five hundred years) and Noah (said to live nine hundred years), and others, such as Pythagoras himself.[21]

Alchemy became secretive partly because it was often a fraud. But it was also dangerous even if the alchemists truly believed that they were pursuing a noble art, because if someone were to discover how to manufacture gold, that person could then gain immense power and even undermine established rulers. If gold became increasingly available, the values of old riches would diminish; currencies could collapse.

The wealthiest rulers opposed the alchemists' enterprise. The Catholic Church condemned alchemy and denounced it as forgery. In 1404, King Henry IV enacted the Act against Multipliers, a law that prevented anyone from using alchemical crafts to produce silver or gold. Alternatively, other rulers also sought to command the secrets of alchemy. King Henry VI appointed royal commissions to investigate the alchemical arts to try to manufacture more wealth for the crown. He also asked the Catholic clergy to produce gold—after all, they routinely claimed to transubstantiate bread into the body of Christ. They refused, offended.

In several countries, accused alchemists who were not supported by a royal license were persecuted, indicted, exiled, excommunicated, imprisoned, hanged, burned at the stake, or immersed in boiling oil.[22]

Gold, what substance has caused more bloody theft, deception, and murder? Ironically, it is one of the most useless metals: too soft for tools and construction, inconveniently heavy, and chemically un-reactive. Yet gold not only looks good, it lasts; other metals such as iron and copper tarnish and rust, but not gold, as if somehow it holds onto youth, just as humans wish they could. Beauty, youth, and wealth were represented by gold.

In the Renaissance, Bernard Trevisan sought alchemists' secrets. One autobiographical text explains his struggles for the Philosophers' Stone. His family was wealthy, and Trevisan invested much money to study alchemy. He praised Hermes and Pythagoras, but complained that impostors deceived him. He worked with minerals, metals, vegetables, blood, hair, and excrement, praying for God's help. With a collaborator, he spent over a year trying to use salt to make the Stone. At age forty-six, he bought two thousand eggs, separated yolks and whites, putrefied them in horse manure (later distilled thirty times to extract a white liquid and red oil). He worked two years on those residues, then gave up and spent eight years on other experiments. Trevisan worked with silver, mercury, sulfur, olive oil—again with no results. Relatives ridiculed him. He stopped eating and drinking and became emaciated. He traveled, spent more money, sold properties. Poor by the age of sixty-two, he retired to the isle of Rhodes, still seeking men who claimed to know the Stone. Taking loans, he mixed gold, silver, mercury, horse manure, fire, and urine for months, but still saw no results. Sleepless, old Trevisan suspected that Nature cannot be altered, that those writing about transmutations were damnable cruel thieves.[23]

But apparently, not all alchemists were so unlucky. One intriguing alchemist wrote under the name Basil Valentine. Allegedly he was a Benedictine monk, but his identity and when precisely he lived remain a mystery. He wrote about various occult and alchemical topics, such as the use of certain poisons as medicine.[24] One of his short treatises, first published in 1599, discussed secret powers and virtues which, he claimed, God had put into metals and minerals. It became known as *The Twelve Keys of Basil Valentine*.

The author claimed to have spent years at a monastery, where through his devotion to God and by studying old texts he eventually progressed toward an earthly treasure. He claimed that one of the brothers in the convent suffered from a severe disease that physicians had failed to cure. To help him, Valentine worked with vegetable substances for six years. But he too failed, so he proceeded to labor with minerals and metals. He claimed to have managed finally to concoct a colorful mineral substance from which he extracted a "spiritual essence" that then served to completely cure the ailing brother. Allegedly, it was the Stone of the Ancients.

Valentine argued that to receive the knowledge of the Stone, one first had to show devotion and gratitude to God, help the needy and the poor, and truly repent of one's sins. Valentine added various other requirements involving discipline and dedication that would help one to attain the ancient secret knowledge of the Stone. He also explained that he was forbidden from plainly revealing some secrets by the law of God, whose wrath he feared. Yet he did express a desire to share his findings with other practitioners of the secretive art.

One example of Valentine's keys, the second, depicts two swordsmen lunging toward one another. On one sword clings a serpent, and on the other an eagle. Between the two stands a naked man, readily identifiable as the god Mercury, because the symbol for that god is over his head and because he has wings on his back and at his feet. In each hand, Mercury holds a caduceus, that is, a staff entwined by a pair of serpents facing each other. In the background landscape, the sun and moon appear near the ground. What did this image mean?

Valentine accompanied this image with several paragraphs alluding to a palace, the sea, and the marriage of the gods Apollo and Diana. He argued that although such divine bride and groom be gloriously adorned by their wedding garments, they must be naked and clean in the bridal night. He explained that the King Apollo should be cleansed by two "hostile" substances, and then wrote:

Figure 4.1. The Second Key of Basil Valentine, as illustrated in 1624.

> But if you cast the Eagle onto the icy Dragon who for a long time dwelled in the rocks and has crawled out from the caverns of the Earth, and you place both of them together on the infernal chair, then Pluto will blow wind, and, from the icy Dragon there will arise a volatile and fiery spirit, which by its great heat will consume the wings of the Eagle and produce a perspiring bath so extraordinary that the snow from the highest mountaintops will begin to melt and become a water, with which the invigorating mineral bath may be prepared, to thus give to the King fortune and health.[25]

Such cryptic expressions, simultaneously intriguing and silly, drew the attention of practicing alchemists for centuries. Were there really any alchemical secrets hiding in Valentine's keys?

In the early 1980s, an undergraduate student at the University of

Delaware named Lawrence Principe became fascinated by Valentine's keys. Principe spent a lot of time deciphering texts and reproducing alchemical processes in a laboratory. Proceeding to graduate school to study organic chemistry at Indiana University, he continued to painstakingly study the old emblems and language of alchemy and to recreate the experimental procedures that the alchemists obliquely described.

Looking at the words and image for the second key, Principe realized that the moon and the sun represent silver and gold. The sun was a traditional symbol for the god Apollo and the moon represented Diana. But what about the two fighters wielding swords with an eagle and a snake? And what about the winged figure of Mercury standing between them? In alchemical emblems, weaponry often designated a philosophical fire, a way to kill one substance to make it become another.[26] Principe reasoned that the clash of the swordsmen represented the mixture of two substances, which in turn would yield something symbolized by the winged Mercury. The snake entwined on a sword recalls "serpentine powder," an old kind of gunpowder that includes the substance saltpeter (crystals that grow like brushes in dank caves, produced by bat guano or urine and rotting manure). On the other sword, the eagle was the traditional symbol for a rare, volatile substance, sal ammoniac (also found in bat guano and volcanic vents).

Principe realized that these two substances, saltpeter and sal ammoniac, joined, produce a volatile acid, depicted by the winged figure of Mercury. This acid would serve to corrode silver and dissolve gold.[27] In present-day terms, the second key describes a reaction between potassium nitrate and ammonium chloride:

$$KNO_3 + NH_4Cl$$

Does that look more obvious than the image of the two fighters? In the right proportions, the two substances can make the potent acid that is actually a mixture of nitric acid and hydrochloric acid:

$$HNO_3 + 3HCl$$

It's a dangerously corrosive yellow liquid that spews toxic fumes as it quickly expends its potency. This acid, which the alchemists called *aqua regia* (royal water), dissolves gold.

Through further analysis, Principe came to think that Valentine's third key discussed how to use the royal water and the fourth showed how to volatize gold. But that seemed unlikely: could alchemists possibly have known the subtle procedures by which gold actually can be made into gas? In his ensuing research, Principe found an article from the late 1800s that discussed the volatility of gold chloride, and that cited an old work by the famous chemist Robert Boyle.

Robert Boyle is famous for advancing chemistry into what we consider a science. His many experiments involved clever techniques and important instruments such as the air pump. In schoolbooks, he shows up for having formulated "Boyle's law," $PV = k$. This law, that the product of the pressure and volume of a gas has a constant value $k$ (if the temperature is kept constant), was confirmed by Boyle, but not discovered by him.[28] Boyle devised useful classifications of substances, stressed the importance of securing experimental evidence that is repeatable and publicly witnessed, advanced the use of quantified analysis of chemical transformations, and advocated the value of mechanical explanations in order to make sense of puzzling phenomena. Boyle believed that material bodies are all composed of atoms in various arrangements.

In 1661, Boyle published a book titled *The Sceptical Chymist,* in which he critically analyzed and shattered old theories of matter, including the ancient theory that all was made of earth, water, air, and fire. Thus Boyle became famous for having separated science from nonsense by debunking the phony quackery of alchemy. He became known as "the father of modern chemistry."[29]

Lawrence Principe found that Boyle had some definite knowledge of how to volatize gold. And when Principe read Boyle's notes, he found that the method that Boyle described for volatizing gold was the very same that Principe had deciphered from Basil Valentine's keys. Boyle called the acid "the *Aqua pugilum,* aenigmatically describ'd by *Basilius.*" *Aqua pugilum* was one of the names sometimes given to *aqua regia,* the royal water that could dissolve the noble metals such as gold. Principe noted that *aqua pugilum*, or "water of the fighters," referred to what Valentine called a "most precious water made cunningly from two fighters."[30] Thus Boyle alluded directly to Valentine's image of the two swordsmen. Why was the famously skeptical chemist citing the alchemical literature?

Digging through Boyle's private papers, Principe found multiple manuscripts, letters, and notes showing that Boyle was intensively reading the writings of alchemists, deciphering their symbols, and reproducing their experiments. Principe also found fragments of a manuscript by Boyle titled *Dialogue on the Transmutation of Metals*. Meanwhile, another researcher, Michael Hunter, found an interview of Boyle taken by his confidant Bishop Gilbert Burnet. These two documents converge on this odd story: that Boyle once visited a foreigner who introduced him to a stranger who claimed that he could convert lead into a material supple as butter. Boyle's own assistant provided the lead and the crucible on which to perform the conversion, so that there would be no fraud. By fire, the lead melted in the crucible, and then the stranger cast a small dash of a bright powder into the crucible. Soon they removed the crucible from the fire, waited for it to cool, and then Boyle, reportedly, saw that the lead had actually become true and fine gold.[31]

Impossible? Chemists now teach that it is impossible to make gold by any chemical procedure. Yet in the 1600s, Boyle supposedly witnessed a transmutation. Thus Boyle concluded in his *Dialogue* that, despite all reasons to the contrary, the production of gold might actually be possible.

Also, by 1666, Boyle became convinced, by his own experiments, that gold could be destroyed and even transmuted into silver, but he lacked the opportunity to repeat such experiments because he was driven out of his laboratory by the bubonic plague.[32] Furthermore, Boyle believed that the Philosophers' Stone would facilitate communication "with good spirits." It would demonstrate the existence of rational spirits, angels. Thus it might even prove the existence of God. The Philosophers' Stone would not only transmute metals into gold, it would transmute atheists into believers. At the time, some adepts eagerly searched for the power to communicate with God or gods, a skill that allegedly had been mastered by Pythagoras.[33]

Lawrence Principe further found that Boyle had many interactions with alchemists and that he took many steps to join a secret society of alchemists in France which soon dispersed when King Louis XIV allegedly managed to kill many of its members in an explosion at the castle where they hid.

Robert Boyle provided monies to fund the works of alchemists in England and Europe. And in 1689, Boyle and the Bishop of Salisbury testified to the Parliament to have actually *seen* the transmutation of metals into gold, which helped repeal the Act against Multipliers, the old law against the production of silver and gold in England. It became legal to perform such transmutations. But Boyle did not get an opportunity to try to benefit from that ruling—he died on 31 December 1691. To organize his stacks of manuscripts, he had appointed three friends, including the philosopher John Locke. Promptly after Boyle died, Locke received some odd letters from, of all people, Isaac Newton.

For over twenty years, Newton had privately studied alchemy, performing countless experiments and writing thousands of notes. It was he who wrote the lines quoted at the beginning of this chapter about the wind bathing the sun and moon, about Apollo and Diana and the perfection of the world. Like Boyle, Newton read Valentine and others and pursued the art of transmutations. Newton wrote to Locke that he knew that Boyle had a process involving a "red earth and mercury," the process for which Boyle had repealed the Act against Multipliers. Newton asked for a sample of that red earth along with a copy of Boyle's written prescriptions.[34]

Locke complied by mailing the prescriptions along with a bit of the mysterious red earth. But afterward, Newton became dismissive, as if he had failed to make any gold, as if it were a sham. Chronically sleepless, in 1693 Newton suffered a nervous breakdown. Then in 1696, an anonymous "adept" visited him, claiming to produce a solvent for all metals. Again Newton pondered alchemy. That year he became warden of the Royal Mint at the Tower of London; for years he supervised the production of coins and ruthlessly persecuted forgers.

And what happened to the guy who supposedly demonstrated transmutation before Boyle's own eyes? According to a medical doctor

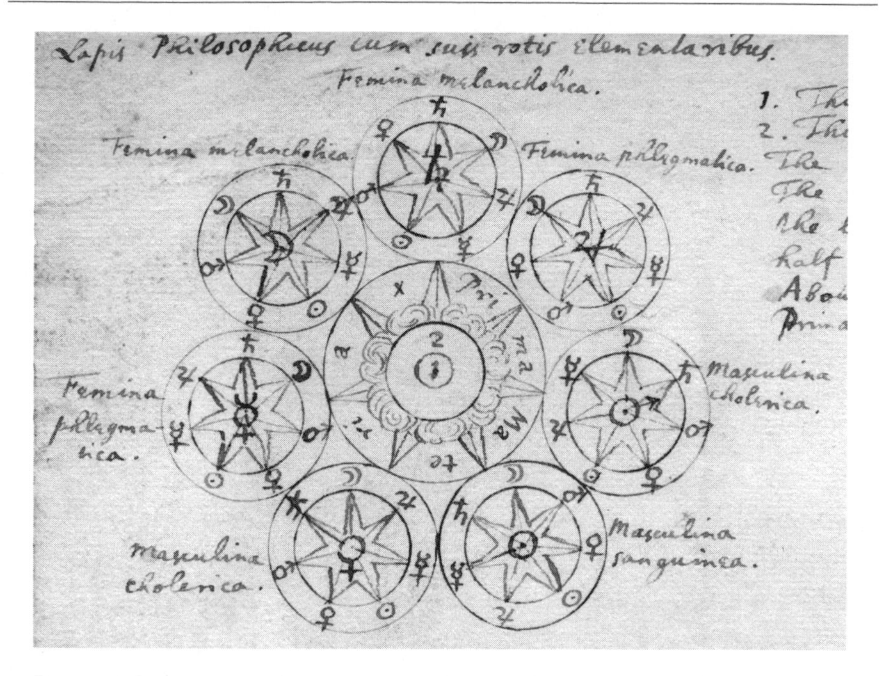


Figure 4.2. Circles upon circles: a manuscript diagram of the Philosophers' Stone, by Isaac Newton.

at Frankfurt am Main, writing in 1706, the alchemist who "while Boyle watched, converted lead into gold," was traveling to France to take Boyle's letters to his master but then he fell off his horse and died.[35]

The secret of transmutation eluded Boyle and Newton, or maybe there never was any such secret. Chemists concluded that certain substances, the elements, are unalterable. During the next centuries, some individuals occasionally claimed to make gold, but few people believed them. Chemists rose in prominence while alchemists sank in disrepute. Alchemy as a whole drew disdain and ridicule. A chemist turned historian, Henry Carrington Bolton, remarked: "We imagine it will be hard to discover in the whole range of literature writings having scientific pretensions more senseless than the aphorisms of the disciples of Pythagoras, collected in the 'Turba Philosophorum,' so often quoted by the alchemists of the sixteenth and seventeenth centuries."[36] Distinguished historian Edward Gibbon had also complained that "these ancient books, so liberally ascribed to Pythagoras, to Solomon, or to Hermes, were the pious frauds of more recent adepts."[37]

Paracelsus, another alchemist, complained that the secret of transmutation had been hidden by enigmatic expressions: "But since the supercelestial operations lay more deeply hidden than their capacity could penetrate, they did not call this a supercelestial arcanum according to the institution of the Magi, but the arcanum of the Philosophers' Stone according to the counsel and judgment of Pythagoras. Whoever obtained this Stone overshadowed it with various enigmatical figures, deceptive resemblances, comparisons, and fictitious titles, so that its matter might remain occult. Very little or no knowledge of it therefore can be had from them."[38]

In Ovid's *Metamorphoses*, the character of Pythagoras claimed that all is subject to change: "Even the things we call elements do not persist. Apply your concentration, and I will teach the changes they pass through."[39] But the red tincture and its medicinal powers remained elusive and occult, the talk of secretive groups.[40] No such teachings were found in any ancient texts. Such exalted tales became myths, buried in the coffin of scientific impossibilities. But sometimes, the dead come back to life.

By the 1890s, some strange substances had emerged in chemistry. Henri Becquerel found that uranium, a heavy and puzzling metal, emitted invisible rays. Gold didn't emit rays, neither did silver. Yet uranium had the power to develop photographic film, even in total darkness. And its invisible rays also electrified the air around it. Metals can become electrified, of course, but uranium was itself a source of electric charge.

In 1898, a Polish student of chemistry, Maria Sklodowska, also known as Marie Curie, began studying uranium as part of her doctoral research in Paris. Her husband, Pierre Curie, had constructed a device to detect the invisible rays of uranium by measuring faint electrical charges in the air. Marie used that device to analyze the effects of various factors, such as light and moisture, on the radiation. She found that the rays remained unaffected, which led her to conclude that the "radioactivity" was an intrinsic property of uranium.

Marie also tested whether elements other than uranium also emitted invisible rays. At first, no other element seemed to have this property. Then she tried thorium and found that it too emitted invisible rays. Having searched for such rays in all the elements, she tested many compounds too. Uranium was commonly extracted from pitchblende, a heavy black ore from mines between Germany and Czechoslovakia. Marie found that even after the uranium had been extracted, the pitchblende residue continued to emit rays. Quite surprisingly, it emitted *more* rays than the pure uranium. A few other minerals also emitted rays. She also found that such rays were independent of conditions such as heating, light, and acid baths. In 1898, she announced that the invisible rays were "atomic properties" of yet undiscovered elements. After more months of grueling work, Marie Curie managed to identify two new elements that she named polonium and radium. She estimated that these elements were hundreds of times more radioactive than pure uranium.

While Pierre Curie studied the radioactivity of radium, Marie worked to isolate radium entirely from the pitchblende. Only then would she be able to definitely measure radium's chemical properties. Some chemists remained skeptical of the very existence of this element. Trying to work in a suitable laboratory, Pierre and Marie only obtained access to a large but cold and shabby wooden shed where medical students used to dissect corpses. The ceiling leaked, and the building looked like a stable. Still, Marie, Pierre, and several hired workers came to labor upon piles of pitchblende residue to slowly extract minute traces of radium. They boiled the residue and repeatedly washed it with acids, alkaline salts, and water. It was backbreaking work. Marie Curie recalled: "Sometimes I had to spend a whole day mixing a boiling mass with a heavy iron rod as large as myself. I would be broken with fatigue at the day's end."[41]

After two years of work, the Curies and their assistants had processed eight tons of pitchblende, using four hundred tons of rinsing waters and thousands of chemical treatments and distillations.[42] When Pierre and Marie entered the shed at night, they saw in the darkness that the distilled substances in various containers glowed, in her words, "like faint fairy light." By 1902, Marie finally had a roughly pure sample of radium, a few tiny metallic bits—about one-fiftieth of a teaspoon—painstakingly extracted from about ten tons of pitchblende.

Most people today hardly know why Marie Curie became an international celebrity. She found new elements, coined the word "radioactivity," became the first woman professor at France's elite university the Sorbonne, and she was awarded two Nobel Prizes. Yet such achievements do not fully explain why journalists, photographers, and autograph-hounds came to stalk her.

To better understand why her contributions were extraordinary, one should list some of the astonishing properties of the elusive substances that she found and isolated. Tiny portions of radium affected glass and even diamonds by giving them colors. The substance also electrified the air around it, and its effects penetrated through solid objects. It became known as the most potent kind of poison. Radium did not need to be ingested or even touched to transmit its effects: it poisoned at a distance. A tube containing only a pinhead bit of radium placed over the spine of a mouse caused paralysis in just three hours, followed by convulsions, and then death. It killed even microbes, and its presence sterilized seeds. Moreover, the radium extract was self-luminous, it shone like electric blue light bulbs—but without requiring an influx of energy. It continually emitted heat, about 250,000 times more than the heat produced by burning an equal amount of coal. Calculations showed that one ton of radium would suffice to boil a thousand tons of water for an entire year. The energy output of radium was so great that it implied the possibility of unbelievably terrible new weapons. Frederick Soddy commented that radioactivity led one to envision the planet "as a storehouse stuffed with explosives, inconceivably more powerful than any we know of, and possibly only awaiting a suitable detonator to cause the earth to revert to chaos."[43] Likewise, Ernest Rutherford remarked that "some fool in a laboratory might blow up the universe unawares."[44] But radium also had positive effects: it cured skin cancer. Just years earlier, the mere idea of a metal having even just one of these properties would have been dismissed as a *ridiculous* alchemical dream.

Radium became a consumer sensation. In 1903, two patients in St.

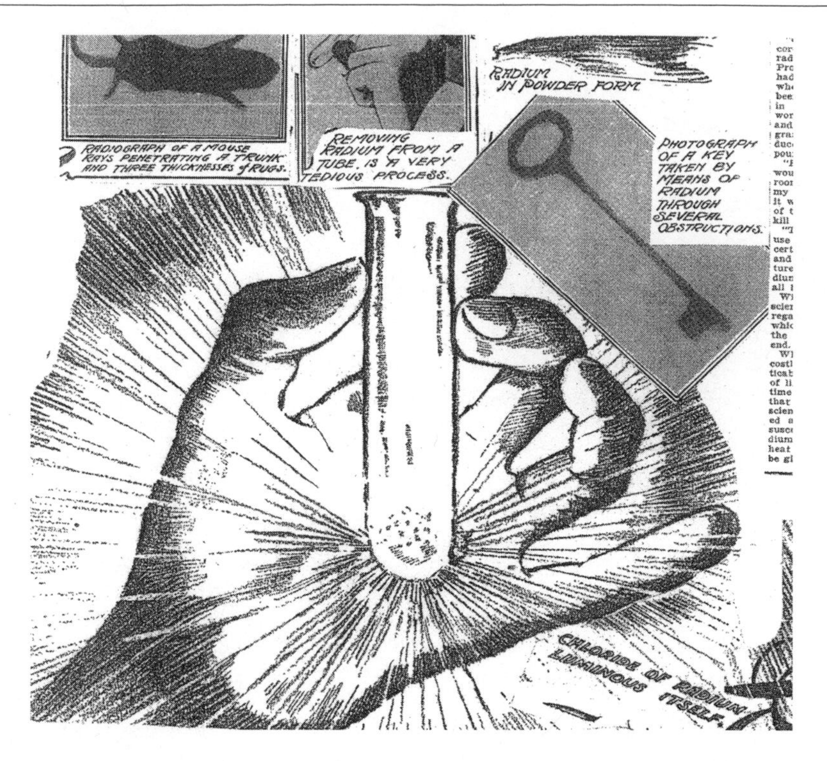


Figure 4.3. Illustration from a *Philadelphia Press* news article on radium, early 1900s.

Petersburg were actually cured from facial cancer. More were cured subsequently, giving rise to the myth that radium was the ultimate cure for cancer.[45] Pierre Curie hoped that it would cure blindness and tuberculosis.

Moreover, the rare metal powder had an even stranger property. In 1898, Marie Curie had speculated that radioactivity might be a kind of "disintegration of the atom,"[46] but Pierre had convinced her that it wasn't. Then, in 1902, at McGill University in Montreal, Ernest Rutherford and Frederick Soddy analyzed experimentally the emissions of radioactive elements. Soddy detected a strange gas emanating from thorium, so he stood there transfixed "stunned by the colossal import of the thing," and he blurted in shock: "this is transmutation: the thorium is disintegrating and transmuting itself into argon gas"—but then Rutherford shouted: "For Mike's sake, Soddy, don't call it *transmutation*. They'll have our heads off as alchemists."[47]

In hindsight, years later, Soddy remarked: "Nature can be a sardonic jester at times, when you come to think of the hundreds of thousands of alchemists in the past few thousand years toiling and broiling over their furnaces, spending laborious days and sleepless nights trying to transmute one element into another, a base into a noble metal, and dying unrewarded in the quest, whilst we at McGill, by my first experiment, were privileged to see, in thorium, the process of transmutation going on spontaneously, irresistibly, incessantly, unalterably!"[48]

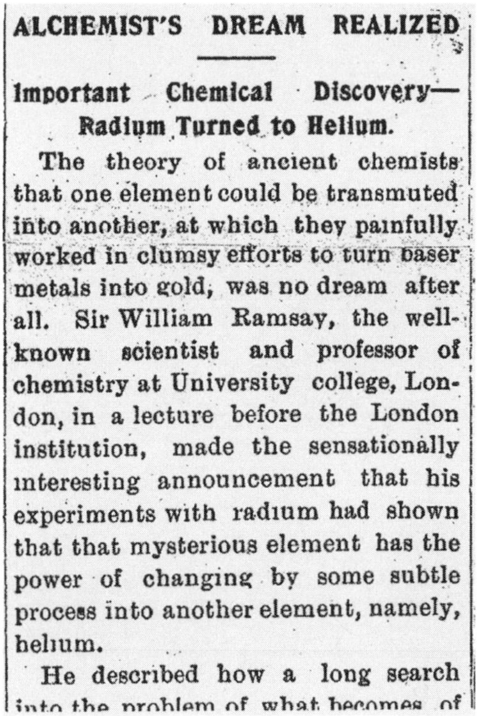

**ALCHEMIST'S DREAM REALIZED**

**Important Chemical Discovery—Radium Turned to Helium.**

The theory of ancient chemists that one element could be transmuted into another, at which they painfully worked in clumsy efforts to turn baser metals into gold, was no dream after all. Sir William Ramsay, the well-known scientist and professor of chemistry at University college, London, in a lecture before the London institution, made the sensationally interesting announcement that his experiments with radium had shown that that mysterious element has the power of changing by some subtle process into another element, namely, helium.

He described how a long search into the problem of what becomes of

Figure 4.4. The *Arizona Blade* and *Florence Tribune*, 12 December 1903.

In 1902, Soddy inferred that radioactive elements emit helium (named after the mythical Greek name for the sun, *Helios*, because its distinctive light had been detected earlier around the sun), and in 1903, Soddy and William Ramsay proved that radium emits helium gas.[49] Again, this meant that radioactive elements do change. Marie Curie acknowledged in 1904 that such findings would finally prove "that the transmutation of elements is possible."[50]

Still, chemists believed that this sort of transmutation was distinct from the alchemists' dream. For one, it happened by itself. In 1903, Rutherford and Soddy noted that radioactivity was a process that lay "wholly outside the sphere of known controllable forces."[51] Likewise, in 1906, Pierre Curie argued that if Rutherford and Soddy were correct, theirs would be "a veritable theory of the transmutation of the elements, but not as the alchemists understood it. Organic matter would necessarily evolve across the ages, following immutable laws."[52]

But one day in 1906, as he was rushing across a street in Paris, Pierre Curie was trampled by a horse-drawn wagon, he died. Marie mourned

ALCHEMY, LONG SCOFFED AT, TURNS OUT TO BE TRUE

Transmutation of Metals, the Principle of the Philosopher's Stone, Accomplished in the Twentieth Century.

in the hand of man the power to harness nature as was never dreamed of before, and change civilization to a fairy tale.

Hitherto chemists have dealt with atoms as architects have dealt with bricks. Bricks have been built up into different forms, and produced widely different results, but they were always bricks. Nobody ever got back of that fact. What would be the surprise of a builder if he suddenly discovered that a brick would pull a wagon or keep him warm, or go off with a bang like gunpowder? His surprise would be no greater than that of the chemist when he discovered that atoms of matter of their own accord were prepared to do things quite as startling.

"Science," says Prof. Soddy in his interpretation of radium, "has broken fundamentally new ground, and has delved with distinct step further down into the foundations of knowledge." Into the foundations, he says. Science has been building up and spreading out, but now it has dug deeper and is getting near the bed rock of things.

Radium is an ordinary looking affair, differing in no respect, so far as the eye can see, from any one of the ordinary common elements known to everybody.

time you will find that the Welsbach burner has photographed itself on the plate through the light proof paper. This is because thorium is radio active, in a far less degree than radium, but still quite enough to be highly interesting.

Radio activity was discovered by M. Henri Becquerel in 1896, the year after Prof. Roentgen discovered the X-ray. It came about by an accident. He had been experimenting in the sunlight with uranium and a photographic plate. The sun failed, and he put things away in a dark drawer for several days. When the sun came out again and he went for his plate he discovered that the uranium had made a photograph of itself in the dark. Thus, by accident, although it was an accident which must soon have happened, the key was given to a mighty secret of nature.

A few years later M. and Mme. Curie were able to isolate the element radium from a piece of pitchblende.

Now comes the transmutation part of the wonder. Everybody knows that radium is giving out energy at a tremendous rate. It gives out three hundred times as much as the same weight of burning coal. Nothing in nature can work like

If radium would change altogether in 2,500 years, a piece of uranium the same size would require 7,500,000,000 years to change altogether. If lead, copper, gold, and silver are changing, which seems likely, they are certainly not doing so at the rapid rate of uranium, and you can add to this 7,500,000,000 about any number of ciphers that you like. We used to think of these things as unchanging, and when we get into the billions of years they might as well be so so far as we are concerned at present; but the changes of radium have been calculated in a very narrow space, and science has determined what they are. It may be possible to determine the same thing for other elements, and of course in the far distant future the dream of the old alchemists may come true and we may be able to transmute one into the other at will.

The singular fact about all this experimenting has been that radium does not always give the same product. If the conditions under which the experiment is carried on differ radium seems willing and able to turn into different things. The Alpha particles were identified with

Figure 4.5. Excerpt from the *New York Times*, 19 February 1911.

his death and became increasingly reclusive. People increasingly realized that radium is extremely toxic; many were poisoned by it, including the Curies, and many dozens died. Still, it really did have some remarkable properties. Newspapers and advertisements hailed radium as a "miracle drug." Entrepreneurs developed ways to mechanize the process of extracting the substance from pitchblende for commercial purposes, including it in products for arthritis, lupus, insanity, birthmarks, and various other conditions and diseases, as well as soaps, hair tonics, facial creams, tea, toothpaste and glow-in-the-dark paint.

Marie Curie became internationally famous. In 1921, one American interviewer and fundraiser hailed her as "The Greatest Woman in the World," comparing her to Julius Caesar, Buddha, and Jesus Christ for having "reached into the bowels of the earth for one of the healing secrets of God." The interviewer raved that radium was "the most priceless stuff in the world" and "the strongest force in the world. The power contained in a gram is enough to raise a battle-ship of twenty-eight thousand tons one hundred feet in the air."[53]

Meanwhile, speculators anticipated that the alchemists' dream was at hand: the manufacture of gold. In 1914, the novelist H. G. Wells wrote a fictional account, *The World Set Free*, pondering the consequences of atomic transmutation. He dedicated the book "To Frederick Soddy's Interpretation of Radium." In the novel, scientists managed to convert bismuth into gold, causing the collapse of the world's economies and even a nuclear war. Wells speculated: "What chiefly impressed the alchemists of 1933 was the production of gold from bismuth and the realisation, albeit upon unprofitable lines, of the alchemist's dreams."[54] In reality, people were increasingly dreading the implications of transmutation. In the 1920s, financial experts discussed the possible impact of "modern alchemy." Representatives of the U.S. federal government issued public comments in the *New York Times* trying to reassure the public that even if modern alchemists succeeded in making gold, the monetary standards would not collapse.[55]

**WAY TO TRANSMUTE ELEMENTS IS FOUND**

**Dream of Scientists for a Thousand Years Achieved by Dr. Rutherford.**

**NEW AGE, SAYS RICHARDSON**

**Remarkable Result of Bombarding Nitrogen Gas With the Alpha Rays of Radium.**

The transmutation of elements, the dream of both charlatans and scientists for nearly a thousand years, has actually been accomplished by the recent work of Sir Ernest Rutherford, and his results are generally accepted by scientists and physicists, according to Dr. James Kendall, Associate Professor of Chemistry at Columbia, who said, on the other hand, that there was not the slightest reason to believe that the Germans had accomplished their reported feat of making synthetic gold.

Nitrogen, sodium, aluminum, chlorine, oxygen and carbon have been transmuted, or broken up by Rutherford into hydrogen and helium, according to Dr. Kendall.

This was first accomplished, according to the claims of Rutherford, by bombarding nitrogen gas with the

Figure 4.6. Excerpt from the *New York Times*, 8 January 1922.

Soon, chemists in various countries reported having synthesized gold. Repeated efforts showed such claims to be mistaken, yet the struggle continued for decades more. Part of the problem was that minute quantities of gold can be extracted from mercury by distilling it in a vacuum. Yet researchers at Berkeley, by bombarding platinum with neutrons, detected radioactive gold isotopes, at least.[56] Also, scientists at the University of Michigan studied radioactivity induced artificially in gold.[57]

In 1941, scientists at Harvard University used a cyclotron to achieve the "transmutation of mercury."[58] They shot fast neutrons at about 350 grams of mercury and found that it produced three kinds of radioactive gold (isotopes that decay in hours or a few days). Hence chemists and physicists were actually making gold, but unfortunately, a kind of gold that vanishes in a few days. And in fitting fulfillment of the alchemists' conjectures, they were producing it from, of all metals, mercury.

But the imminent possibility of artificial gold did not undermine the financial markets, as economists realized that it would be preferable not to ground money on metals anyhow. In 1971, the United States ruled that dollars would no longer be valued on the basis of convertibility into gold. This ruling reduced the recurring fears that the world's economies would be wrecked by transmutation.

Finally, in 1980, scientists in California at the Lawrence Berkeley Laboratory converted a tiny little bit of the rare pink metal, bismuth, into gold.[59] They shot ions of carbon and neon at extraordinary speeds against the bismuth, knocking out minute fragments to produce gold, which in turn was extracted into ether, washed with hydrochloric acid, and plated thinly onto platinum. The entire procedure cost about $10,000. It produced a nearly negligibly small quantity of gold, equivalent then to less than a penny, about one billionth of one cent. Thus the alchemists' old dream was undermined by the discovery that it cost much more money to make gold than to dig it from the earth. Worse, making gold cost more than just buying it at a store.

Still, as chemists had anticipated decades earlier, the monetary value of the energy released in a transmutation is immensely greater than the value of gold.[60] Of course, the "transmutation" of the chemists was not the same as the alleged transmutations of the alchemists. Similar, but not the same. Yet the early alchemists would have been pleased with what scientists did find: immensely valuable elements that glow and have spectacular properties that act at a distance to energize or destroy, and the ability to convert elements into one another. As in the writings of Basil Valentine, there are poisons that can be used as medicine. Reflecting on the nature of science, Marie Curie once wrote "A scientist in his laboratory is not only a technician, he is also a child placed before natural phenomena, which impress him like a fairy tale."[61]

Thus ends the story on transmutation which I set out to tell, from the legendary Pythagoras to the meticulous chemists. But let me add an afterthought. In the end, the alchemists' gold surprisingly was found not by a chemist but by a writer. As was hoped and desired, eventually the "Philosopher's Stone" *did* generate unimaginable wealth. But it was not from a glittering red powder or an actual piece of metal, it turned out to be the idea itself, the ink on paper. The key to wealth was not literally a stone, instead it was words, like on this page this ink: *the Philosopher's Stone.*

In 1995, an unknown writer in Edinburgh, Scotland, finished a manuscript titled *Harry Potter and the Philosopher's Stone.* After numerous rejections, it was published in 1997, and then it quickly made its author, Joanne Rowling, very, very rich. It began a series of books that became the fastest selling books *in history.* They earned their author hundreds of millions of dollars, *far surpassing the wealth ever earned through writing by any author on Earth.* It made her richer than the Queen of the United Kingdom. Anyone who has read such books can see that they're fun, good, but why the unprecedented scope of the success? Are Rowling's stories so immensely superior to most everything ever written in human history? I did not think so, so this puzzle stood in my mind for years. I called it "the Harry Potter problem."

Was it just a coincidence that the original title of this book involved the Philosopher's Stone? (I say the original title just because in the United States it was re-titled *Harry Potter and the Sorcerer's Stone,* because of the publisher's belief that the title would not be as effective in America.) Many readers immediately liked it, but why so many millions? In 2009, someone emphasized to me that those stories borrow a lot of material, images, and myths from older stories. A lot, "it's *all* in there, everything." Even the Stone. Then it hit me: that's it, by incorporating many of the most outstanding elements of good old stories, the author made a work that could well provide rich mythical images that were otherwise *lacking* in readers' lives. Maybe Rowling's books thoroughly conveyed the uplifting power of myths, secrets of the ancients, to readers who were practically starved for myths. This idea seemed striking because it matched an issue that was raised years ago, by the inspirational scholar Joseph Campbell, who analyzed the value of myths. He complained that because classic myths are nowadays seldom taught and told, people are losing their way, losing their connection to meaningful epic experiences of the past. "Our society today is not giving us adequate mythic instruction of this kind, and so young people are finding it difficult to get their act together."[62] Storybooks that revive such myths are helpful. And the apparent and unnecessary rupture between sciences and myths is a related problem to solve. There, the study of chemistry should begin with alchemy, with myths.

# 5

## *Darwin's Missing Frogs*

MANY old books claim that when Charles Darwin visited the Galápagos Islands, he was inspired to think about evolution by seeing variations in finches' beaks. Most people believed that species were as unchangeable as the chemical elements. Just as no metallurgist could ever make gold, fish or lizards could never become birds. As the story goes, Darwin changed all that when he theorized the "Transmutation of Species." Allegedly, he found that each species of finch belonged to a particular island and had developed distinct feeding habits that matched their evolving beaks, for cracking small or big seeds or for

Figure 5.1. Illustration of finches from the Galápagos, from Darwin's book of 1845.

eating insects. That's what many people still think, and so, one of the most widely reproduced pictures in history is that of Darwin's finches.

However, in sterling historical studies, Frank J. Sulloway of Harvard University showed that, really, Darwin was hardly influenced by finches and scarcely observed their feeding habits.[1] He did not correlate their diets and beaks; in fact, Darwin collected too few specimens to determine whether any finch species was unique to each island. He did not even keep track of where he picked up every specimen. Really, no finch species was unique to any one island. Unfortunately, some teachers and writers remain unaware of Sulloway's historical findings.[2]

The popular myth that the Galápagos finches crucially inspired Darwin to think about evolution arose because in the second edition of his *Voyage of the Beagle* he added one sentence about finches: "Seeing this gradation and diversity of structure in one small, intimately related group of birds, one might really fancy that from an original paucity of birds in this archipelago, one species had been taken and modified for different ends."[3] But that brief comment was foreign to Darwin's travel books and thousands of research notes; there is no evidence that it represented his thoughts during his voyage in 1835.[4] When he added that comment, in 1845, he had already believed in evolution for eight years. Yet the finches acquired fame partly because editions of his *Voyage* include an illustration of finches that, together with the quoted sentence, created the illusion that Darwin construed the finches as compelling evidence for evolution. Actually, Darwin's observations of finches were so scant that his thoughts on them were inconclusive guesswork—so much that he did not refer to the Galápagos finches in his *Notebooks on Transmutation* or use them as evidence for evolution in his *Origin of Species* of 1859.

Still the legend spread, partly through the book *Darwin's Finches* by David Lack, published in 1947. In the Galápagos Islands, Lack extensively collected and analyzed data on the finches that showed that evolutionary processes could account for such species and varieties as well as their beaks, habits, and geographical locations. But along the way, Lack seemed to attribute some of those scientific findings to Darwin himself.[5] Textbooks decorated by pictures of finches echoed such claims. Lack's brilliant insight, that the finches' beaks could be understood as evolu-

**Table 5.1** Darwin's manuscript records on the Galápagos birds

| | | |
|---|---|---|
| 1835 | Darwin's notes about Galápagos birds, written during the HMS *Beagle* voyage | Abundant species of tame finches exhibit "an inexplicable confusion," and "a gradation in the form of the bill," with "no possibility of distinguishing the species by their habits, as they are all similar, & they feed together . . ." |
| | | Two of four specimens of mockingbirds are distinct kinds *exclusive* to two islands; and some Spaniards claim to know from which island came any tortoise—such remarks might "undermine the Stability of species," and should be examined. |
| 1837–1838 | Darwin's notebooks on transmutation | No comments on the finches of the Galápagos. |
| 1857 | Darwin's big manuscript on natural selection | "I suppose that nearly all the birds had to be modified, I may say improved by selection in order to fill as perfectly as possible their new places; some as Geospiza [finches], probably the earliest colonists, having undergone far more change than the other species; Geospiza now presenting a marvelous range of difference in their beaks . . ." |

tionary adaptations, was substantiated by Robert Bowman in 1961.[6] Thus the insight that is routinely attributed to Darwin in the 1830s was only established scientifically more than a century later.[7]

Why did the legend spread? Why did the story about the finches propagate so widely? Sulloway argued that perhaps the story about the finches spread because it matched the format of traditional hero myths: a man departs from home on a bold adventure, encounters and overcomes hardships, and returns with a deep truth. Another reason is that the picture of four species of finches often seemed attractive as the only illustration in Darwin's books that could be construed as a portrayal of evolution, and the story of discovery evolved as a fitting complement.

Nowadays, science textbooks continue to highlight the imagery of finches, and authors finesse the story by noting that Darwin saw the finches but that only later did they turn out to be great examples of evolution. Other writers speculate about factors that "may have been" of decisive importance. For example, Stephen Jay Gould argued that perhaps five years of arguing against his ship's authoritarian captain led Darwin to turn toward materialism and evolution: "Who knows what 'silent alchemy' might have worked upon Darwin's brain during five years of in-

**Table 5.2** Other writers expound on Darwin and the finches

| | | |
|---|---|---|
| 1835 | Charles Darwin visits the Galápagos and collects animals, including birds. | |
| 1837 | John Gould | Darwin collected thirteen species of Galápagos finches having different beaks while their plumage is nearly identical, by contrast to finches on the continents. |
| 1839 | Darwin's *Narrative of the Surveying Voyages* | "All the species [of finches], excepting two, feed in flocks on the ground, and have very similar habits. It is very remarkable that a nearly perfect gradation of structure in this one group can be traced in the form of the beak. . . . I very much **suspect**, that certain members of the series are confined to different islands." |
| 1845 | Darwin's *Journal of Researches* | Illustration of four finches. "Seeing this gradation and diversity of structure in one small, intimately related group of birds [finches], **one might really fancy** that from an original paucity of birds in this archipelago, one species had been taken and modified for different ends." |
| 1859 | Darwin's *Origin of Species* | No comments on the finches of the Galápagos. "**I felt fully as much difficulty in believing** that they [pigeons] could ever have descended from a common parent, as any naturalist could in coming to a similar conclusion in regard to the many species of finches, or other large groups of birds, in nature." |
| 1944–1983 | David Lack | "...in 1835, Charles Darwin collected some dull-looking finches in the Galapagos Islands. They proved to be a new group of birds and, together with the giant tortoises and other Galapagos animals, they started a train of thought which culminated in the *Origin of Species*, and shook the world." |
| 1977 | Stephen Jay Gould | "tortoises and finches have always received the nod as primary agents in the transformation of Darwin's world view." |
| 1982 | Frank Sulloway | Finches did not inspire Darwin to his theory. |
| 1999 | Raven and Johnson *Biology* | "the correspondence between the beaks of the 13 finch species and their food source immediately suggested to Darwin that evolution had shaped them." |

*Note:* Boldface has been added to emphasize wording that reflects uncertainty.

sistent harangue."[8] The question is, what really led Darwin to evolution? We can tell the history fairly, if briefly, and explain it by highlighting a neglected story about reptiles and frogs in hell.

In his book of 1691, *Wisdom of God Manifested in the Works of Creation,* John Ray argued that the parts of all animals are perfectly fitted to their use because God is good. Thus natural science lent support to theology.

People believed that God had created the Garden of Eden as a beautiful and harmonious place where every being fit comfortably in its particular environment. Similarly, for cold places, God created furry animals, while in any desert, there were placed appropriate animals as well. Moreover, some plants and animals were perfectly suited for one another, showing clear evidence of the Designer's plan. For example, some plants have seeds bearing little hooks, and by snagging onto animals' fur, they get transported to distant places, thus disseminating beyond the spaces that are already crowded by other plants.

In the early 1800s, in Paris, Georges Cuvier studied vertebrates. Like most people, he believed that all species were created at the same time, all coexisting with one another. He too believed that species remained constant over time, unchanged. Yet some naturalists, such as Jean Lamarck, believed that all species evolved progressively over time. Many people ridiculed such ideas. Yet naturalists found increasingly various odd fossils that did not seem to correspond to any known creature. People dismissed such anomalies by assuming that such creatures were alive somewhere in the vast continents.

Yet Cuvier also pieced together some bizarre old bones that constituted what seemed to be monstrous kinds of elephants: the Siberian wooly mammoth and the American mastodon. Problem: if such huge animals lived somewhere, then travelers or hunters should have encountered them; yet nobody had reported any such sightings. Such bones and other bizarre fossils seemed to imply that the world used to have a different kind of population in the distant past. Therefore, Cuvier acknowledged that he had been wrong—not all animals coexisted at the same time. He conjectured that the ancient population of creatures was replaced at some point by a modern population of species.

But another problem arose: geologists increasingly found older layers of sedimentary rocks bearing distinct kinds of fossils. It seemed that each geological period had its own distinctive animal population. And the more ancient the layers, the more bizarre the fossils. Rather than two distinct populations in Earth's history, there seemed to be a varied sequence. Many species were no longer alive. What horrible disasters could kill all the animals of a kind?

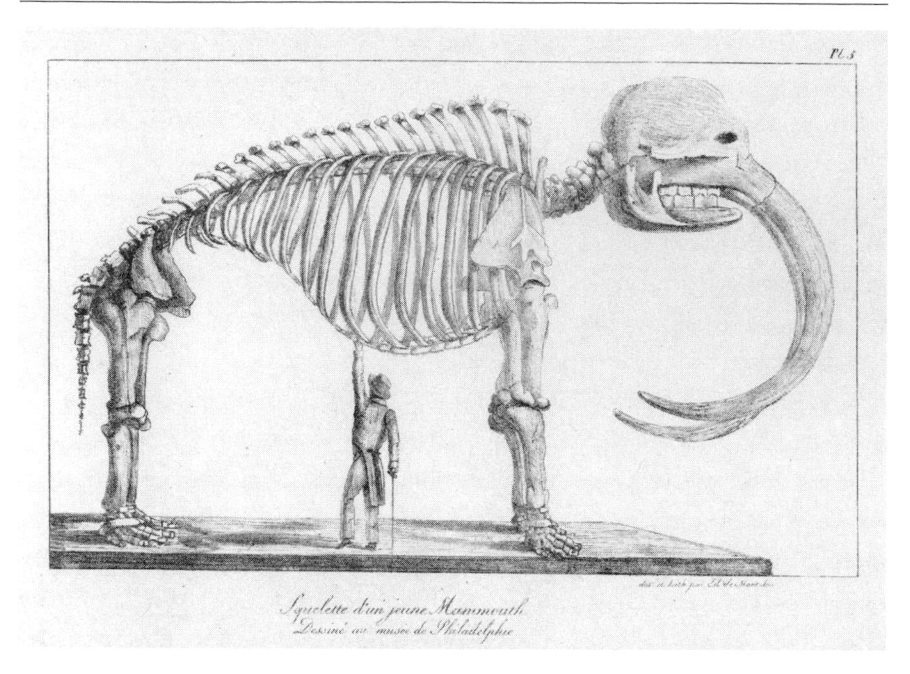


Figure 5.2. Mastodon skeleton found in America; mistakenly reconstructed with the tusks curving down. Cuvier categorized it as "extinct."

Cuvier conjectured that major catastrophes had caused extinction. He argued: "Numberless living beings have been victims of these catastrophes; some have been destroyed by sudden floods, others have been laid dry in consequence of the bottom of the sea being instantaneously elevated. Their races even have become extinct, and have left no memorial of them except some small fragments which the naturalist can scarcely recognize."[9]

There were more problems. Fossils found in Europe showed animals that presently did not live in Europe, while the older sedimentary rocks did not seem to contain fossils of species currently living in Europe. Accordingly, Cuvier conjectured that the past catastrophes, which had caused fossil deposits, were localized in particular continents. And so, animal species that survived, in Europe for example, had emigrated from the zones of upheaval. If animals currently in Europe had originated elsewhere, then we would expect to find fossils of European animals in other continents. But no such fossils were found.

Stranger things were buried in the rocks. Cuvier identified fossil remains of giant lizards, larger than elephants. William Buckland found the fossil remains of another monstrously large reptile, which he called "Megalosaurus" in 1824. Other large extinct reptiles emerged, and by 1841, Richard Owen proposed the term "Dinosauria" for such creatures. There seemed to be an age in the history of Earth when reptiles large and small dominated the land. Scarcely any mammals seemed to exist at that time, whereas in human history, mammals are the predominant predators on land. The age of mammals was preceded by an age of reptiles, and in deeper layers of rock, geologists found that previously there had been an age of fish and invertebrates, where there were no fossils of mammals or reptiles at all. Moreover, at even older layers of sedimentary rock, they found not even fossils of fish, but only of stranger, smaller beings. Why were there no human remains in the older layers of Earth? Naturalists speculated that maybe the temperatures and other conditions in the early Earth were not yet fit for humans.

In the late 1820s, the young Charles Darwin studied to become an Anglican clergyman. But he became increasingly interested in studying nature. At Cambridge, he studied William Paley's *Natural Theology* of 1802 and became impressed by how well the parts of animals suit their functions, showing God's benevolent and intelligent design.[10] He had also learned, from his professor Adam Sedgwick, that major catastrophes had deformed and reshaped the surface of Earth, killing many past species.

In 1831, at the age of twenty-two, Darwin had the opportunity to travel in a voyage around the world on His Majesty's Ship, the *Beagle*. Its main mission was to survey the coasts of South America. He would be able to explore and study many exotic places and species, and he would provide company for the ship's young captain, Robert FitzRoy, since the captain was not supposed to socialize with his underlings. FitzRoy wanted company also because he feared that the isolation of a long voyage could drive him insane. He worried that he might have inherited the insanity of an uncle who had killed himself. FitzRoy was well educated in the sciences; he even followed the study of physiognomy, that a person's facial features correspond to their character. When he met Darwin, they got along very well, although FitzRoy, whose nose was long, thin and aquiline, doubted whether Darwin's rather thick nose implied the necessary energy and determination for the long voyage.[11] Still, he extended the offer, and Darwin accepted to join the journey.

The HMS *Beagle* sailed along the coast of South America, carrying out a detailed geographical survey of the coasts. Often the ship docked, and Darwin studied the animals in various regions. Traveling southward, he was impressed by how particular species of animals were replaced by similar but distinct variations and species. For example, a large species of flightless birds, the rhea, occupied much territory, and farther south, a distinct species of rhea shared the territory with the northern kind. Even farther south, the southern rhea were present, but there were none of the northern kind. Their territories overlapped, but why did one species of rhea give way to the other?

Later, traveling in the Pampas region, Darwin found the impressive fossil remnants of a huge armored animal. It resembled the small armadillos of South America. He also found the ancient skull of a toothy rodent as big as a hippopotamus. the bones of an "ant eater" as big as a horse, and bones that seemed to belong to a mastodon. He increasingly sensed how much extinction had obliterated species. How to explain it?

Darwin collected hundreds of samples of unusual and bizarre specimens and analyzed the soils and geological formations of the continent. For reading during the trip, FitzRoy had gifted him a copy of *Principles of Geology* by Charles Lyell, published in 1830. At Cambridge, Darwin had learned that ancient major catastrophes had shaped the world. But Lyell denied any such catastrophes and argued instead that ordinary environmental processes—rain, wind, rivers, erosion, earthquakes, and volcanoes—had gradually shaped all land. Lyell contended that progress in geology was made mainly by scientists who cited only known causes rather than speculating about hypothetical spectacular events. He praised the ancient Pythagoras, who according to the poet Ovid (ca. 8 CE) allegedly declared: "Nature, the renewer of all things, continually changes every form into some other shape. . . . I have seen what had been solid earth become salt waves, and I have seen dry land made from the deep; and, far away from ocean, sea-shells strewn."[12] Lyell claimed that the Pythagorean doctrines confirmed "a principle of perpetual and gradual revolution inherent in the nature of our terrestrial system," and he noted that Pythagoras, had he applied such ideas to geology, would be further admired—just as astronomers admire him as an ancient Copernicus.[13] Lyell's theory was later construed as a key development that led to Darwin's evolution, and therefore, one writer gifted the credit to the ancient wise man: "By Pythagoras, who resided for more than twenty years in Egypt, these ideas were introduced to the Greek philosophers, and from that time 'Catastrophism' found a rival in the new doctrine which we shall see has been designated under the names of 'Continuity,' 'Uniformitarianism' or 'Evolution.'"[14]

Lyell believed that the rate of all natural changes was absolutely uniform through time. To explain rock formations, even the most extraordinary mountains, canyons, and cliffs, he appealed only to observable intensities of natural processes. Therefore, he acknowledged, such processes must have acted for immense periods of time. Strangely, atop some of the Cordillera Mountains, Darwin found fossils of seashells and maritime beings. While some might have imagined that spectacular catastrophes had uplifted such soils out of the ocean, Lyell required that such matter had gradually risen over millions of years.

On 20 February 1835, Darwin and company were at the coastal city of Valdivia, Chile, when suddenly, a great earthquake struck the region. Darwin wrote: "An earthquake like this at once destroys the oldest associations; the world, the very emblem of all that is solid, moves beneath our feet like a crust over a fluid; one second of time conveys to the mind a strange idea of insecurity, which hours of reflection would never create."[15] Unsettling—remember the supposed words of Pythagoras, that "Earthquakes were nothing else but Conventions of the dead."[16] Houses were shaken violently, and people ran in terror; but the wooden constructions at Valdivia survived. Yet the port of Talcahuano, where the *Beagle* docked afterward, and the adjoining town of Concepción, were utterly destroyed by the earthquake. Nearly every house was shattered to rubble, dozens of villages were devastated, and the port and ruins were wiped out by three terrible tidal waves.

For Darwin, the most remarkable effect of the earthquake was that the land had risen. The land at the Bay of Concepción was actually lifted by two or three feet; rocks that used to be underwater were now exposed. Moreover, FitzRoy noted that at the island of Santa María, one shore had risen by ten feet, such that beds of mussel shells for which the inhabitants used to dive underwater were beginning to rot in the sun. Darwin saw that hundreds of feet uphill, one could find the old remains of similar shells, and at the city of Valparaiso he had found seashells even at an elevation of 1,300 feet above sea level.

Such evidence, appearing in successive layers of differently weathered strata, showed that Lyell was right: major features of the landscape, such as the Andes Mountains, had not arisen abruptly in a single ancient catastrophe, but gradually, through many events such as the severe earthquake.

In 1835, the *Beagle* arrived at dark islands known to the British as "the Enchanted Islands," just below the equator and about 550 miles west of South America. The Spaniards called them Galápagos. These towering volcanic islands were the last series of stops before completing the mission and then sailing around the world back to England. Far from anywhere, the foreboding shores of ugly black rocks, broken and jagged, swarming with red crabs and huge, hideous crested iguanas, reminded Captain Robert FitzRoy of the legendary capital of hell: Pandemonium.[17]

Since the Galápagos are volcanic islands, their geological makeup is thoroughly different from the continental coast. The jagged black rocks, arid soils, high elevations, and the unusually temperate climate, are all distinct. Darwin encountered many craters, including some more than three thousand feet high. The vegetation was ugly and nearly leafless, it smelled bad, the stunted trees seemed dead, and only some tall cacti were large enough to cast shade. Darwin agreed with FitzRoy that the deathly landscape seemed hellish, like "the cultivated parts of the Infernal regions."[18]

The expedition members encountered massive tortoises and some odd birds. Only Charles Island was populated, mostly by people who had been banished there for political crimes. Chatham Island housed pigs and goats; the settlers had introduced them. There were mice at

Chatham Island and rats at James Island, these being the two islands most visited by sailors, and Darwin surmised that these mammals likely came from ships that sailed there for over a century. Darwin collected samples of plants, fish, seashells, and insects, many showing unusual traits. For example, although tropical regions usually have many insects, many of them colorful, the Galápagos had very few kinds, of very small sizes and dull colors.

The most striking native animals were reptiles: sea turtles, giant tortoises, countless ugly lizards, and several kinds of snakes. The massive tortoises were nearly deaf, and to Darwin they seemed like ancient survivors of the biblical flood: "they appeared most old-fashioned antediluvian animals; or rather inhabitants of some other planet." One kind of hideous and strong lizard, "imps of darkness," lived exclusively on the rocky sea-beaches, their claws admirably fitted for crawling about, and their flat tails suited for swimming. Remarkably, it was the only known lizard that subsists by eating marine vegetables. Another species of lizards was not aquatic at all, but remained inland, moving sluggishly and looking "singularly stupid."[19] Darwin recorded strange behaviors for all of these reptiles. For example, the large aquatic lizards entered the sea often to eat submarine seaweed, but when Darwin confronted one on them, as if threatening it, cornering it on the rocks, he found "one strangle anomaly," namely, that:

> when frightened it will not enter the water. From this cause, it is easy to drive these lizards down to any little point overhanging the sea, where they will sooner allow a person to catch hold of their tail than jump into the water. They do not seem to have any notion of biting; but when much frightened they squirt a drop of fluid from each nostril. One day I carried one to a deep pool left by the retiring tide, and threw it in several times as far as I was able. It invariably returned in a direct line to the spot where I stood. It swam near the bottom, with a very graceful and rapid movement, and occasionally aided itself over the uneven ground with its feet. As soon as it arrived near the margin, but still being under water, it either tried to conceal itself in

> the tufts of sea-weed, or it entered some crevice. As soon as it thought the danger was past, it crawled out on the dry rocks, and shuffled away as quickly as it could. I several times caught this same lizard, by driving it down to a point, and though possessed of such perfect powers of diving and swimming, nothing would induce it to enter the water; and as often as I threw it in, it returned in the manner above described. Perhaps this singular piece of apparent stupidity may be accounted for by the circumstance, that this reptile has no enemy whatever on shore, whereas at sea it must often fall a prey to the numerous sharks. Hence, probably urged by a fixed and hereditary instinct that the shore is its place of safety, whatever the emergency may be, it there takes refuge.[20]

Overall, the Galápagos were unusual compared to other places visited by Darwin because instead of there being large quantities of herbivorous mammals, there were large quantities of herbivorous reptiles. As for birds, Darwin collected sixty-four specimens, including water-birds, unusual finches, mockingbirds, flycatchers, a weird hawk, and owls.

Despite such oddities, there were obvious major similarities among the animals of the Galápagos and those of South America. There were close resemblances in their appearance, voices, gestures, and habits. Darwin wrote: "almost every product of the land and water bears the unmistakable stamp of the American continent," adding that the "naturalist feels as though he's standing on American land."[21]

A common story about the birth of evolution is that Darwin based it on the unique giant tortoises of the Galápagos. The vice governor of the islands told Darwin that he could identify the origin of any giant tortoise by the shape of its shell. The story goes that Darwin then reflected on that statement, analyzed the tortoises' shells, and realized that those unique tortoises had adapted to distinct environments.

But this myth too was dispelled by Frank Sulloway.[22] Actually, back then, Darwin did not heed that one might distinguish tortoise species based on their shells: dome or saddle shaped (*galápago* meant saddle). He assumed that the tortoises he saw in the Galápagos were of same

Figure 5.3. Darwin and the "imps of darkness" on the lava rocks of the Galápagos, October 1835; the HMS *Beagle* at bay.

species found in islands in the Indian Ocean. Hence, he did not care to collect the shells of any tortoises to analyze them. FitzRoy did take some thirty giant tortoises as their ship departed the islands, but just as food. Darwin and company ate every one of those giant, tasty tortoises before they returned to England, and they threw out their huge shells and bones in the oceanic waters. Darwin kept two young tortoises as pets, but all of the giant tortoises brought on board were destroyed.

On the way back to England, Darwin began to organize his notes. While writing about his four specimens of Galápagos mockingbirds, he noted that two of them were distinct kinds exclusive to two islands. At that point he noted that some Spaniards claimed to know from which island came any tortoise by looking at the animal's size, shell, and scales. He briefly remarked that such statements "undermine the Stability of species," and should therefore be examined.[23] This does not mean that he had a sudden conversion to transmutation, but rather, as he later recalled, that during the voyage he experienced some "vague doubts" about the immutability of species.[24]

What was so puzzling? The oddity was that Darwin had not expected that islands so close together, having the same rocks, climate, and elevations, could yet have different inhabitants. Species were supposed to match their environments, so why did different kinds inhabit equivalent environments? He had not pondered this question early enough to label and segregate all specimens by location. Therefore, he could hardly make any inferences about finches in particular, and about tortoises he only had hearsay. At least for the four specimens of mockingbirds, he had noted the four islands where he obtained them, and moreover, he had collected more mockingbirds on the continent—east, south, and west—in Uruguay, Patagonia, and Valparaiso.

Trying to identify specimens, Darwin was plagued by confusions. Were diverse specimens different species, genera, or merely different varieties? Where had he obtained some of them? Which animals were new species? He had not visited the northwestern coast of South America, above Lima, so he could not judge whether animals in the Galápagos were unique to the islands. He could hardly identify the large fossils from South America, either.

Finally, he delivered his specimens to specialists in London. Richard Owen analyzed the fascinating collection of mammal fossils. John Gould analyzed the collection of birds. Owen found that the mammal fossils resembled animals living presently in South America. Darwin's presumed mastodon was actually a kind of giant llama, big as a camel.

The apparent ant eater resembled also an armadillo, as did the huge fossil with the massive armor shell. Another giant skeleton, larger than a bear, resembled the sloth.

Other naturalists, too, were concerned with the puzzle of how new species replace old species. Did God repeatedly create new species whenever bad weather wiped out an entire species? In 1836, John Herschel had written to Charles Lyell, "Of course I allude to that mystery of mysteries, the replacement of extinct species by others. Many will doubtless think your speculations too bold, but it is as well to face the difficulty at once."[25] Did God operate instantaneously or through a series of intermediate causes? By abrupt sporadic miracles or by systematic laws?

Lyell became eagerly interested in Owen's findings on Darwin's specimens, because they confirmed a "continuity of types": that animals presently living in South America resembled ancient extinct species. Darwin also was impressed. As for the birds, John Gould painstakingly realized that the specimens from the Galápagos were especially interesting, because most of them were *absent* on the continent. Darwin's inferences in attempting to classify his birds were mostly erroneous. Whereas he identified only one species of mockingbird in the Galápagos specimens, Gould distinguished three. Whereas Darwin presumed that his mockingbirds from the continent were all the same species, Gould concluded that they were not. Likewise, Gould organized the finches as species with various affinities, whereas Darwin had thought that they were more diverse. Gould also found that the weird hawk was an "intervening link" between rather disparate groups: common buzzards and the South American caracaras.[26]

By 1837, Darwin had privately inferred that species can change in time. In July he began a notebook on "Transmutation of Species." He began to collect and organize facts about many species and their relations in places and times. By 1840, he was fully convinced that species were mutable.[27] He had been struck by the overall character and distribution of the South American fossils and especially the Galápagos species. The distribution of animals in space and time, geography and fossils, suggested that species evolved. The differences and similarities between the ancient extinct animals and the living were stunning. There was a "continuity of types." But why would God replace species with similar but distinct species? Having designed all species to live in harmony and to fit in perfect balance with their environments, why would the benevolent Creator allow entire species to die?

The environment did not always suit all species. Thinking about what he had seen regarding the distinct species of rheas in southern regions, Darwin considered the possibilities: perhaps the climate *favored* the more southern species, but one could also infer that it *hurt* the more northern species. If each species was perfectly suited to its territory, Darwin thought, then neither was perfectly suited to the region in between. The two species might be *competing* to occupy territory. How would this match the theory that every creature fits perfectly in its environment? The assumption that ecology is perfectly balanced became doubtful.

Some animals did not seem to match their environment. Animals at the Galápagos resembled American animals, yet the soils and environment of the Galápagos were quite distinct from those of the continent. Darwin realized that geologically the Galápagos islands resemble instead the Cape de Verde Archipelago, volcanic islands off the coast of West Africa, near the equator. He exclaimed: "there is a considerable degree of resemblance in the volcanic nature of the soil, in climate, height, and size of the islands, between the Galapagos and Cape de Verde Archipelagos: but what an entire and absolute difference in their inhabitants!"[28]

Instead, the Galápagos animals resembled the American animals, and the animals of Cape de Verde resembled the African animals. Why? This seemed to be an "almost universal rule": island populations resemble those of nearby continents. Darwin reasoned that "this grand fact" could not be explained by the usual theory of multiple independent creations.

As Darwin thought about transmutation, he reflected on the words of the Galápagos Spaniards and the vice governor about the tortoises. Darwin tried to compare his two pet tortoises to other specimens, but their shells only become distinctive when they grow big—so unfortunately, he could infer nothing. Darwin trusted the words of the Galápagos settlers, yet they had exaggerated; even experts on tortoises today can scarcely guess the island origin of a Galápagos tortoise by just examining it. There are gradations between the dome and saddleback shells. When zoos returned fifty tortoises to the Galápagos in the 1970s to breed them with locals, specialists failed to identify the likely origin of any but *one* such tortoise. In short, Darwin just did not reach his theory of transmutation by inspecting tortoise shells.

Thus far I have mentioned various animals that Darwin encountered in the Galápagos. But what about those that he did *not* find? Those also had a puzzling significance. Presumably, God had placed the many unique species of birds and reptiles on the islands. But what about amphibians? Darwin knew that a French naturalist and cartographer, Bory de St. Vincent, had pointed out that on volcanic islands off the coast of Africa, strangely, there was a complete absence of Batrachians—the whole order of animals consisting of frogs, toads, newts, and the like.[29] Yet some oceanic islands' conditions are ideally suited for such animals. In particular, certain areas of the Galápagos were appropriate. Therefore, Darwin painstakingly searched for such creatures. He concluded, "Of toads and frogs there are none. I was surprised at this, considering how well the temperate and damp woods in the elevated parts appeared adapted for their habits."[30]

Likewise, he noted that there were no frogs in the Canary Islands near North Africa, none at the Sandwich Islands (Hawaii), none at St. Jago in the Cape de Verde, none at St. Helena Island (between Africa and South America).[31] But why not? Why did God choose *not* to place any frogs in the oceanic islands? Darwin commented: "The absence of the frog family in the oceanic islands is the more remarkable, when contrasted with the case of lizards, which swarm on most of the smallest islands. May this difference not be caused, by the greater facility with which the eggs of lizards, protected by calcareous shells, might be transported through salt-water, than could the slimy spawn of frogs?"[32] And Darwin was right. Frogs, similar animals, and their spawn have semipermeable skins that let their inner fluids seep out in saltwater, which kills them. Such animals *cannot* float across a stretch of salty oceanic waters without promptly dying. By contrast, the eggs of some lizards can well be transported by oceanic currents or in floating tufts of plants and dirt.

Darwin also contemplated the lack of native mammals in the Galápagos. On land, only some kinds of rats seemed to be native, and Darwin imagined that they had somehow arrived by ships or otherwise. There were also seals and sea lions. And there was another distinctive kind of mammal on the Galápagos, one kind of mammal found in virtually all islands: bats. The Galápagos had two species, the hoary bat and the red bat. Many other islands, such as New Zealand, Norfolk Island, the Viti Archipelago, Hawaii, Mauritius, also possessed unique bats. Why would the Creator make unique bats but no terrestrial mammals or frogs on most islands? Darwin knew that bats fly across great bodies of water, and Darwin noted that some bats were seen wandering stranded over the ocean in daytime, and some bats were known to visit Bermuda, six hundred miles from the continent.

Darwin realized that all the various kinds of species on the Galápagos were *only* such as could have arrived there by water or air: birds, reptiles, insects, bats. He conjectured that the diverse inhabitants were not created in these islands—*they were colonists.* Their ancestors all originated somewhere else, such as in South America. In sum, the *only* kinds of native animals on the Galápagos were those kinds that could have gotten there by themselves or by accident without an independent miracle.

Yet many of these animals were distinct from those on the continent. So Darwin reasoned: they first arrived at the Galápagos and *afterward they changed.* They somehow adapted to their new environments. But how? Years later Darwin reflected that in the Galápagos, "both in space and time, we seem to be brought somewhat near to that great fact—that mystery of mysteries—the first appearance of new beings on this earth."[33]

Darwin also tested his inferences. Was it possible that eggs and seeds could really float across hundreds of miles of saltwater and still come to life on land? Darwin found that timber drifting in the ocean, as well as icebergs can carry clusters of trapped soil and seeds in such a way that the seeds remain able to produce vegetation. Further, he tested whether a dead pigeon, floating on saltwater for thirty days, could bear seeds in its stomach that would afterward germinate. He found, "to my surprise," that indeed such seeds could then germinate. Also, by studying pigeons' excrements he found that some kinds of seeds passed undigested and could then germinate. Darwin also tested whether seeds can germinate after being immersed in saltwater for weeks. He found that out of 87 kinds, 64 did germinate after being immersed in water for 28 days. A few germinated even after being immersed for 137 days. Most seeds sink, but fruits float for a long time. Dried hazelnuts floated for 90 days, and when Darwin planted them later they did grow. He found that 18 of 94 plants with ripe fruit floated for more than 28 days. He estimated that 14 out of 100 plants from a given country might float for 28 days and retain the power to germinate.

One of the examples of the seemingly perfect match between plants and animals in their environments was that some plants have seeds with hooks that are useful because they snag onto the wool and fur of mammals to be carried to fertile soils. Yet Darwin knew that some such plants and seeds are also found on islands where there are no furry mammals.[34] It hardly made sense in the theory of independent creations, but it made sense given the notion that seeds simply arrive at places by natural transport.

Such transport did not seem random because it followed definite oceanic currents. He also knew that the average speed of Atlantic currents was about thirty-three miles per day (some as high as sixty miles per day), and so he estimated that seeds might travel nine hundred miles in open ocean in a month. After similarly considering how the eggs of certain animals travel, Darwin's findings converged on this conclusion: life on the Galápagos was not created there, it arrived by natural processes.[35]

So organisms thus spread to environments to which they initially were *not* perfectly suited, and there they changed. *But how?* Being acquainted with the theory of Lamarck, Darwin believed, wrongly, that as individual organisms change their habits in response to their environments, they acquire traits that are subsequently inherited by their offspring. Still, this mechanism seemed insufficient to explain the great complexity of many species.

Elsewhere, back in 1755, worrisome observations about population growth had been published by an anonymous author in Pennsylvania. He had analyzed demographic data on births, deaths, and marriages, and he concluded that owing to America's extensive resources and land, population there would continue to grow nearly unchecked, to thus double at least every twenty or twenty-five years. America seemed destined to eventually have more Englishmen than Great Britain, and someday, even all the land in America would not be enough to sustain the population. The article was widely reprinted, even in a book on electricity. The anonymous author was a printer and civic activist in Philadelphia: Benjamin Franklin.[36] He further argued that Englishmen should therefore seize Canada for the sake of the future.[37] This finding, that human populations growing unchecked would increase exponentially, became an issue of political concern.

In 1798, another anonymous author, in London, also published his concerns about population expansion. He included his name in the subsequent editions of his essay, it was Reverend Thomas Malthus, an Anglican clergyman who became a prominent political economist. In late September and October of 1838, Darwin was reading the 1826 edition of Malthus's *Essay on the Principle of Population*.[38] Malthus had argued that because the sexual attraction among people is so strong, populations have a gross tendency to expand. In principle, the rate of expansion would be geometrical (he noted that in the United States the population had continued to double every twenty-five years), but expansion was limited by lack of resources to feed everyone, given a limited terrain. For Malthus, poverty and starvation were not results of the unfair distribution of wealth, they were natural and nearly inevitable. Although he earned the contempt of optimistic social reformers who claimed that society could flourish by redistributing wealth, Malthus contended that redistribution (as when the wealthy tried to help the poor) usually failed because people continued to breed too much, again creating poverty and starvation.[39] Wars would ensue unless people exercised moral restraint to avoid giving birth to children they could not feed. Malthus believed that God allowed such apparently harsh conditions so that people would strive for a moral purpose. Misery, starvation, and war could teach humans the virtues of labor and moral behavior. He explained that population expansion exerts constant pressure against the local means of sustenance, thus acting to drive people to migrate to inhospitable lands, to invade peaceful countries, and to fight against other restless tribes like barbarians in a "perpetual struggle for room and food." He wrote: "And the frequent contests with tribes in the same circumstances with themselves, would be so many struggles for existence, and would be fought with a desperate courage, inspired by the reflection, that death would be the punishment of defeat, and life the prize of victory."[40]

As Darwin read Malthus, he realized that such struggles for existence happen among animals, too. Lacking morals, animal populations expand until there are not enough resources to support them all. Darwin reasoned that animal populations could "increase at a geometrical ratio FAR SHORTER than 25 years," and hence, that species compete for resources, and individuals compete against one another.[41] Those less suited to obtain enough food and comforts die. Darwin realized that this pressure would effectively change a species by favoring particular traits.

But could it be, really, that species just were not in perpetual balance with their environments? Again, Darwin tested his conjectures, partly with experiments. In one plot of soil he planted native English weeds, from which there germinated 357 seedlings, and he found that, soon, 295 of them were killed, mainly by insects. He also found that in a small section of turf subject to mowing or grazing, 9 out of 20 species of plants perished; the more vigorous ones survived. Such experiments, plus many observations in the wild, led Darwin to conclude that plants and animals compete constantly against one another to survive. Plus, the climate affected all living things, chiefly by reducing the food supply. Darwin noted, for example, that just one winter season killed about four-fifths of all the birds on his grounds. Observing nature, he found that there are tremendous rates of death.

When we briefly walk in the woods, nature seems harmonious. But all beings, Darwin reasoned, are subject to severe competition. Some trees scatter thousands of seeds to subsist, as organisms feed on one another. He estimated that just *one pair* of elephants, apparently being the slowest breeding animals (by birthing only about six offspring in ninety years), would produce fifteen million *living* descendants in just five centuries.[42] How many more would they produce in a thousand years or more? The absence of any such huge population of elephants showed that an *im-*

*mense* number of offspring perish continuously. Darwin wrote: "The face of nature may be compared to a yielding surface, with ten thousand sharp wedges packed close together and driven inwards by incessant blows, sometimes one wedge being struck, and then another with greater force."[43] And he realized that the struggle is greatest among members of the same species, especially among siblings; the young suffer most.

All offspring exhibit certain variations compared to their parents. Some of those variations are advantageous whenever there are hazards or a lack of resources. Thus, individuals having even slight advantages tend to survive in greater numbers and thus to reproduce and transmit such traits in greater proportion. Over time, the population changes as disadvantageous traits diminish. Thus Darwin realized that nature affects species analogously to how breeders make "pure" varieties of pigeons, dogs, and cattle. By selecting which individuals mate, breeders or nature shape future animals. The continued selection of specific traits in an isolated population gives it a distinctive look. And if no barriers prevent intermixing among populations, there arise new mongrel varieties. Darwin carried out experiments that showed that, indeed, distinct types of plants mix with their neighbors; they do not keep their "purity" in the wild. The environment determines whether species evolve or become extinct. Darwin called this process *natural selection*: how environments shape species. He expected that it proceeds very gradually, requiring no sudden large changes, no major catastrophes. Species change would be retarded by intercrossing, but it would also be facilitated in places where populations are isolated for immense periods of time.

Thus, Darwin changed his mind. He had believed that species appear by miracles in their native lands, perfectly suited to live in constant harmony with the environment. But many species had ceased to exist, just as landscapes and climates had changed. Lacking permanent balance, nature is in flux, and some species survive in environments where they don't comfortably fit. Animals and plants struggle, and through that process, Darwin thought, many populations gradually evolved and adapted exquisitely to their dynamic environments.

Such thinking seemed to suggest that species had not been instantly created by God. Or had God made this very mechanism so that populations adapt to changing conditions? Darwin dreaded religious controversies, so he kept his theory secret except to some close friends.

In 1844, an anonymous book was published, arguing that species evolve by a progressive law of development.[44] Scientists rejected it, but it became a bestseller and generated much controversy. Darwin continued to work on his own theory in relative secrecy. Fifteen years later, he published *Origin of Species*. But all that is another story.

To return to the myth about finches, we can summarize that old story as follows: While visiting the Galápagos Islands, Charles Darwin noticed that various species of finches had beaks of different shapes and sizes. Observing their eating habits, he noticed that the shapes of their beaks corresponded to their diets. He also noticed that some species were distinct to some islands. Hence he inferred that the various species were related: they were descended from common ancestors that had populated the islands and had adapted variously to the distinct island conditions. Species evolved.

This short story works because it fits in the space allotted by a science textbook. And it works because, as Sulloway argued, it fits into the form of a classic journey of discovery: man departs from home on a bold adventure, encounters and overcomes hardships, and returns with a deep truth. But the story is false, so the challenge is how to replace it with something better. Selecting elements from the longer account above, I suggest that we can well write: Halfway around the world, the young traveler Charles Darwin arrived at foreboding towering volcanoes, the "Enchanted Islands." Their dark jagged terrain held swarms of hideous reptiles, "imps of darkness," and tame birds. Yet Darwin found no frogs or toads on the islands. He found *only* the kinds of animals that could cross the salty waters from the continent. All resembled American species, but oddly distinct. He later concluded that such island species descended from *colonists*, but somehow evolved.

The latter story is just as short as the old, just as appropriate for a textbook. And it's better, because it involves mythic imagery but is actually true.

# 6

## *Ben Franklin's Electric Kite*

LIKE Darwin's finches, other images have greatly influenced popular ideas about the history of science. One that is etched in our minds is that of stocky Benjamin Franklin flying a kite in a thunderstorm. It shows up in books, stamps, art, and currency, even on a U.S. silver dollar issued in 2006. It enchants because it shows a self-made American using a child's toy to make a major contribution to science: to prove that there is electricity in the clouds, that the awesome force of lightning involves the same stuff as electricity. But that image has lost some of its apparent certainty. Some commentators have faced a perplexing lack of evidence that Franklin ever conducted such an experiment.[1]

The young Benjamin Franklin published and printed a newspaper called the *Pennsylvania Gazette*. Long before he investigated the nature of electricity, he published several reports about lightning. In 1731, Franklin reported: "From Newcastle we hear, that on Tuesday the 8th Instant, the Lightning fell upon a House within a few Miles of that Place, in which it killed 3 Dogs, struck several Persons deaf, and split a Woman's Nose in a surprizing Manner." One year later, in 1732, Franklin described an incident in which a house in Allenstown "was struck by Lightning. It split part of the Chimney," melted butter, and caused a fire.[2] In 1736, Franklin's newspaper reported this bizarre anecdote:

> We hear from Virginia, that not long since a Flash of Lightning fell on a House there, and struck dead a Man who was standing at the Door. Upon examining the Body they found no Mark of Violence, but on his Breast an exact and perfect (tho' small)

> Representation of a Pine Tree which grew before the door, imprest or printed as it were in Miniature. This surprizing Fact is attested by a Gentleman lately come from thence, who was himself an Eye-witness of it; and 'tis added that great Numbers of People came out of Curiosity, to view the Body before it was interr'd.[3]

Franklin, the printer, actually believed the story—that lightning had imprinted the image of a tree on the man's chest.[4] In 1742, another report appeared in his *Gazette*: "two labouring Men (standing under a Sawyers-Shead, on Society-Hill, to shelter themselves from the Rain) were struck down by a Flash of Lightning: But one of them recovering, found his Companion, Thomas Smith, dead; his Hat was much torn,

Figure 6.1. An engraving from the 1870s: Benjamin Franklin and son; leisurely flying a kite, awfully close to a bolt of lightning.

and part of one of his Shoes torn off; on his Head, Neck, Breast, and the Inside of one of his Thighs were spots which appear'd as if burnt. The Survivor had most of the upper Leather of one of his Shoes torn away, and was burnt several Parts of his Body."[5]

Since ancient times, philosophers wondered about thunder and lightning. For example, Ovid claimed that Pythagoras knew the origin of lightning, whether it was caused by the god Jupiter, by storm winds, or by colliding clouds.[6] Centuries of speculations transpired before scientists finally reached up toward the clouds and found electricity there. The story about Franklin using a long string to capture lightning from the sky resembles an ancient myth. In epic Greek poems, the god of sky and thunder, Zeus, hid fire from humans, yet Prometheus managed to steal it with the long stalk of a fennel plant.[7] Prometheus stole fire from the god of thunder; Franklin caught "electric fire" from the thundering sky.

If so, he was not the first. In May 1752, a few individuals at Marly-la-Ville, France, used a sharply pointed, forty-foot-tall iron bar to test whether storm clouds transmit electricity. As a cloud passed overhead, they extracted sparks of electric "fire" from the iron bar. This group, usually led by Thomas Dalibard (although he was absent during this first successful trial), was roughly following a proposed experiment by Franklin, although they worked independently.[8] So Franklin himself was not the first to draw electricity from storm clouds. (Years later, in 1768, Franklin praised Dalibard by writing that Dalibard was "the first of Mankind that had the Courage to attempt drawing Lightning from the Clouds.") Just a few months later, in July 1752, another French experimenter, Jacques de Romas wrote a letter to a scientific academy, stating that he planned to use "a child's toy" to explore the electrification of clouds.[9]

On 27 August 1752, Benjamin Franklin's newspaper, the *Pennsylvania Gazette*, included a letter that summarized the lightning rod experiments of Dalibard and others.[10] Franklin did not add the slightest note to the effect that he had carried out any comparable experiments. In October of that year, Franklin published in his newspaper a brief description of a kite experiment. In contrast to other experiments he had described, his account of this one was vague, and he did not specify when or where in Philadelphia it was carried out, he did not allude to any witnesses, and he did not actually state that *he* had conducted it.

To clarify, Franklin did not describe what one might imagine: that lightning struck the kite to run electricity down the string—that would be deadly and the string would vaporize, as a bolt can have a temperature of 50,000° F, about five times hotter than the surface of the sun! Franklin wrote that during a thunderstorm, the experimenter should fly a kite bearing a metal spike, letting rain wet the twine to transmit electricity from a cloud, which would then charge a metal key attached to the twine before an insulating silk ribbon (kept dry because the experimenter should stand under a cover, holding the line out from a window or doorway). Then the experimenter could detect the "electric fire" on the key, using an instrument or his bare knuckle. Franklin wrote that the experiment is "easy"—that anyone could carry it out—but that just isn't so. The danger involved is so horrifying that few people dared to try anything of the sort. Besides, the difficulty of flying a kite out of a window during a rainy thunderstorm seems considerable. And details of Franklin's account seemed puzzling. For example, how long was the string?[11]

Still, Franklin claimed: "the sameness of the electric matter with that of lightning completely demonstrated."[12] If so, why did he publish, just months later, a request asking readers of his newspaper to please send him any information about the effects of lightning "on Wood, Stone, Bricks, Glass, Metals, Animal Bodies, &c. and every other Circumstance that may tend to discover the Nature, and compleat History of that terrible Meteor"?[13] Promptly after the account of 1752, a fellow scientist and friend wrote to Franklin that having read it, "I hope a more perfect and particular account of it will be published in a manner to preserve it better and give it more Credit than it can gain from a common News paper."[14]

The vagueness of Franklin's brief report is especially frustrating because if *ever* an experiment begged for detailed prescriptions, this is it. One does not play with lightning, a discharge that descends from about five miles (twenty-five thousand feet) in the sky, at a speed of about sixty

Figure 6.2. Franklin and son, as portrayed in 1884.

thousand miles *per second* with heat and force that can split a tree, burn a house, melt an iron rod, and fuse silica sand into glass. As Tom Tucker has rightly commented, when we're merely reading we might not care much about details, but if you are going to actually try this experiment, then you are risking your life, and in that context, *every word counts.*

In 1753, Jacques de Romas in France reported his own successful attempts to collect airborne electricity with the twine of a large kite.[15] (De Romas did not know whether Franklin or anyone else had conceived the same experiment.) Failing to collect electricity with a scarcely wet string, de Romas ran a thin copper wire along the hemp string. In daytime, he raised his kite to roughly 550 feet high (780 feet of string), and tied three and a half feet of silk at the bottom, anchored to a heavy stone pendulum. Near the silk, on the string, he tied a one-foot-long tin tube. De Romas then drew sparks by approaching a metal-tipped glass wand toward the hanging tube. He and several assistants and bystanders also used their fingers to draw sparks. The sparks ceased when the dark clouds overhead drifted away. When more clouds arrived, the witnesses used their fingers, keys, canes, and swords to feel the electricity, but then de Romas used a knuckle and received a painful shock—in his fingers, wrist, elbow, shoulder, abdomen, knees and ankles. After more experiments, the storm approached, growing violent, but no rain. So, fearing a deadly accident, de Romas then used only the glass-metal wand to draw sparks. Everyone took steps away. Straws on the ground stood upright and danced under the hanging tin tube. De Romas felt an electrical effect, like spider webs on his face. Then a long straw jumped from the ground to the tin tube and caused a terrifying explosion that cracked like thunder and caused a bright spark of electric "fire" roughly eight inches long. Crackling sounds happened, along with more sparks, and the string became luminous. The experiment ended because wind and rain made the kite fall; fortunately, nobody was injured.

Unlike Franklin, de Romas gave abundant detailed observations about various procedures, measurements, precautions, times, conditions, findings, observations, sounds, and even smells. Franklin had briefly described only one experiment, which was oddly cryptic and prescriptive, whereas de Romas described the execution of multiple specific experiments. Soon, some individuals wondered whether de Romas should well be credited as being the first to carry out the kite experiment and make the effect clearly visible.[16] De Romas complained to the Paris Academy of Sciences that he was really the first to succeed and that aspects of Franklin's report seemed questionable. Nevertheless, scientists began to compare Franklin with Prometheus: "Among all the phenomena of electricity it is difficult to find one as marvelous as that which Mr. Franklin has discovered, if it is true that this new Prometheus managed to draw the fire from the Sky."[17]

A committee of the academy concluded in 1764 that de Romas should indeed be regarded as having priority, *unless* Franklin or anyone provided evidence to the contrary. The committee included Franklin's scientific nemesis, Abbé Jean Nollet. The contentious Nollet had experience in criticizing and exposing electrical frauds and charlatans in France and Italy. But Franklin did not respond, just as he remained oddly quiet toward admirers who independently asked him about that particular experiment.[18]

Later, in 1767, Joseph Priestly published an account that added some details to Franklin's original brief article. Priestley wrote (presumably echoing Franklin) that Franklin had conducted the kite experiment in a field, under a shed, allegedly with one witness: his son. Priestley's account dated Franklin's experiment to *June* 1752—just a month after experimenters in Marly had already shown that lightning consists of electricity, but before Franklin had heard about it.[19] Yet if this was the case, why did Franklin wait more than three months to first report his extraordinary and brave experiment in the August *Gazette*?

But why would we doubt Ben Franklin? Looking at Franklin's achievements prior to 1752, we find that he invented the lightning rod, he operated a newspaper, he founded a library and a volunteer firefighting company, he implemented techniques to prevent the counterfeiting of currency, he founded the Academy and College of Philadelphia, he worked to establish the Pennsylvania Hospital, he was elected to the Pennsylvania Assembly, and he served as a Justice of the Peace for Philadelphia.

Aside from his many positive contributions to science, politics, diplomacy, and printing, Ben Franklin was also good at hoaxes. In 1732, he began publishing a comic almanac that started with the hoax of a (fictional) editor who predicted the death of the real editor of a prominent almanac. In 1742, he wrote a "Plain Truth" essay pretending to be a Scottish Presbyterian trying to incite military actions. In 1747, he invented and published a story about a woman who was put on trial by Puritans for the sin of having many babies out of wedlock.[20] The story gained much international attention, but twenty-two years later, Franklin admitted that it was entirely a prank. In other publications, he concocted plausible but fake messages from other characters: a Jesuit, the King of Prussia, and a Muslim who supposedly relished slavery. He also published a humorous letter stating that dead flies come back to life when submerged in wine. He published plenty of words that were simply not true.

Benjamin Franklin was not a Presbyterian, Jesuit, or Muslim; he was an active member of a secretive brotherhood: "that most ancient and Right Worshipful Fraternity," the Freemasons. In 1734, when he served as Grand Master of their Pennsylvania Lodge, Franklin reprinted "The

Constitutions of the Freemasons," a treatise from 1723 that noted, among other things, that freemasonry, all of it—civil, military, and sacred—was based on the geometrical theorem authored by Pythagoras.[21]

Did Franklin really carry out the electric kite experiment? How can we fly a kite out of a window or doorway during a rainstorm? One would need to prevent the line from touching the window or door frame, so that the electric charge is not lost. How can the rain wet the line while keeping a silk ribbon tied to it dry? Tom Tucker, having researched the matter, even trying to fly kites out of a window and through a wooden frame, concluded, "*He-really-didn't-do-this.*"[22]

I used to agree with Tucker, but now I'm not sure. I had studied this topic in books, and I became convinced that a compelling argument against the story was that if Franklin were to fly a kite out of a building it would be terribly awkward to constantly try to prevent the line from touching or snagging against the sides of the window or doorway, as Tucker described. Therefore, one day, I went out to a field by El Morro castle, in Puerto Rico, and I flew a kite with my father from under the archway of an old structure, and also from under the entrance of another tall building with columns. The line was remarkably stable, it did not touch the sides of these entrances, not for dozens of minutes, not at all. So I no longer think that flying a kite from a building and keeping the line steady is a difficulty. Still, I did not fly the kite in a thunderstorm, and I did not fly it under a drizzle of rain. These factors in Franklin's account still make his story seem implausible: first, one would have to elevate the kite before the rain starts, then move to the covering provided by a building; then hopefully a drizzle would start, and finally, hopefully, the rain would not wet the portion of line beneath the silk ribbon. (Why not simply soak the string before elevating the kite?) Personally, I choose not to guess the past, I just don't know whether Franklin managed to do this.

Still, most writers and historians duly credit Franklin as having carried out the experiment. For example, in a classic article, I. B. Cohen accurately discussed much evidence, addressed various ambiguities, and he concluded that Franklin did fly the kite in June 1752. But at the end of his article, Cohen admitted that he had begun by assuming that what Franklin and Priestley had written was true, and subsequently had tried to interpret all the evidence in ways that would square with their account. Having overtly used a plethora of reasonable conjectures, he admitted that several more were implicit: "I am fully aware that many more statements in this article should contain such words as 'very likely,' 'possibly,' 'probably,' 'may well have,' etc."[23] By contrast, I have used no such speculative arguments, and therefore I have not reached the same conclusion as Cohen.

The television show *MythBusters*, in an episode aired in 2006, tried to replicate Franklin's experiment. The show's cast members constructed several kites, following Franklin's vague specifications, but they did not manage to get them to fly until they modified them. Instead of testing how a kite flying under a storm cloud can gather electricity, they tested the obvious myth that lightning struck the kite. Using a Van de Graaff generator, they transmitted an electrical discharge onto the wet string, which created a tiny spark from the key. A much greater discharge set it on fire. Next, they raised the generator to 480,000 volts and flew the kite—a spark of electricity jumped to the kite, down the wet string to the key, and onto the hand of a dummy made to look like Franklin. A heart monitor showed that this shock would have killed him.[24]

Meanwhile, in 2006, an electrician in Belize was flying a kite with a niece. But he wanted a longer line so, not having any more string, he added some long copper wire. Flying, the kite approached some high tension power lines. Reportedly, the kite broke loose and the copper wire fell onto the electrical lines which then electrocuted the man, knocked him down, and burned him severely; he died soon after.[25] Following this incident, Belize Electricity Limited issued a press release giving several kite safety tips, saying that one should never use wire, and also: "Never fly kites in wet or stormy weather."

I don't know whether Franklin flew the kite, but even if he didn't, the story still works. That one of America's founding fathers crowned his real contributions to science with a captivating bluff, that the mere *idea* of his experiment was extraordinarily influential. The rags-to-riches commoner who boldly used a child's toy to draw down the awesome power of lightning from the sky—this image insinuates equality: that self-educated, lone amateurs can contribute to science just as much as the intellectual elite. Even if Franklin did not actually carry out the experiment, some brave individuals managed to pull it off with various contraptions and safety precautions. Besides, there are other experiments in the history of electricity that were similarly extraordinary and clearly did happen. Whereas in the kite experiment we may well be fearful and hesitant to test its truth by carrying it out, there are other perplexing experiments that we can well try.

# 7

# *Coulomb's Impossible Experiment?*

In many schoolbooks, electricity shows up as just another dry, boring, difficult, and monotonous subject. But it wasn't always that boring. Electricity seemed to hold the secret of life after death. In 1802, an Italian experimenter, Giovanni Aldini, performed demonstrations before French scientists, using dead animals and prongs to transmit electrical currents. Witnesses reported: "Aldini, after having cut off the head of a dog, passed the current of a strong battery: the mere contact triggers truly frightful convulsions. The mouth opens, teeth rattle, eyes roll in their sockets; and if reason did not deter the agitated imagination, one would almost believe that the animal is again suffering and alive."[1]

Aldini also had the nerve to conduct this kind of experiment on human bodies, publicly. And audiences had the courage to witness the results. Back then, punishment for criminals did not always end at death. In London, for example, a man could be sentenced to death followed by a public dissection. Bodies were cut and flayed and organs were pulled out for the edification and education of the masses, and also as a continuation of penal torture. On 18 January 1803, George Foster was executed by hanging at Newgate Prison, London, for murdering his wife and child by drowning them. His lukewarm corpse was then taken to a house where Professor Giovanni Aldini would "galvanize" it. The proceedings were reported in the *Newgate Calendar: The Malefactor's Bloody Register*.[2] In front of a medical audience, Aldini applied an electrical rod to the cadaver's mouth and another to an ear, whereupon the face grimaced horribly, his jaw quivered, and an eye opened. Aldini also applied the electrical rods to the corpse's rear end, causing the entire body, legs and arms, to convulse.[3] Some spectators feared that murderer George Foster was coming back to life. One old official of the Surgeons' Company was so alarmed that soon after he left he died of fright.[4] The *Bloody Register* noted that if a convict were revived, he would have been killed by hanging again.

Electricity was a high-stakes field in the early 1800s: Aldini's ultimate aim was to learn to "command the vital powers," and disturbed by the implications of electrical experiments, Mary Shelley wrote the horror novel *Frankenstein*. Electricity seemed to hold not only the power of life and death, but also of sanity, as Aldini reportedly cured mentally ill patients by inflicting electric shock. Even decades earlier, in the 1740s, when experimenters had found a way to store electrical fluid in a so-called Leyden jar, electricity was used to amuse and perplex, by administering it even to dozens of men in a chain, all of whom would then scream and contort. Moreover, electricity could be used to really do the presumably impossible—to move things without touching them.

Since ancient times, people had seen that when certain materials are rubbed, especially amber, small things such as hair move toward them. Amber could pick up small bits of stuff from the ground. According to Diogenes Laertius, the ancient philosopher Thales of Miletus believed that there was soul or life in inanimate objects such as amber.[5] This precious yellow substance is the fossilized resin of extinct coniferous trees, and since it was yellow, the Greeks called it *elektron*, as it resembled the pale gold metal that had the same name. That metal, electrum, is a natural alloy of silver and gold, and its root word, *elektor*, meant "beaming sun."

By the late 1700s, numerous devices harnessed the divine powers of electricity. It became increasingly important to understand electricity. How does it work? Could it be understood in terms of natural laws? If anyone could find a mathematical order in electrical effects, then electricity would be understood not as an occult wonder, but as a natural phenomenon. A solution to this problem was advanced, some fifteen years prior to Aldini's works, by a retired French engineer: Charles Augustin Coulomb.

What Coulomb claimed to find amounted to saying that electricity obeys the following algebraic law:

$$F = k\frac{q_1 q_2}{d^2}$$

It states that two electrical charges ($q_1$ and $q_2$) attract or repel one another with a force that increases as the square of the distance ($d$) between them decreases. The remarkable thing about this equation is that it is formally identical to Newton's law of gravity:

$$F = g\frac{m_1 m_2}{d^2}$$

That is, the force of gravity between masses varies with the square of the distance in the same proportion. Newton's law seemed to meet the reputedly Pythagorean ambition to understand phenomena numerically. And likewise, Coulomb's law exhibited the same form, submitting the

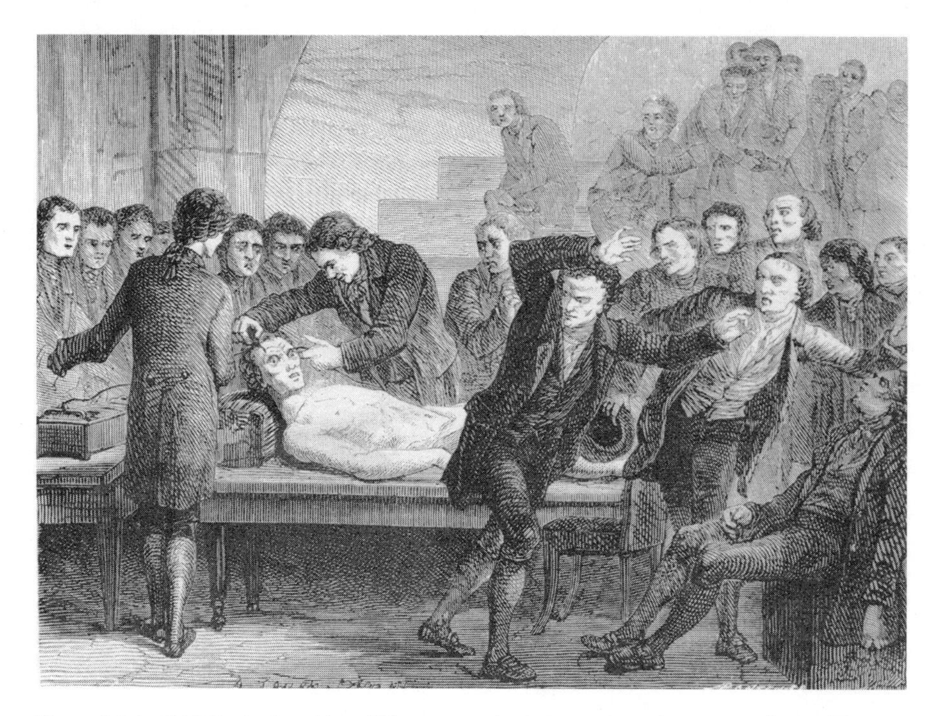

Figure 7.1. In 1818, Dr. Andrew Ure of Glasgow applied electricity to the corpse of a murderer, and then the dead man apparently grimaced in anguish and anger.

mysterious phenomena of electricity to the harmonious rule of numbers.[6]

Why did the two laws have the same form? Gravitational effects seemed quite different from electrical effects. For example, gravity seemed to be universal; according to Newton all bodies have gravity, whereas not every object seemed to have electrical charge. And charge flows in and out of bodies such that one object can have a lot of charge at one moment and apparently none at another, whereas gravity never seems to vary in any one object. Moreover, electrically charged objects can attract or repel one another, whereas gravity only attracts. So why would such different forces follow the same mathematical equation?

One might imagine that electrostatic attraction is weaker than gravity, because we usually see it in minor things such as hair and sweaters, whereas the sun's gravity holds the immense planet Jupiter in orbit, millions of miles away. Actually, electrical force is stronger than gravity. By contrast, the gravity on a comb is much too weak to attract anything as large as a flake of dandruff. Electrical force is billions of times stronger. Roughly put, electrical force is $10^{39}$ times *stronger* than gravity. It does not instantly kill us because we live in a mostly delicately balanced, nearly neutral electrical environment. But lightning kills. So, scientists struggled to figure out how electricity operates.

Following Newton's successful account of the dynamics of the heavens, people in Europe tried to replicate his mathematical method in various fields. The members of the Paris Academy of Sciences, for example, were convinced that reason aided by mathematics could comprehend the most puzzling physical phenomena. Charles Coulomb was a member of the Paris Academy, and he worked to analyze the electrical force. In 1785, he announced to the academy that he had proven that electrical forces behave mathematically like Newton's gravity. Whereas the academy had cast doubt on Franklin's brief and vague report, they accepted and praised Coulomb's experiment. How did Coulomb do it?

First, let's look at the inverse square equation. Until we have made some experimental tests, it is only a conceivable theoretical account. So, to test the equation, one might find some of its numerical consequences and check to see if such numerical values, or others, actually show up in some experiments.

One simple property of the equation is that it entails that if the distance between two bodies decreases by half then the force between them should become four times greater. This might not be obvious to everyone, so it may be explained as follows. Consider two electrically charged bodies, $q_1$ and $q_2$, separated by a distance *d*. If the electrostatic repulsion between them really can be described by an inverse square law, then we expect that their force of repulsion is:

$$F_1 = k\frac{q_1 q_2}{(d)^2}$$

Now, if these two bodies, while keeping their charges, are brought closer together, so that their separation is now *half* as large, then we may express their force of repulsion as:

$$F_2 = k\frac{q_1 q_2}{(d/2)^2}$$

Again, only the force and distance have changed. This equation may be rewritten as:

$$\frac{1}{4}F_2 = k\frac{q_1 q_2}{d_2}$$

which together with the equation for $F_1$ yields:

$$4F_1 = F_2$$

This means that *at half the distance*, there is *four times the force* between the two bodies. (As with the alleged finding of the Pythagoreans: the same tension upon a string half as long acts four times more.) Coulomb devised a way to test this simple relation. If electrical repulsion really acted as the equations suggest, then wherever the distance between two charged bodies is reduced by half, the force between them should quadruple.

Coulomb designed an instrument for measuring electrical forces. Experimenters knew that bodies repel if they carry the same kind of electrical charge, moving away from one another. One might measure the force by measuring their mutual repulsion. Coulomb had studied the properties of wires under torsion. He had found that when a wire is twisted by a certain amount, it exerts a proportional force against that torsion. For example, if you twist a wire to a given angle, it reacts by pushing back with a certain force, and if you twist it to an angle twice as large, then it pushes back with a force twice as strong. Coulomb realized that he could use this property of wires under torsion to measure electrical forces.

Coulomb constructed an apparatus that he called the "torsion balance." He provided an illustration, which, like the image of Ben Franklin flying a kite, has been reprinted countless times—in fact, it has been characterized as the one diagram of any experimental device that has been "reproduced more often than any other diagram."[7] Inside a glass cylinder, Coulomb hung a thin, silver wire. Onto it he attached a thin vane, horizontally, carrying a small ball of pith on one side and a paper counterweight on the other.[8] Nearby, a rigid stick held another small pith ball at the same height. By imparting electricity onto the two balls, Coulomb could make them repel. One ball remained fixed in place, while the other ball (suspended from the vane and wire) moved away. In turn, the wire, being twisted by this repulsive force, exerted a reactive force in the opposite direction. At some point, the force of repulsion reached a balance against the wire's reactive force, and the movable ball stopped moving.

Once the balls were separated by a constant distance, held in place by the balance of the two forces, Coulomb could twist the top of the device to force the suspended ball to come closer to the stationary ball. Then he could measure how much any given twist affected the separation between the balls.

Now, Coulomb wanted to test the force equation. Once the bodies separate, to measure the force of electrical repulsion means to assign a number to it. The farther away the two balls are, the greater the force pushing them apart, and the farther the balls separate, the more the wire is twisted. Therefore, the torsion of the wire expresses the force. If the balls separate by, say, 30 degrees, we expect that the wire is twisted by 30 degrees, and so we may well say that the force is "30." That number represents the force, quantified in terms of how much the wire is twisted. Now, in this case, the number also represents the distance (the angular separation) between the two balls.

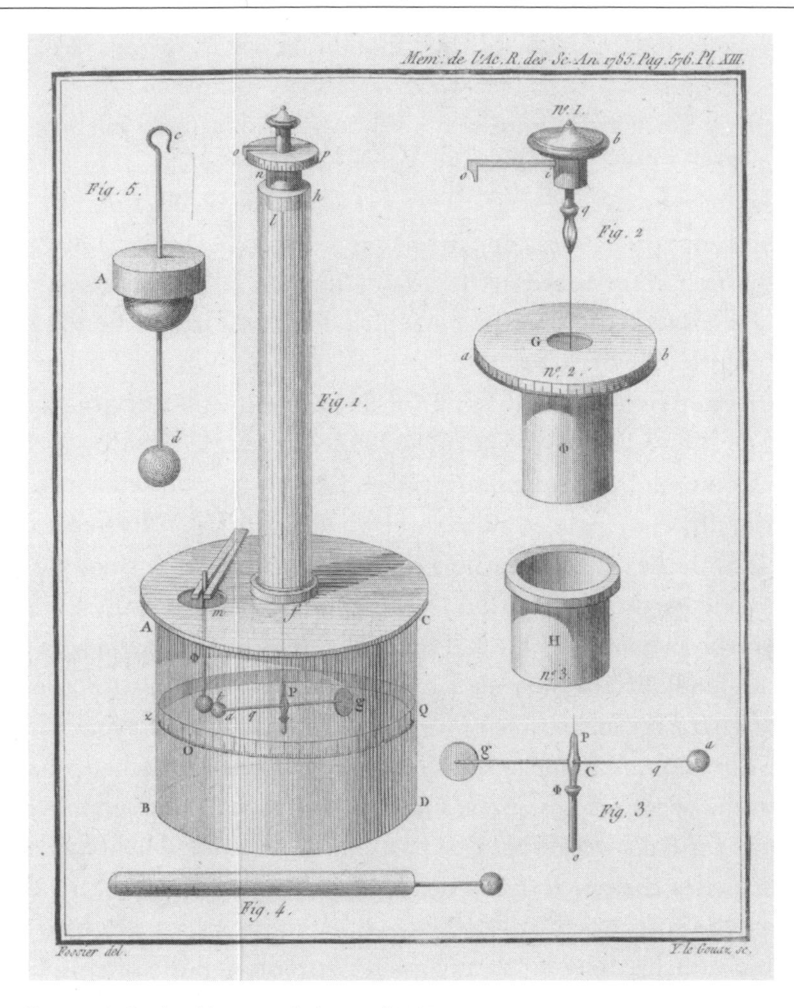


Figure 7.2. Coulomb's torsion balance of 1785.

Coulomb could increase the force between the balls by forcing them closer together. And he did so by turning a knob at the top of his instrument, where the wire was attached, bringing the one ball closer to the other. Thus the force between them increased, and it could be quantified by taking note of the total torsion on the wire. Coulomb had marked degrees all around the knob, to show how much torsion he had added to the wire. Meanwhile, the two balls continued to repel, thus adding extra torsion onto the wire. Thus, in his device:

$$\textit{total wire torsion} = \textit{angle on the dial} + \textit{angular separation between the balls}$$

Meanwhile, the distance between the two balls is measured easily, approximately, by the angular separation between the two balls. And to measure that separation, Coulomb used a strip of paper, divided into 360 degrees, pasted all around the large glass cylinder, at about the height of the two pith balls.

Now, since he could measure the distance between the balls, plus the force acting between them, he had two terms, $d$ and $F$, with which to test the equation in question. One way to do so is to use the algebraic property we identified: that at half the distance the force is quadrupled.

So, Coulomb compared how much the separation changes when the balls are forced closer together. First, he imparted electricity onto the two balls. He reported that the suspended ball moved away from the stationary ball until the two were separated, steadily, by 36 degrees. Thus, the total torsion on the wire, at that stage, was 36 degrees as well.

Next, Coulomb, turned the knob enough to add four times as much force as before, to see what separation would result. He expected that if he quadrupled the force, the balls should become separated by half the initial distance. The balls should then rest at 18 degrees. So how much should one twist the knob to quadruple the force of repulsion? The total torsion of the wire should be $36 \times 4 = 144$. But the total torsion of the wire, as we pointed out above, consists of the angle on the upper dial plus the separation between the two balls. (The knob pushes the ball in one direction while the repulsion pushes it away in the opposite direction, causing a greater total torsion on the wire.) To have a *total torsion* of 144, we need to include the separation of the balls, 18. So Coulomb knew that one should twist the knob to: $144 - 18 = 126$, in order to get a total force of torsion 144.

Next, one would expect that if the force were quadrupled again, 144 $\times 4 = 576$, then the separation should become half as large, $18/2 = 9$. So, to obtain that separation, one would need a total torsion of 576 on the wire, consisting of 567 degrees on the upper dial plus 9 degrees on the lower ruler.

In short, the inverse square equation of electrostatic repulsion predicts that if the initial separation between the two balls is 36, then one would obtain the following results by forcing the balls closer together, as we have described:

| | | | |
|---|---|---|---|
| *knob dial* | 0 | 126 | 567 |
| *separation* | 36 | 18 | 9 |
| + | | | |
| *force* = | 36 | 144 | 576 |

These are purely abstract, theoretical results, assuming the simple algebraic equation that expresses the expectation that electrical force varies inversely with the square of the distance. Compare the theoretical numbers with the numbers that Coulomb reportedly obtained in his experiment:

| | | | |
|---|---|---|---|
| *knob dial* | 0 | 126 | 567 |
| *separation* | 36 | 18 | 8.5 |
| + | | | |
| *force* = | 36 | 144 | 575.5 |

These numbers, the only data that Coulomb published on the matter, are remarkably close to the predicted series. Coulomb concluded that he had found that the fundamental law of electrostatics was an inverse square law. In his paper, Coulomb did not allude to any witnesses to his experiments. Yet he presented his apparatus to the Paris Academy of Sciences and wrote that his results were "easy to repeat, and immediately disclose to the eyes the law of repulsion."[9] Just as Ben Franklin said about his kite experiment: *easy*.

In order for the experiment to work at all, it was essential that the electric charge be imparted only onto the pith balls and that it remain there rather than spread onto the other parts of the apparatus. For that purpose, it would help to use plastic as an insulating material separating the pith balls from their supports. But in 1785, synthetic plastic was not available. Instead, Coulomb used natural thermoplastic polymers manufactured from the sticky resin secreted by various species of lac insects. Such insects live mainly in soapberry and acacia trees in India, wherefrom their sticky resin was collected and exported by Venetian merchants to Spain and France. Thus, Coulomb used a compound, so-called Spanish wax (though it contains no wax) to coat the single strand of silk into a thin, rigid, horizontal vane for the suspended pith ball, and he further insulated it with an additional coating of gum-lac on the tip. Likewise, the stationary pith ball was also supported by a rigid stem of Spanish wax.

As far as I know, in 1785, Coulomb's torsion balance was the most sensitive instrument ever devised for measuring forces. To turn the thin wire a full 360 degrees would require merely a force so tiny that it would lift just one 40 millionth of a pound.

The French soon became convinced that Coulomb had indeed proven a fundamental law of nature. Over the next two hundred years, Coulomb's experiment became known as an excellent example of how experiments establish physical laws. Many physics textbooks reprinted Coulomb's diagram of his torsion balance. Also, various instrument makers produced versions of the torsion balance, which were purchased by physicists in Europe and America.

Still, books seem to have referred to Coulomb's data alone, rather than adding any other data whatsoever. Eventually, this lack of reports corroborating Coulomb's results raised some questions.

In the early 1990s, Peter Heering, at the University of Oldenburg, in Germany, carried out historical research to find out whether past scientists had operated the torsion balance as successfully as Coulomb. But Heering found no evidence that any other physicists had ever obtained results as good as Coulomb's single report. Instead, he found that several German physicists in particular, in the 1800s, had reported difficulties operating the torsion balance. Of course, maybe there were others who had no such difficulties obtaining results similar to Coulomb's. But if so, why does it seem that no such results were ever reported in print?

Accordingly, Peter Heering decided to replicate Coulomb's experiment. Heering and his colleagues at Oldenburg constructed a replica of the torsion balance, following Coulomb's original prescriptions such that the dimensions of every part of the instrument were copied closely. Still, not all materials were duplicated; for example, the wire was not pure silver but copper, it was not clamped but soldered, and the needle was not made of silk and Spanish wax but of PVC. Nevertheless, the properties of each part were presumably equivalent to those prescribed by Coulomb, so it all should have yielded comparable results. Once all the components were in place, it took six more months of daily work to stabilize and calibrate the device before being able to take meaningful measurements. Heering carried out many experiments. But in none of the experiments did results similar to Coulomb's emerge. Finally, Heering reported: "in none of the experiments was it possible to obtain the results that Coulomb claimed to have measured."[10]

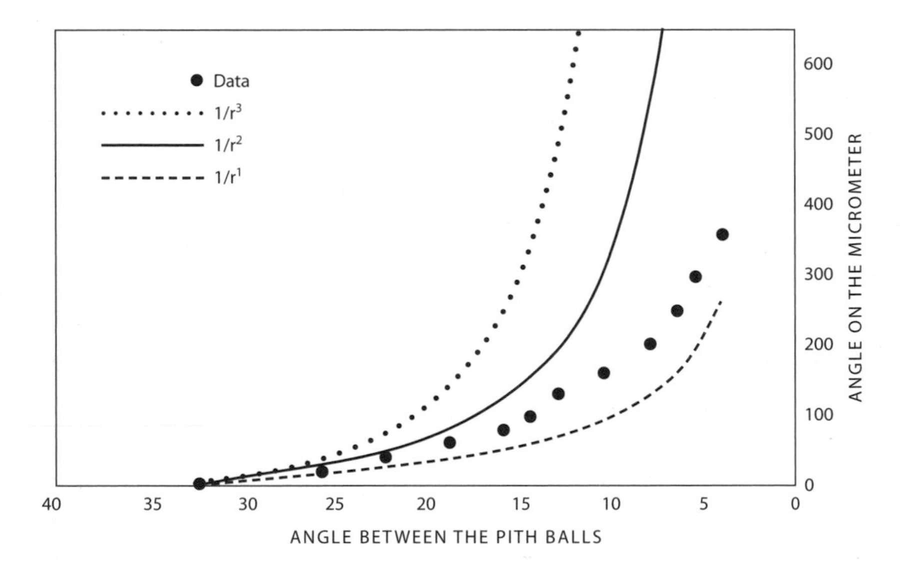


Figure 7.3. Results from one of Peter Heering's attempts to replicate Coulomb's experiment.

As shown in the graph above, in Heering's results, the data does not confirm the central line, which stands for the results predicted by the inverse square law of electrostatics. Even though Heering gathered much more data in this trial than Coulomb reported in his, Heering confirmed that the delay in taking more readings did not affect the result (because the pith balls did not lose much charge in that time). Using this data and some trigonometry to calculate the exponent $n$ in a presumed inverse law $1/r^n$, we get a result of $n = 1.28$ instead of the theoretical $n = 2$.

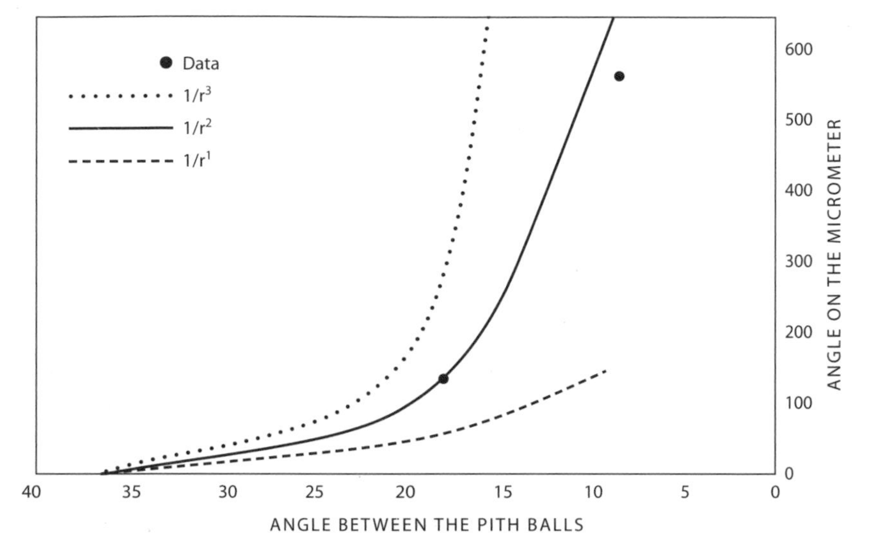


Figure 7.4. Results from Coulomb's experiment of 1785.

Compare that to Coulomb's reported results, as graphed here, which give an exponent of $n = 1.91$. Clearly, Coulomb's reported numbers seem to very nearly select the curve that belongs to an inverse square law. The only way that Heering managed to obtain results nearly as "good" was by altering the experimental setup.

The suspended needle behaved erratically. Heering reported that "it was absolutely impossible to measure the exact position of the movable pith ball because of these oscillations."[11] Heering noticed that whenever the experimenter, himself, approached the torsion balance, the movable pith ball oscillated. It became evident that electrical charge on the body of the experimenter affected the behavior of the experiment. So, to shield the device, Heering surrounded it entirely with a wire mesh, known as a "Faraday cage." Only then did Heering obtain results that were almost comparable to Coulomb's. But Coulomb described no such metal insulation around his device. Moreover, such a metal cage was devised only decades later by Michael Faraday, in the 1830s, so it seemed unlikely that Coulomb would have a comparable arrangement, especially since he did not mention anything of the sort.

Therefore, Heering argued that "Coulomb did not get the data he published in his memoir by measurement."[12] Another historian, John Heilbron, remarked that "It appears from Heering's careful and resourceful labor that Coulomb either faked his numbers altogether or obtained them under experimental conditions materially different from those he reported."[13]

Meanwhile, Christian Licoppe argued that in Coulomb's various papers, his rhetoric of purported instrumental precision was tailored to his primary audience, mathematicians of the Paris Academy who were ready to believe that electricity obeyed a simple mathematical law, like Newton's law of gravity, and to accept the rather unique and private results of a fellow member.[14] There seemed to emerge consensus that a disparity exists between Coulomb's elegant and simple narrative of 1785 and complications that arise in actual practice. In particular, Coulomb's wonderfully accurate numbers seem unlikely in light of Heering's results as well as Coulomb's admission in his original memoir that his balance suffered from defects that he planned to correct later. Accordingly, some historians have advanced conjectures about tacit procedures or practical knowledge that Coulomb might have exercised but left unreported. If the movable pith ball, once charged, oscillated for a while before settling, then how could Coulomb possibly make his three readings in only two minutes? Heilbron conjectured that Coulomb may have guessed where the moving ball would come to rest and reported that guess as an actual reading of its position. Heilbron also proposed: "Probably Coulomb put some torsion on the wire by twisting the knob even for the first data pair, although in the narrative he says that in its first position the needle balanced between the electrical force of the balls and the torsion given the wire by their repulsion alone. Coulomb's convenient round number for his first measurement ($f = q = 36°$) suggests that he had his hand on the micrometer knob for the first setting and that he had a special way of estimating the equilibrium position of the needle."[15]

The apparent discrepancy between the text and the practical subtleties led to various speculations about the actual operational procedures involved as well as past practices of scientific reporting. It also led to perceptive questions about possible unidentified material factors in the laboratory environment. Christine Blondel wondered whether Coulomb glazed his apparatus with an insulating varnish. Jed Buchwald wondered whether Coulomb used a short-range telescope to make his readings while staying a distance away from the apparatus. Maria Trumpler, and Heilbron, even commented that perhaps factors such as a wig and a silk shirt could affect the charge on the experimenter.

Furthermore, students at the Massachusetts Institute of Technology, under the guidance of Jed Buchwald, attempted to replicate Coulomb's experiment. But their results were not even as good as Heering's. And in France, Bertrand Wolff operated a torsion balance and observed unstable needle behaviors similar to those reported by Heering, which rendered the device utterly inconclusive for testing Coulomb's law. A documentary video concluded: "Since the results announced by Coulomb were in such good accord with the inverse square law, one may think that from among numerous measurements, he selected those which confirmed the previous hypothesis of a law analogous to Newton's law of universal gravity."[16] A plausible conclusion, but speculative. Strangely, in over two centuries, and despite hundreds of discussions about the torsion balance in textbooks and journals, no documents showing actual data comparable to Coulomb's three data pairs have surfaced.

Today, a widespread and disturbing habit exists in physics classrooms. Students in many lab classes try to reproduce one or another basic experiment, but often, they do not get the results predicted by the physical theory. Then, when writing the report that will determine their grade, they proceed to "fudge" the results. Students discreetly adjust data to make it seem as if they were closer to the theory than was actually the case. Wouldn't it be disturbing to find that, originally, famous scientists also fudged the data in their experiments? Don't we hope that, rather than the theory determining whether the results are correct, experiments decide whether a theory is valid?

So whom should we believe: Coulomb and the French Academy of Sciences and the many old teachers and books that parroted Coulomb's report? Or should we believe the physicists and historians who, two centuries later, having the resources of high-tech experimental physics and the benefit of critical hindsight, argued that Coulomb's report was suspicious? It might seem that the apparent triumph of Coulomb's law was artificial and socially constructed.

But there's an alternative. Belief in science should not have to be decided by appeals to authority or by popularity votes. Instead, we could well carry out the experiment and see what results. Maybe then one's judgment would not be an opinion.

In light of Heering's account, Coulomb's experiment seemed so fascinating and puzzling that once I actually tried to carry it out. I'm not a physicist, and I had no prior significant experience replicating old scientific experiments, but teaching history sometimes leads to such attempts. In spring 2005, at the California Institute of Technology, I put together a torsion balance following Coulomb's specifications and benefiting from advice from historian of science Jed Buchwald. I tried various materials for some of the components, starting with foam balls and relatively thick wires, as a metallurgist told me that it would be "impossible" to make a wire as thin as Coulomb claimed to use. Failing also to follow some of Coulomb's prescriptions, I fitted components in various ways. For example, I could not find a way to clamp a very thin copper wire (thinner than Coulomb's; the metallurgist was wrong), so instead I tied it at the top, where it hung. Also, I attached the fixed pith ball to a wooden rod instead of a rod made of sealing wax. Later I used rods made from a white plastic coat hanger. To insert electricity into the system, I attached a metal pin with a round tip onto a plastic rod and rubbed it on a bundle of cat hair. For months, I did not obtain results similar to Coulomb's. It was quite an unsuccessful effort.

Gradually, I made a series of adjustments, identifying components that seemed responsible for the erratic effects. For example, I managed to clamp the wire firmly to flattened metal, inserted in turn into the tip of a mechanical pencil. Also, for a while I used a thread covered in sealing wax as the needle that held the suspended pith ball. And it exhibited erratic oscillations such as Heering had described. (By this point, my original skeptical doubts increased, for how could such a crude tabletop instrument test a fundamental law of physics, given that it does not even use a vacuum?) But in my case, it turned out that the oscillations were caused by a leak of electric charge from the pith ball onto the waxed thread, as I *finally* found the waxed thread itself was attracted by the other charged pith ball. So, I replaced the waxed thread with a thin stick of blue plastic that I made by melting a disposable Gillette razor.

Figure 7.5. The dimensions and properties of each component closely approximate Coulomb's prescriptions of 1785.

Thus, I refined the parts of the device one by one until each component other than the pith balls was effectively insulating against the charge, and until the fittings among the parts were just as Coulomb had described them. I was also delighted to obtain and use wire nearly as thin as Coulomb's and made of 99.99% pure silver. And I found that it was important to ensure that the bare dehydrated pith balls were polished very smoothly.[17]

Experiments are complicated by constraints: in my case, I was running out of time. I soon had to leave California to move to a new job in Texas. But by August, once all the components were in place, and all resembled closely Coulomb's prescriptions, the balance behaved very stably. I carried out another series of experiments. The experiments were digitally recorded, including having a camera follow the manual adjustment of the dial knob to a given position and the ensuing stable position of the movable ball, held in electrostatic balance, as it aligned with a particular angular number on the tape measure around the large cylinder.

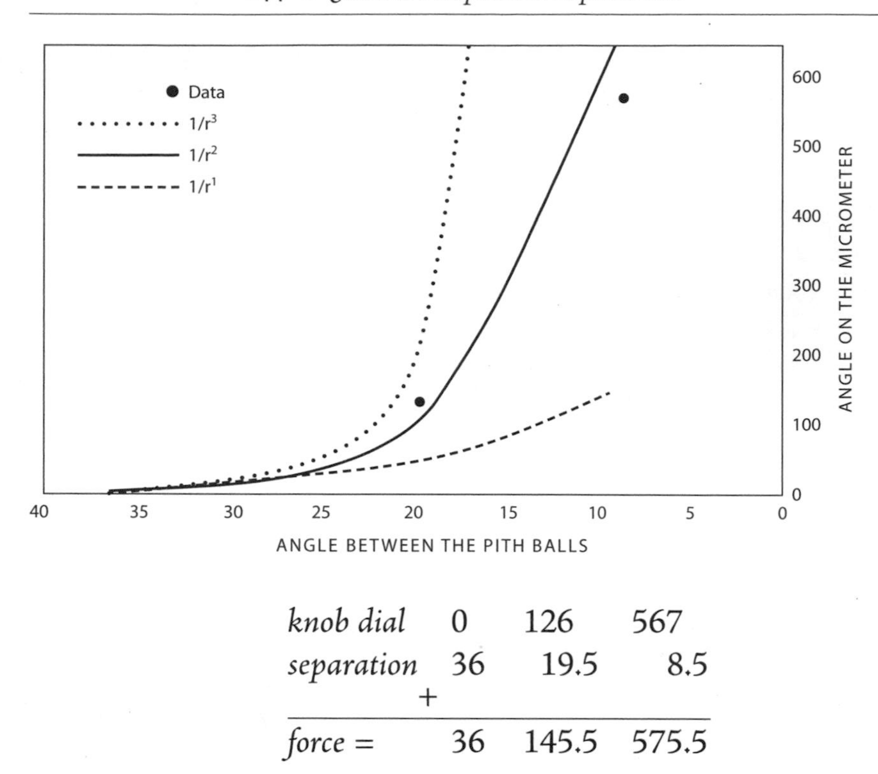


| | | | |
|---|---|---|---|
| *knob dial* | 0 | 126 | 567 |
| *separation* | 36 | 19.5 | 8.5 |
| + | | | |
| *force =* | 36 | 145.5 | 575.5 |

Figure 7.6. Results from a replication of Coulomb's experiment.

I was stunned when I saw these results! This experiment, and others like it, showed that one can obtain a neat separation of 36 degrees at the start, just as Coulomb reported. It also showed that when one next moves the dial knob as Coulomb noted, it is readily possible to obtain numbers similar to his. Moreover, these results showed nice agreement with the expectation that at four times the force, the separation is reduced to about half. And, in the results shown above, the data points fall closely near the curve for the inverse square law. This experiment gave an exponent of $n = 1.96$. Consider another example, graphed here.

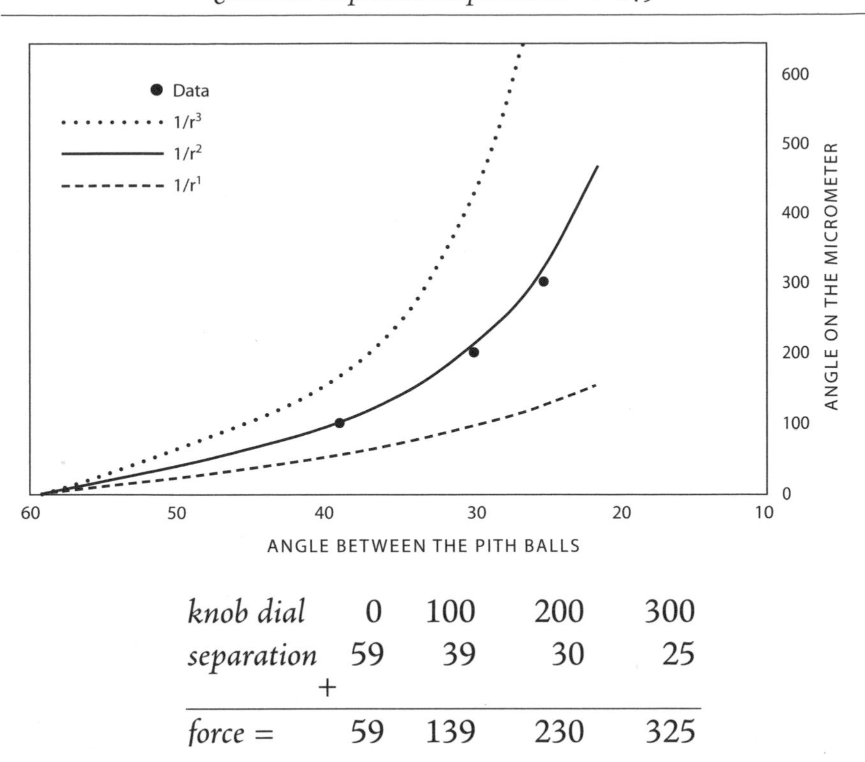


| *knob dial* | 0 | 100 | 200 | 300 |
|---|---|---|---|---|
| *separation* + | 59 | 39 | 30 | 25 |
| *force =* | 59 | 139 | 230 | 325 |

Figure 7.7. Another replication of Coulomb's experiment.

Here, four readings were made instead of three. Consider the initial force on the wire, a twist of 59 degrees. Multiply that by four, and we have 236 degrees of torsion. Thus at 236 degrees, one should observe a separation of 29.5 degrees, half of the initial separation. And indeed, this is supported by the data: at 230 degrees of total force the separation between the balls became 30 degrees. Again, these numbers nicely confirm the expectation that at four times the force, the distance is reduced by half. Moreover, the data supports the inverse square equation.

This experiment gave an exponent of $n = 1.92$. The similarity to Coulomb's results was striking. My materials were not identical to Coulomb's, for example, I used a plastic needle from 2005, whereas he used Spanish wax manufactured in the 1780s—but they need not be the same, because what we were testing was a "law of nature," a general property of electricity.

Following Coulomb's prescriptions, the behaviors of my device matched his claims to a degree that I had not anticipated as probable.

Such results and more, together with various considerations, led me to conclude that there was no fudging in Coulomb's original report of 1785. Coulomb could well obtain his numbers from actual experimental measurements on the very torsion balance that he described in his report, using the procedures he described. The hundreds of textbooks and teachers who for centuries parroted Coulomb's claims were at least parroting a reliable source. Again, I am not a physicist or an experimenter, so it felt wonderful to participate in one of those processes in which persistent curiosity leads to surprising results. It strengthens a conviction that anybody who sets their mind to a problem has a chance of solving it: the ideal that is conveyed so well by the image of Ben Franklin flying a kite.

Coulomb showed that it is possible to design elegant experiments that neatly test fundamental physical theories with stunning numerical results. Moreover, the history of electricity reminds us that in science, the truth of propositions should not be decided by the authority of books or by living or dead great physicists—but by experiments. The case of Coulomb's experiment also tells us something about how carefully we should measure our words. In hindsight, we can see that despite the best intentions in reconstructing historical conditions, one might exaggerate the implications of particular findings. For example, it was an overstatement for Heering to claim, "Coulomb did not get the data he published in his memoir by measurement."[18] Likewise, we should be careful in using words such as "impossible."

Finally, there's a big difference between *algebraic accounting* and *causal explanations*. Indeed the motions of electrically charged bodies can be described by an inverse square equation, but we deceive ourselves if we think that by itself that's the best explanation we can obtain: that bodies "obey" the mathematical law. Why did the pith balls bearing equal charges move away from each other, rather than, like gravity, attract? What was electricity, really? What was this invisible thing that could move objects at a distance, cause thunder in the clouds, and even reanimate a headless corpse? About those things, Coulomb's experiment said nothing at all.

# 8

## *Thomson, Plum-Pudding, and Electrons*

We learn in school that electricity is a current of subatomic particles that are negatively charged: *electrons*. But who discovered that and how? Just as the question of whether Pythagoras discovered anything is complex, so is the issue of what constitutes authorship of a scientific discovery. In the case of electrons, fortunately, we have plenty of documentary evidence. That evidence involves a story about the British physicist J. J. Thomson.

Incidentally, I once had a problem with Thomson, when I was growing up in Puerto Rico. When I was about sixteen years old, I participated in a television quiz show with my school, and it came down to one last question, directed at me: "Which great physicist, from the beginning of the twentieth century, made breakthroughs that led other physicists to formulate theories and discoveries that later led to nuclear fission?" I puzzled over this question as the seconds ticked by, and I asked that it be repeated. Time ran out; I hesitated, and replied: "J. J. Thomson."

Wrong. The moderator expected the answer: Albert Einstein. I immediately argued that I thought that the question was ambiguous, but a producer in the shadows stopped me flat, yelling: "This is *television!*" Then they asked a student from the rival school: "Who was the first scientist to discover nuclear fission?" He replied: "Enrico Fermi." And they won.

Afterward, my teachers sympathized as I argued that Einstein was one of the physicists who "formulated theories that later led to nuclear fission," and that he lived and worked until 1955, and hence that he was not "from the beginning of the twentieth century." So I had thought that the answer should be someone who contributed to the discovery that the atom had structure, parts, and could therefore be split. (In hindsight, my guess was not good, partly because the question asked about splitting the nucleus, and J. J. Thomson did not seem to have contributed to that.) I also complained that the other student was wrong, because actually, Otto Hahn and Fritz Strassmann had split the atomic nucleus (as explicated by Lise Meitner, as I later learned). But to no avail, we still lost the game. That game-show incident now seems typical of a common disdain to check facts about the history of science. What difference does it make? This attitude shows up plainly in science classrooms. Consider the case of J. J. Thomson.

In countless schoolbooks, J. J. Thomson is known for two things. He discovered the electron, and he formulated the "plum-pudding" model of the atom. One achievement was a breakaway success, the other a flop. In the early 1900s, apparently, the atom seemed to be like a faint glob of positive charge with small negative electrons stuck at random places—like raisins in that bready old English treat (plum-pudding has no plums). But schoolbooks tell us that Ernest Rutherford and his assistants falsified Thomson's plum-pudding atom by spraying alpha particles (positively charged helium atoms) against thin gold foil. Sort of like shooting bullets at a fluffy slice of pastry. Surprisingly, some of the bullets bounced back, suggesting that the atom was really not a soft glob of fuzz but that it had a dense, hard nucleus. It's a classic story of progress in physics.

For years, Rubén Martínez (no relation to me) researched the history of the plum-pudding model of the atom. He analyzed hundreds of books and documents. Yet he found no evidence at all that J. J. Thomson ever advanced a plum-pudding model of anything.[1]

None of Thomson's models of the atom resemble the pictures or accounts of the plum-pudding model that famously shows up in science books. Instead, Thomson theorized, for example, that the atom consists of a series of stacked planes on which negatively charged particles rotate in circles, around the axis of an immaterial positive sphere. The earliest appearance in print of the plum-pudding tale, found by Rubén Martínez, was in a physics textbook of 1943. If anyone did formulate an atomic model that roughly looked like plum-pudding, it was actually an older Thomson. In 1899, William Thomson (no relation to J. J.), better known as Lord Kelvin, described the atom in a way that closely resembles the plum-pudding cartoon that decades later showed up in so many schoolbooks.

Picking up this search, I have found several early mentions of a pudding atom, even in a book of 1919, in which the author attributed it mainly to Lord Kelvin.[2] More generally, and irrespective of any Thomson, an analogy was published in the 1890s by someone who argued that perhaps matter is not made of particles, but is instead roughly continuous, "like a plum-pudding."[3] Thus, J. J. Thomson's plum-pudding is just a tasty myth. If ever you're trying to remember whether there's a *p* in J. J. Thom(p)son, just say to yourself these lines:

> *There's no P in J. J. Thomson,*
> *no plum-pudding in his atom.*

Textbook writers should rewrite their books. But there are better things that they can keep—right?

"In 1897 J. J. Thomson discovered the electron." That's what countless teachers, textbooks, websites and encyclopedias routinely say. But here, too, ever since the late 1980s, increasingly many historians have cast a cloud of doubt on that claim.[4] Before we consider their complaints, let's look at what J. J. Thomson actually did.

In the 1870s, when the young Joseph John Thomson studied physical science at Trinity College, Cambridge, there was a growing feeling that physics was finished. He later recalled that there existed "a pessimistic feeling, not uncommon at the time, that all the interesting things had been discovered and that all that was left was to alter a decimal or two in some physical constant."[5] In Germany, likewise, a professor of physics, Philipp von Joly, advised his student Max Planck not to pursue a career in the field, on the grounds that physics was essentially complete.[6] Not everybody shared this pessimism, but it persisted into the 1890s. Another student of physics, Robert Millikan in New York, similarly recalled: "In 1894 I lived in a fifth-floor flat on Sixty-Fourth Street, a block west of Broadway, with four other Columbia graduate students, one a medic and the other three working in sociology and political science, and I was ragged continuously by all of them for sticking to a 'finished,' yes, a 'dead subject,' like physics, when the new 'live' field of the social sciences was just being opened up."[7] But then, as the usual story goes, new things were discovered: X-rays, radioactivity, electrons, and so forth.

In April 1897, J. J. Thomson announced findings to the Royal Institution concerning his analysis of the mysterious "cathode rays," and he elaborated his arguments and results in a paper published in October.[8] Cathode rays were produced by sending an electrical current through a wire ending in a metal bit (the cathode), inside a glass tube containing a rarefied gas, such that a hazy colorful glow appeared inside the glass tube. These rays transmitted electricity to the opposite end of the tube.

In 1895, Jean Perrin had shown that cathode rays seemed to be stuck to their electrical charge.[9] He showed that when cathode rays enter a receptacle, they convey charge, but when the same rays are deflected by a magnet, such that they don't enter the receptacle, no charge is collected in the receptacle. Thomson repeated Perrin's experiments in a modified arrangement to check whether when the cathode rays were deflected by a magnet, the charge was deflected equally. He concluded that the charge was inseparably attached to the rays.

Thomson also tested whether the cathode rays could be deflected by an electric field. In 1883, Heinrich Hertz had tried to deflect the rays in an electric field but found no such effect and thus had concluded that cathode rays could *not* consist of negatively charged particles. Yet Thomson inferred that residual gas in the "vacuum" tubes affected Hertz's experiments. Thomson managed to better empty his vacuum tubes for the same kind of test, and accordingly he found that when the cathode rays passed between electrified metal plates, the rays were clearly deflected. This meant that the rays might well consist of negatively charged bodies.

Thomson also analyzed the curving paths of cathode rays deflected by magnets while crossing through various gases. He found that the path of the rays was independent of the kind of gas used.

In sum, since the rays behaved in the same way as negatively charged material particles, Thomson concluded that the rays consisted of minute particles of matter. Still, he wondered, were they molecules, atoms, or something even smaller?

Thomson analyzed the properties of such negatively charged "corpuscles." He estimated their velocity, charge and mass. First, the cathode rays traveled across the vacuum tube to a collector at the opposite end. An electrometer there measured the total quantity of charge $Q$ received at the collector. By assuming that this charge was composed of the sum of a number $N$ of individual corpuscles, each having an electrical charge $e$, Thomson wrote:

$$Ne = Q$$

Another device at the collector measured the rise of temperature there, and from that, Thomson calculated the energy of the cathode rays. By definition, for each corpuscle the *kinetic energy* $= 1/2mv^2$ (where $m$ and $v$ are its mass and velocity). Thus, the total energy $E$ transported by all the corpuscles would be:

$$E = N(1/2mv^2)$$

Thomson also placed two magnetic coils of wire along the sides of the vacuum tube. And when he ran an electrical current in those wires, their magnetic field $M$ deflected the cathode rays. The stronger the field, the more sharply the rays were deflected. Thomson described the ray's trajectory by its "radius of curvature." To explain, imagine a small bullet flying while deflected by a powerful magnet away from a straight path. The stronger the magnet, the more the path of the bullet curves. If we extend that curved path to sketch a circle, that circle has a radius; the smaller the circle, the smaller its radius. So, if the bullet has a lot of momentum, or if the magnetic field $M$ is weak, the bullet will not be deflected much, so its radius of curvature will be large. Now, if the bullet is about to hit a target, it will hit harder depending on it being deflected least, therefore, the bigger its radius of curvature $R$, the greater the momentum. Thus, *electromagnetic momentum* $= MRe$.

Thomson used various vacuum tubes to measure the cathode rays: for their electrical charge, the temperature rise they caused, and their radius of curvature.[10] Using these measurements, and converting temperature into kinetic energy, Thomson calculated the velocity, mass, and charge of the negative particles as follows. Since the momentum of projectiles is given by mass times velocity, Thomson had: $mv = MRe$. This equation includes the velocity of the particles, $v = MRe/m$ which can then be entered into the equation for the total energy $E$, above, which together with $N = Q/e$ gives:

$$\frac{QM^2R^2}{2E} = \frac{m}{e}$$

The remarkable thing about this equation is that it translates macroscopic measurements into microscopic invisible quantities: the mass and charge of the negative particles. Likewise, Thomson estimated the velocity of these invisible particles, with the equation:

$$v = \frac{2E}{QMR}$$

Again, this invisible property, the velocity of the negative particles, was inferred on the basis of four measurements: the total energy (temperature), the charge transported, the magnetic strength, and the radius of curvature.

In addition to these experiments, Thomson carried out another experiment, substantially distinct, to measure the quantities $m/e$ and $v$. Whereas his first method used the deflection of rays in a uniform magnetic field, the second method employed also an electrostatic field instead of the accumulation of charge at a collector.

By comparing the rays' mass to charge ratio $m/e$ against the mass to charge ratio known from the motions of molecules and atoms in a magnetic field, Thomson conjectured that the cathode ray particles were much smaller than atoms. The $m/e$ of the cathode ray corpuscles were a thousand times smaller than the $m/e$ of a charged hydrogen atom.

Now, during his experiments, Thomson found that the cathode rays behaved pretty uniformly. He used various gases in the tubes: air, hydrogen, carbonic acid. He also tried different cathode materials: aluminum, platinum, and iron. In all, the values $m/e$ of the cathode rays were practically unaffected by such different factors.

Thomson argued that his experiments led to the conclusion that gas atoms can split into smaller "primordial atoms," which he called "corpuscles." He argued that this "new state" of matter (not solid, not liquid, not gas) was all of one kind, and that it was the substance that makes up all the known chemical elements.

That all summarizes Thomson's work of 1897. He was right to conclude that cathode rays consist of negatively charged particles smaller than atoms. And his estimates of their velocity, mass, and charge were not bad. He was also right that the "corpuscles" are constituents of atoms. So, did Thomson discover the electron?

Well, a key question is: How do we know that any physical effect involves particles? One traditional kind of evidence was to ask: Do any such effects travel in straight lines? Like bullets? Physicists such as Newton used to argue that light consists of particles because it travels in straight lines, casting sharp shadows. Likewise, in 1869, Johann Hittorf showed that cathode rays too cast shadows. His student William Crookes also carried out experiments that showed sharp shadows. For example, Crookes placed an iron cross on the path of the cathode rays and found that they cast a pretty sharp shadow.[11]

Crookes also found that the cathode rays all have identical proper-

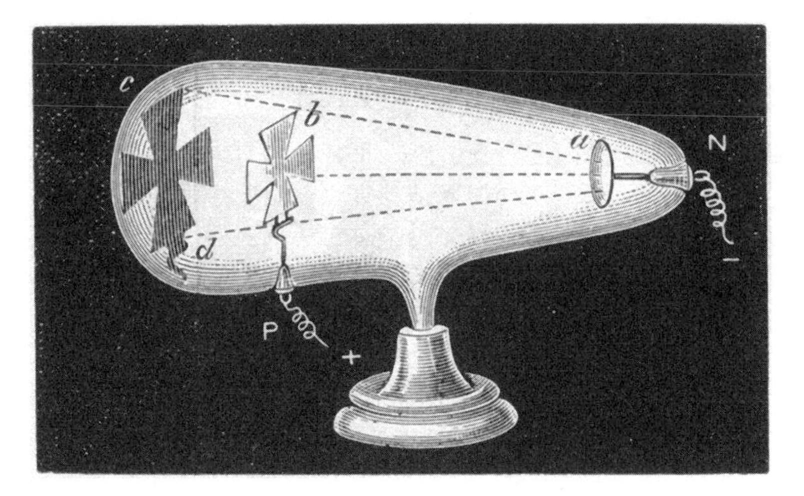


Figure 8.1. Crookes's experiment showing the shadow cast by cathode rays.

ties, independent of the various materials involved. Furthermore, another way to test whether an effect consists of particles would be to determine whether it can *push* things, whether it imparts momentum. Accordingly, Crookes devised a small metal wheel with paddles, on glass rails, such that the paddles would be hit by the cathode rays. He found that when the rays hit the paddles, the wheel spun and moved, showing that the rays impart momentum. Crookes also showed that the rays could be deflected by a strong horseshoe magnet, and that two parallel streams of rays repel each other like electrified bodies (like the pith balls in Coulomb's experiment).

In 1879, Crookes concluded, on experimental grounds, that cathode rays were "a fourth state of matter" (not solid, liquid, or gas), a kind of "Radiant Matter." He argued: "The molecules shot from the negative pole may be likened to a discharge of iron bullets," bullets extremely minute and moving at extraordinary speed and encountering resistance from gases. He described the rays as "the little indivisible particles which with good warrant are supposed to constitute the physical basis of the universe."[12]

In 1884, Arthur Schuster argued that cathode rays consist of negatively charged particles produced when molecules near a negative cathode were broken up.[13] He inferred that scarce but significant experiments seemed to suggest that the particles carried a constant charge. By experimenting with magnetic deflections, by 1890, Schuster estimated the

Figure 8.2. Crookes's experiment showing the momentum of cathode rays.

upper and lower bounds of $e/m$, the charge to mass ratio.[14] He quoted a statement by Hermann von Helmholtz, which said "we cannot avoid concluding" that electricity behaves as consisting of "atoms of electricity." But Schuster did not think that atoms were divisible or had removable pieces; that might sound too much like the old and supposedly crazy alchemy. Schuster later recalled that "The separate existence of a detached atom of electricity never occurred to me as possible, and if it had, and I had openly expressed such heterodox opinions, I should hardly have been considered a serious physicist, for the limits of allowable heterodoxy in science are soon reached."[15]

Since the 1870s, George Johnstone Stoney had argued that there are material units of electricity, positive and negative, permanently attached to atoms. He imagined that they rotated around atoms, and in 1891 he called these tiny units "electrons."[16]

Still, another way to argue that cathode rays were particles smaller than atoms would be to prove that they can penetrate materials that are opaque to atoms. This was demonstrated in Bonn in 1892 by Hertz and his student Philipp Lenard.[17] But they argued that since cathode rays penetrated matter, they were some sort of waves, like sound passing through a wall, not particles. Lenard showed that as the rays passed through a metal foil, they deviated, fanning out. (This result could instead be interpreted like bullets ricocheting through a fence.) Lenard also manipulated rays outside of glass tubes, showing that they were not an effect of rarefied gas, but were self-standing phenomena.[18] Lenard also studied the rate at which the rays diminish in brightness as they travel distances in gas. He found that regardless of the gas in which they moved, the rays exhibited the same magnetic deflection.[19] Thomson interpreted Lenard's findings as suggesting that corpuscles were smaller than atoms because they crossed between many molecules without hitting them.

In 1895, Jean Perrin showed experimentally that the negative charge invariably accompanies the cathode rays, a result that, he said, supports the claim that the rays consist of particles, not waves.[20] Thomson acknowledged that these results influenced his work. Furthermore, in April of 1896, Gustav Jaumann published experimental results showing the electrostatic deviation of cathode rays.[21] So contrary to widespread claims, Thomson was not really the first to do that.[22]

Yet none of these scientists identified all the properties of electrons (or "ions" or "corpuscles"); they made mistakes in some of their characterizations and conjectures. Then again, so did J. J. Thomson. For example, he claimed that the corpuscles were the *only* constituents of atoms—he was wrong.

Thus, prior to 1897, various physicists had demonstrated that cathode rays seemed to consist of negatively charged particles smaller than atoms because: the rays traveled in straight paths, cast shadows, penetrated thin foils, imparted momentum, and could be deflected magnetically. Some of those physicists also estimated the charge to mass ratio of such particles, and they conjectured that these were components of atoms.

What's left of the claim that J. J. Thomson discovered the electron? One might argue that Thomson was the first to *measure* the charge to mass ratio. That argument was propagated by several influential physicists including Oliver Lodge, Norman Campbell, and Robert A. Millikan. But they were mistaken.[23] By 1890, Schuster had published upper and lower estimates. In 1896 and in March of 1897, prior to Thomson's work, Pieter Zeeman measured and published the charge to mass ratio of negatively charged particles and concluded that they were about a thousand times smaller than charged atoms.[24] Zeeman described his findings as "direct experimental evidence" of the existence of the charged material particles.[25] And, in January 1897, also prior to Thomson's work, Emil Wiechert too showed that cathode rays consist of charged particles, "electric atoms," that are smaller than ordinary molecules. He also calculated upper and lower bounds of $e/m$. Wiechert declared: "Here we are not dealing with atoms as we know them in chemistry, because the mass of the moving particles turns out to be 2000 to 4000 times smaller than the mass of a hydrogen atom, the lightest of the known chemical atoms."[26]

One might imagine that maybe Thomson's measurements of $e/m$ were just the most accurate at the time. But in 1897, Walter Kaufmann obtained values that were considerably closer to the value that we now recognize.[27] Moreover, Thomson's values did not clearly seem to converge toward a constant.[28] His largest reported value was about five times greater than his smallest. And the results from his electric-magnetic deflection method compared to his magnetic method differed by about 20 percent. This discrepancy was so large that one of Thomson's biographers commented that it would have been unacceptable even in a grocery transaction.[29]

What about the Nobel Prize? After all, Thomson was awarded the prize in 1906, so it might seem that physicists agreed that his contribution was outstanding. However, his prize was not awarded for having discovered any subatomic particle. It was awarded for his work on "the conductivity of gases"—cathode rays. Moreover, Lenard was awarded the Nobel Prize in 1905, for *his* work on cathode rays. (And, Lenard later claimed that his early experiments proved the existence of electrons.)[30]

Regardless, another way in which a few physicists and historians have defended Thomson's reputed discovery is by claiming that even if Thomson's work of 1897 did not prove the existence of the electron, his later work did. In 1899, Thomson showed two more ways of measuring the charge to mass ratio, roughly matching the earlier results.[31] Moreover, he managed to measure the charge alone of the corpuscles, which in turn led to a calculation of their mass.[32] Those were immensely valuable contributions. Still, as noted by historian Theodore Arabatzis, those are just two properties of electrons, among many others, so why should the identification of those properties in particular give someone the title "discoverer of the electron"? (Moreover, the values of mass and charge were estimated earlier by Joseph Larmor and H. A. Lorentz.[33]) Physicists who already recognized the existence of electrons were impressed by the accuracy of Thomson's new measurements, but they did not construe them as a qualitative discovery. For example, Arthur Schuster recalled about Thomson's important lecture of 1899: "It at once carried conviction, and though to those who had followed the gradual development of the subject, it only rendered more certain what previous experiments had already plainly indicated, the scientific world seemed suddenly to awake to the fact that their fundamental conceptions had been revolutionised."[34] In 1900, Pierre and Marie Curie did not characterize Thomson as having discovered the electron; instead they esteemed him as having "completed" the ballistic theory of William Crookes.[35]

One more way to defend Thomson's alleged role is to argue that his work, at least, was the key contribution that finally convinced most scientists to accept electrons. This approach implies that the credit for being the discoverer of something consists not merely in being the first to make the claim, but in being the one to actually succeed in persuading the majority of the community of peers.

For example, Charles Coulomb was not the first to propose "Coulomb's law," as others such as Joseph Priestley proposed it earlier. Likewise, Charles Darwin was not the first to propose that species evolve by natural selection. In a book of 1831, on the breeding of trees to build boats, Patrick Matthew argued that the superabundance of offspring against pressing competition and circumstances, makes species change.[36] His conjectures seem to have been ignored. But after Darwin's success, Matthew claimed credit, and he described himself as the "Discoverer of the Principle of Natural Selection." Yet he did not get the credit. In Darwin's opinion, speaking generally, "all the credit" for discoveries goes to whomever succeeds in convincing readers.[37]

Yet there were several physicists who helped to gradually convince the community, or ascertain phenomena, regarding electrons. Some showed that cathode rays were rectilinear; others showed that the rays transfer mechanical momentum, like bullets; still others showed that the rays consist of particles smaller than atoms; and so forth. The problem with claiming that J. J. Thomson convinced everyone is that, again, this claim does not hold up historically. Plenty of physicists remained unconvinced. In 1897, for example, George FitzGerald considered Thomson's claim that the corpuscles were the fundamental constituents of all matter, and noted the dramatic implication that physicists were now on "the track of the alchemists," moving toward "a possible method of transmutation of matter."[38] FitzGerald also considered other plausible hypotheses, including that negative corpuscles were just charges devoid of matter. Similarly, John Zeleny, a young physicist who worked in a laboratory in Berlin at the time, reported that "nobody in Berlin" believed the announcement that the electron was a corpuscle.[39] Max Planck too, in 1900, still disdained material electrons as a dubious hypothesis. Only by the 1910s, after the works of many experimenters, did nearly all physicists accept the notion that matter is made of particles such as atoms and electrons. As late as 1914, one of Thomson's own students, Owen Richardson, who intensively investigated electrons, commented that a series of discoveries over "the last fifteen years" had finally established their existence.[40] Ernest Rutherford had a similar view.

Just as some chemists doubted the existence of Marie Curie's radium before she finally isolated it, physicists could doubt electrons until one was isolated. Thus in 1899, for example, the chemist Henry Armstrong criticized Thomson's work as not showing that corpuscles were separable from atoms; and further, even the physicists who then did appreciate Thomson's work did not construe it as the discovery of a new particle.[41] The isolation of a single electron was accomplished finally by Robert A. Millikan in 1911. Incidentally, Millikan did not even believe that Thomson alone discovered the electron.[42]

Looking back at the "discovery of the electron," Theodore Arabatzis proposed to construe that expression properly as describing a complex process that led to the consolidation of physicists' belief in electrons. In that sense, he concluded that J. J. Thomson did not discover the electron, he just contributed to the process of its acceptance.

J. J. Thomson was not the first to believe that cathode rays consist of particles, and he was not the first to experimentally substantiate that conjecture. This was clear to physicists and engineers in the early 1900s. In 1906, Edmund Fournier D'Albe published a historical presentation of the electron theory in which he argued: "It has not been heralded by a flourish of trumpets, nor has it been received with violent opposition from the older schools. No one man can claim the authorship of it. The electron dropped, so to speak, into the supersaturated solution of electrical facts and speculations."[43] Yet decades later, it became commonplace in textbooks of physics and chemistry to claim simply that "in 1897 J. J. Thomson discovered the electron." What was it about that claim that made it so appealing, so common, and easy to echo? I really don't know, but one might hazard to guess a few reasons. Maybe Thomson's image suited textbooks: a proper Englishman with eyeglasses and trimmed mustache, hunched over his experimental apparatus, a flattened flop of black hair over his prominent forehead. Perhaps this image of a serious-looking, beady-eyed gentleman who paid attention to the smallest details might insinuate to young students that they too, in their own disciplined way, could someday discover important things? And the name—*J. J.*—does its phonetic repetition help people to remember it? Marginal human tidbits to gradually shape students into physicists? Maybe Thomson seemed appropriate for being regarded as a discoverer because he was acknowledged to be a gentleman. (By contrast, William Crookes seemed more eccentric: he claimed to have witnessed spirits. He claimed to hear rapping sounds, see objects moving by themselves, even see persons levitate while standing or float up on chairs, sometimes in full daylight. He claimed that invisible beings communicated by writing, even with a floating pencil. During séances in which Crookes held onto the hands and feet of a medium, he claimed to have seen small, luminous hands floating in the air, some that even touched him and pulled at his coat. In darkness dimly lit, he claimed to see, speak with, and repeatedly touch, a beautiful apparition called "Katie King." She wore white robes, and Crookes touched her hands, neck, put his ear to her chest to hear her heartbeat and lungs, and followed her into a cabinet.[44])

I think that a stronger reason why the Thomson claim became common is that it gives a simple and definite story of origins: the electron is such an important object in physics textbooks, such a seemingly solid, active, and forceful thing, that it begs for a single discoverer plus a plain date when it was found. By giving a seemingly straightforward provenance to the invisible electrons, teachers can avoid the tortuous intricacies of how such entities were actually identified, how physicists gradually and contentiously became convinced of their existence.

Myths disguise our ignorance. The simple discovery story of electrons conveniently hides the complications of physics and history, to enable students to promptly imagine these invisible objects, to just use them in explanations and calculations. If we were to try, instead, to tell students the actual process by which from the 1870s until 1913 physicists struggled to identify these components of matter and electricity, it would take a fair bit of work. So instead, textbooks portray a simplistic vignette. But as historians have delved into the matter, they have increasingly rejected the attribution of the electron's discovery to any one individual.

Still, some writers have concocted new and subtle ways in which it might still be fair to say that Thomson did discover the electron, at least in a weak sense of the term "discover." For example, the philosopher Peter Achinstein has argued that by elucidating the notion of discovery, on the basis of three main criteria, we might well say that Thomson discovered the electron, because: Thomson discovered something that really exists, he knew that he had done so, and he had some priority in that realization. Yet Achinstein admitted that the question remains complex, and opinions differ.[45] Looking at this issue, Bruce Hunt, a historian of physics and electricity (and my colleague at the University of Texas), commented that even if we do apply Achinstein's own criteria, we would hardly conclude that Thomson discovered the electron. Instead, it would then seem that Zeeman, or perhaps Lenard or Wiechert did so.[46]

As for me, I see no reason at all to regard J. J. Thomson as the discoverer of the electron. And I think that, pedagogically, there are better stories we can tell about how physicists came to know that cathode rays consist of subatomic matter that is negatively charged. In particular, students will more readily understand how physicists came to think of cathode rays as negatively charged material particles thanks to the results on the experiments in which the rays cast shadows and impart momentum (Crookes). And, they came to think that such matter was negatively charged in view of magnetic deflection (Crookes, Schuster, Perrin). And, they came to appreciate the subatomic size of electrons thanks to the experiments of various physicists (including Crookes, Lenard, Zeeman, Wiechert, Thomson, and Kaufmann). And later they isolated such entities (Millikan).

You might still want to know how the legend of J. J. Thomson started. Thomson had some part in the matter, for in old age he described his own achievements in ways that did not fairly acknowledge the contributions of others. But also, historians of physics, such as Isobel Falconer and E. A. Davis, have pointed out that apparently it was Thomson's former students who began to promulgate the narrow, simplistic discovery tale.[47] Graeme Gooday has pointed out that one of Thomson's colleagues wrote an influential article, in a manual of electrical engineering published in 1931, in which he claimed that J. J. Thomson, as a "disinterested seeker after the truth," discovered the electron in 1897 and that from that one key discovery there followed many practical devices.[48] But even this latter claim was false. Technologies (such as television) that are often attributed to the discovery of the electron actually originated from learning how to control cathode rays, rather than knowing what cathode rays are.

I think that the electron discovery story also developed to satisfy an explanatory craving. As readers and students of history, we want to know who discovered what and when. We know that for ages people did not know the invisible nature of electricity, and we expect that with technological improvements someone eventually managed to see what had been hidden for so long: the invisibly small. A similar case was discussed years ago by Thomas Kuhn. He rightly analyzed historical problems of pinpointing the discovery of oxygen, in the 1770s, and he commented: "Though undoubtedly correct, the sentence, 'Oxygen was discovered,' misleads by suggesting that discovering something is a single simple act assimilable to our usual (and also questionable) concept of seeing. That is why we so readily assume that discovering, like seeing or touching, should be unequivocally attributable to an individual and to a moment in time."[49]

It is a myth that J. J. Thomson discovered the electron in 1897, but by "myth" I do not mean to belittle it as a plain falsehood. It is a myth because it functions as a marker, an apparent milestone to punctuate and orient our historical imagination. The electron, many students think, is a solid, discrete particle of electricity, like an invisibly small billiard ball. Such a thing, yes, such a solid natural thing, known for ages for its awesome effects—lightning and static electricity—begs for a story of origins, at least about how we came to find it. Such a definite object seems to demand a location in space, time, and history; like atoms and other subatomic particles, it seems to require a historical junction that pinpoints its entry into the common consciousness. Myths give a neat account of how somebody ingeniously overcame a great difficulty. After setting up the initial ambiguity between wave and particle theories, teachers have traditionally used the story of J. J. Thomson to convey an apparently neat solution, to give a sense of finality, and then move on. But really, no single person deserves sole authorship for this particular discovery.

Moreover, just as its discovery is actually smeared in time, the electron itself seems also to be smeared in space. In 1928, Thomson's own son made a perplexing experimental discovery. George Paget Thomson showed that electrons passing through extremely thin metal films behave not as discrete particles, but like waves smeared in space. Months earlier, C. J. Davisson independently discovered this as well, and they were jointly awarded the Nobel Prize in physics in 1937. In light of their finding, physicists who earlier had stopped short of asserting that electrons are particles now seem more prudent than Thomson. For example, in 1897, Walter Kaufmann had argued already that "the hypothesis that construes the cathode rays to be charged particles shot from the cathode is insufficient."[50] So, instead of saying that in 1897 J. J. Thomson discovered the electron, we can better say that several times since the 1870s, several physicists have found compelling evidence that electricity consists of particles; but time and again, others have encountered evidence to the contrary.

In the end, the textbook writers should just revise the old tasty myths. It's not enough to perpetuate old sayings by writing ambiguities: that so-and-so "is credited with" this, or that some discovery "is attributed to" such and such. Remember these ancient lines:

*There's no P in J. J. Thomson,*
*no plum-pudding in his atom.*
*Fourth or fifth on the electron.*

# 9

## *Did Einstein Believe in God?*

ALBERT EINSTEIN became famous not only for his physics, but for various clever statements that impressed many people. He sometimes spoke of God and religion, with measured words that resonated across a range of beliefs. Many Jews and Christians alike considered him a kindred soul, while many atheists and agnostics celebrated him as a fellow skeptic. When pressed, he sometimes spoke in ambiguous terms. What did the physicist really believe about religion?

He described his parents as "entirely irreligious" Jews.[1] His sister Maja also recalled that their parents did not discuss religious matters or rules. Yet they chose to provide a religious education for their son. At the age of six, they sent him to a public Catholic school in Munich, while also arranging to have someone teach him the principles of Judaism. Consequently, the boy sensed no major conflict between the two religions and somehow mainly harmonized them. He developed deep religious feelings, and he began to observe religious prescriptions in every detail.[2] He stopped eating pork. He read and accepted the Bible. He composed short songs of praise to God and sang them often to himself.

The young Albert also became increasingly impressed by mathematics and enjoyed reading popular books on science. At the age of twelve, in thinking about science, he became convinced that some of the stories in the Bible were just not possible. Right then, he abruptly abandoned religion. At that age, nearly all Jewish boys prepared to carry out their religious confirmation, the bar mitzvah, even in liberal Jewish families, yet the young Einstein refused to do so.[3]

And at the same time, he became fascinated instead with math, thanks to what seemed to him a "holy book" on geometry. It provided the clarity and certainty that he lost in religion. He was not bothered by how the geometrical proofs depended on assumptions that remained unproven and only apparently clear.[4] He lacked a sense of which parts of mathematics deserved critical attention. Accordingly, he did not pursue pure mathematics; he became a physicist instead, because in physics he could pinpoint assumptions that seemed annoying and questionable.

For Einstein, scientific inquiry, not science doctrine, became a sort of religious activity. He was motivated by a "holy curiosity" focused on an immense world that beckoned "like a great, eternal riddle" to a trustworthy "paradise" that would liberate him from the miseries of life.[5] The young guy was an irreligious free-thinker, yet for him, science and math came to function as a substitute religiosity.

In time, his theories acquired a mythical status, and Einstein became shrouded in false myths: that he was a bad student who dropped out of school but eventually became an absent-minded professor who authored perfect theories; that he was an always-old saint, a bleeding-heart sufferer for all humanity, loved by everyone. Such myths have been corrected, so I need not review them here.[6] But the question about his belief in God still requires attention. On one hand, it is quite clear that he entirely renounced religion at the age of twelve. On the other hand, the older Einstein is famously remembered for several moving and oft-quoted religious sayings. In 1919, Einstein received a telegram stating that astronomical observations confirmed his theory of gravity. A student asked Einstein what would have happened if the confirmation had not occurred, and Einstein replied: "Then I'd feel sorry for the dear God. The theory is correct anyhow."[7] In 1921, having heard about experimental evidence that his special theory of relativity might be wrong, Einstein commented: "Crafty is the Lord God, but malicious he is not."[8] Later, he criticized quantum theory by stating that nature, or God, does not act without causes. Einstein wrote to a friend: "The theory yields much, but it hardly brings us closer to the secret of the Old One. Anyhow I am convinced that he does not play dice."[9] In 1940, Einstein pronounced: "Science without religion is lame, religion without science is blind."[10] And moreover, he sometimes described himself as "a deeply religious man."[11]

However, the philosopher most admired by Einstein (as an adult) was

Baruch Spinoza, a Jew who did not believe in free will, nor in any cosmic purpose, nor in the existence of a personal God. Spinoza expressed a kind of religious reverence toward the grandeur of nature and its causal structure. Einstein admired the way in which Spinoza treated the human body and soul as one entity. Spinoza was sometimes described as an atheist, and he also became known as a pantheist: one who believes that nature *is* God. In 1929, Einstein argued that the rationality of the world, involved in the best scientific work, is akin to religious feeling: "This firm belief, a belief bound up with deep feeling in a superior mind that reveals itself in the world of experience, represents my conception of God. In common parlance it may be described as 'pantheistic' (Spinoza)."[12] And also in 1929, Einstein claimed "I'm not an atheist. I do not know if I can define myself as a Pantheist." But he added that "I am fascinated by Spinoza's pantheism."[13]

By "superior mind," did he merely mean to say *nature*? Einstein repeatedly stated that he did not believe in a God who concerns himself with the actions and fates of people, and repeatedly noted that a God who punishes and rewards was inconceivable to him and incompatible with causality. He also argued that there is nothing divine about morality. Einstein did not believe in any sort of life after death, writing: "I do not believe in immortality of the individual."[14] He repeatedly denied the idea that there exists a God that answers prayers. He also denied the idea of an anthropomorphic God, and he said that he could not imagine any will or goal outside of humans.[15]

So—did he believe in God? How many traditionally presupposed aspects may one subtract from the notion of God before one should say that what is left is just not what anyone means by "God"? In 1929, rabbi Herbert Goldstein sent a telegram to Einstein, which read: "Do you believe in God?" He replied "I believe in Spinoza's God who reveals himself in the orderly harmony of what exists, not in a God who concerns himself with fates and actions of human beings."[16]

In the 1930s, a famous actress, Elisabeth Bergner, also asked Einstein whether he believed in God. Instead of just answering yes or no, he said: "One should not ask that of someone who with growing wonderment tries to explore and understand the authoritative order of the universe."

Bergner asked why not, and he replied: "Because he would probably break down when faced with such a question."[17]

At the age of fifty-seven, in 1936, Einstein replied to a letter from a girl who asked whether he prayed. He did not, but he replied that "everyone seriously engaged in science becomes convinced that the laws of nature manifest a spirit which is vastly superior to man, and before which we, with our modest strength, must humbly bow."[18] What did he mean by *a spirit?* He added that this religiosity differs essentially from that of more naive people.

Einstein was interviewed in 1954 by William Hermanns, a professor and veteran who sympathized with Einstein's so-called "cosmic religion" and had interviewed him previously. Among other things, Hermanns asked him to state precisely his views about God. Years later, he reported Einstein's reply:

> About God, I cannot accept any concept based on the authority of the Church. As long as I can remember, I have resented mass indoctrination. I do not believe in the fear of life, in the fear of death, in blind faith. I cannot prove to you that there is no personal God, but if I were to speak of him, I would be a liar. I do not believe in the God of theology who rewards good and punishes evil. My God created laws that take care of that. His universe is not ruled by wishful thinking, but by immutable laws.[19]

This seems, again, a strange and seemingly abrupt contradiction. The first five sentences seem to come quite clearly from someone who seems to be an atheist or a nonbeliever, even a heretic. But suddenly, in two more sentences we suddenly hear the voice of a believer. Here, in contradistinction to many other instances, Einstein seemed to speak about a God prior to nature, who created the laws of physics. Likewise, a student of physics once wrote that Einstein told her: "I want to know how God created this world. I'm not interested in this or that phenomenon, in the spectrum of this or that element. I want to know his thoughts, the rest are details."[20]

On the grounds of these latter statements, Max Jammer argued that

Albert Einstein never renounced his cosmic religion. In his excellent book, *Einstein and Religion,* Jammer argued, however, that Einstein's allusions to God were not a return to the notion of a personal God, but were just a manner of speaking. Accordingly, one of Einstein's assistants in the 1940s, reported that Einstein once said: "What I am really interested in is knowing whether God could have created the world in a different way; in other words, whether the requirement of logical simplicity admits a margin of freedom."[21] If the second half of that sentence was meant to explain the first half, then it just gives an abstract physical meaning to a superficially religious statement. Likewise, when Einstein once said that God is crafty but not malicious, he later added the following clarification: what he meant, really, was that "Nature hides its secret through the sublimity of its essence, but not though trickery."[22]

We might well wonder whether Einstein just did not believe in God, but liked to refer to nature in religious language. He certainly felt a kind of religious reverence for nature. Did he use the name God merely as a convenient name for nature? No, claimed Walter Isaacson, in his number-one-bestselling biography of Einstein. Isaacson argued, "it was not Einstein's style to speak disingenuously in order to appear to conform. In fact, just the opposite. So we should do him the honor of taking him at his word when he insists, repeatedly, that these oft-used phrases were not merely a semantic way of disguising that he was actually an atheist."[23] Isaacson contended, instead, that Einstein believed in an impersonal God reflected in the glory of creation but who does not meddle in daily existence.[24] But what if Einstein was not hiding atheism, but agnosticism?

Einstein denied being an atheist. In his comprehensive book, Max Jammer agreed that Einstein was not an atheist. Jammer also quoted a statement by a rabbi who commented on Einstein's "cosmic religion" by saying that some theologians complained that it "may become a kind of Pantheism almost identical with Atheism."[25] Likewise, an editorial in the newspaper of the Vatican claimed that Einstein's views were an "authentic atheism even if it is camouflaged as cosmic pantheism."[26] Yet Jammer did not label Einstein a pantheist. Jammer also argued that Einstein was not agnostic.[27] Yet Jammer did not quote statements in which Einstein himself accepted the label of agnostic.

Once, writing to an inquisitive sailor, Einstein wrote: "I have repeatedly said that in my opinion the idea of a personal God is a childlike one. You may call me an agnostic, but I do not share the crusading spirit of the professional atheist whose fervor is mostly due to a painful act of liberation from the fetters of religious indoctrination received in youth. I prefer the attitude of humility corresponding to the weakness of our intellectual understanding of nature and of our being."[28] In 1950, he even more clearly explained "My position concerning God is that of an agnostic. I am convinced that a vivid awareness of the foremost importance of moral principles for the betterment and ennoblement of life does not need the idea of a law-giver, especially a law-giver who works on the basis of reward and punishment."[29]

Discrete labels, "-isms," are often inadequate when trying to pin down someone's views in what actually are not discrete categories, but broad spectrums of beliefs. Anyhow, almost ten years after Jammer's book was first published, a surprising letter by Einstein came to light. In 2008, Bloomsbury Auctions in London put up for sale a letter written by Einstein in early 1954, when he was nearly seventy-five years old. The letter had sat in a private collection for fifty years. It is just one and a half pages long, handwritten in ink on slightly browned, folded paper, with Einstein's signature; plus the envelope. In 1952, the philosopher Eric Gutkind published a book titled *Choose Life: The Biblical Call to Revolt*. Gutkind mailed a copy of his book to Einstein, with whom he had exchanged a couple of letters in 1946. In January 1954, Einstein mailed a letter to Gutkind, thanking him for sending him the book, and stating that he had read much of it. But he criticized Gutkind for proudly claiming a privileged position for Jews. And Einstein wrote:

> The word God is for me nothing more than the expression and product of human weaknesses, the Bible a collection of honorable but still primitive legends aplenty. No interpretation, no matter how subtle, can change this (for me). Such refined inter-

> pretations are naturally highly varied and have almost nothing to do with the original text. For me the unmodified Jewish religion, like all other religions, is an incarnation of primitive superstitions. And the Jewish people to whom I gladly belong and with whose mindset I have a deep affinity, have no different quality for me than other people. As far as my experience goes, they are also no better at anything than other human groups, though at least a lack of power keeps them from the worst excesses. Thus I can ascertain nothing "Chosen" about them.[30]

Instead, Einstein reaffirmed his admiration for Spinoza, for having asserted that humans, like the rest of nature, are not at all free from causality. Then Einstein set his intellectual differences aside, and noted that he yet was close to Gutkind "in essential things": their common overall attitude toward the community and their similar appraisals of human behaviors. Einstein died a year later, in 1955.

A spokesman for Bloomsbury Auctions stated that they were "100 percent certain" of the letter's authenticity. They estimated that Einstein's handwritten, brief letter would sell for £6,000 to £8,000 (roughly $11,676 to $15,568 in conversion to dollars at the time). But on May 15, 2008, it drew spectacular bidding. The successful buyer purchased the letter for the staggering sum of £207,600 (that is, $404,000), almost thirty times its presale estimate. The anonymous overseas collector was described only as someone who had "a passion for theoretical physics and all that that entails."

Thus, Einstein had privately admitted that, to him, the word *God* was only an expression of human weakness. The *New York Times* reported: "From the grave, Albert Einstein poured gasoline on the culture wars between science and religion this week."[31] For decades, many believers had gladly argued that Albert Einstein was on their side, quoting any of his many expressions. But regardless of what we believe or want to believe, we should recognize that Einstein did not necessarily share such beliefs. Much of what he said was subjected to public scrutiny. Accordingly, Einstein learned to express himself in ways that many people could find agreeable, even if they disagreed with one another.

In short, Einstein did not believe in Judaism and he did not believe in Christianity. He denied being an atheist. He admired Spinoza's pantheism, and at least once called himself a pantheist. And he sometimes accepted the label of agnostic. Do agnostics believe in God? No, but they also abstain from presuming to know the fundamental ordering of the universe. Early on, Einstein's thoughts about science erased his boyhood faith in Jewish and Christian theologies. But instead of abandoning "religion," he chose to redefine it, so that, for him, religion became the sense of awe and reverence for the harmony and hopefully causal order of the universe.

# 10

## *A Myth about the Speed of Light*

EINSTEIN kept some secrets, and he often shunned attention as if the details of his life were quite inconsequential. He wrote: "the essential in the being of a man of my type lies precisely in *what* he thinks and *how* he thinks, not in what he does or suffers."[1] Although he was a relatively self-isolating person, people flocked to him, reporters and photographers chased him around for decades, converting him into an icon. We see him in posters, magazines, toys, postage stamps, cereal boxes. As he complained in 1949: "my accomplishments have been overvalued beyond all bounds for incomprehensible reasons. Humanity needs a few romantic idols as spots of light in the drab field of earthly existence. I have been turned into such a spot of light. The particular choice of person is inexplicable and unimportant."[2]

Still, there were some fair reasons why he first won attention, one of which is that he managed to convince many people of some seemingly unbelievable ideas. Suppose that you sit on a spaceship traveling in a straight line, say, at 160,000 miles per second. According to Einstein's physics, the length of planet Earth as judged by you will be about *half* of its original length. And, allegedly, this is *not* an optical illusion. Does the entire planet really contract when you move?

Suppose also that as you're zipping away from Earth at 160,000 miles per second (mps), you aim a flashlight in the forward direction and turn it on. Light shoots out at about 186,282 mps away from you. How fast does that same light ray move away from Earth? We might expect that since it moves at 186,282 mps away from the ship and the ship moves at 160,000 mps away from Earth, the light moves at about 346,282 miles per second away from Earth. But according to Einstein's physics, that ray of light moves at 186,282 mps away from Earth. We might think that the light ray *must* move away relative to Earth at a greater speed, and we'd expect that just flying away on a rocket cannot possibly cause the contraction of the entire planet. Otherwise, the world is bizarre. Common sense seems to fly out the window as physics becomes unbelievable and esoteric.

Such notions crystallized in Einstein's theory of relativity and they became accepted by many physicists. His theory was all based on the idea of the relativity of time, the relativity of simultaneity. Yet for centuries, philosophers had theorized various ideas about time, without generating any intense public interest. For example, according to one source, the prominent religious leader and philosopher "Pythagoras believed that time is the encompassing sphere."[3] In the late 1700s, Immanuel Kant denied that time had any absolute or objective reality.[4] In 1902 the prominent mathematician Henri Poincaré echoed a growing conviction that "There is no absolute time."[5] While such various arguments did not generate widespread attention, somehow, the imaginative conjectures of a lowly patent clerk in 1905 eventually generated immense public acclaim.

Einstein reached his theory by a long and complicated trajectory, but instead of tracing a broad overview, let's take a close view of just one of its seminal aspects: the relations between speed, time, and length. I'll discuss the kind of explanation that we don't find in a physics class. Then in the next two chapters, I will discuss some controversial and historical speculations—myths— about the origins of Einstein's theory. How did Einstein come to think about relativity?

For ten years, from 1895 until 1905, Einstein was trying to understand the motion of light.[6] Physicists conceived of light as an electromagnetic wave in an invisible medium, like air, but much more subtle and intangible, which they called "the ether." When Einstein was just sixteen years old, he was puzzled because he wondered what a light wave would look like if one could catch up to it.[7] It's analogous to waves in the ocean; you constantly see them *moving*. Why don't we ever see a stationary mountain of water on the ocean? One would be perplexed to see such a thing, a standing bump of water. Yet if you fly alongside a wave at its same speed, and look at it, then relative to you, it would look like a bump in the otherwise flat water. Likewise, Einstein was puzzled by the idea of what light would look like if one could catch up to it.[8] Nobody had ever reported seeing a flash of light that stands still.

Figure 10.1. A mound of water on a still lake.

While in college, Einstein was fascinated by questions about ether, electrons, atoms, and light. He learned that several experiments on light had given puzzling results.[9] Even though scientists were sure of the existence of the ether (because light behaves like waves; causing effects of interference and diffraction), they failed to measure the speed of Earth relative to the ether. We can measure the speed of a boat relative to the water, so, why not the Earth relative to the ether?

After exhausting and nearly traumatic final exams, Albert graduated college in 1900.[10] He failed to get a university job, though he said that he applied to every job in Europe. Meanwhile, his parents faced immense financial difficulties. His mother came from a wealthy family, but his father's business with his uncle, an electrical company, collapsed, destroying all their capital. So Albert lived in considerable poverty.

In 1902, Einstein moved to Bern, the capital of Switzerland, and took a low-level job at the Federal Office of Intellectual Property, evaluating patent applications. In 1903, he married his college girlfriend. He smoked a lot. He drank a lot of coffee, but rarely alcohol, and not beer, saying that "beer makes one dumb and lazy."[11] He continued his struggles to understand light and electromagnetism. He worked on the puzzles of light and motion for two more years. He recalled: "I was plagued by all sorts of nervous conflicts; I went around confused for weeks."[12] Einstein continually obsessed over the problem, until he "feared for his health," and "wondered if this was the path to insanity."[13] After struggling for years, he had no results. By spring 1905, when he was twenty-six years old, he had reached a roadblock, frustrated.

He then visited his friend Michele Besso, a coworker. Besso was an absentminded mechanical engineer who knew about many topics, including physics, and was often attentive to petty details. Einstein and his shorter, bearded friend discussed every aspect of the light problem.[14] And right then they realized that there was something ambiguous about the measurement of time. It suddenly occurred to Einstein that the reason he wasn't making much progress was because he had taken the notion of time for granted.[15]

In the equations of electricity and magnetism, there appears a term identified as the speed of light, and since speed is distance over time, it raises a question: how is this measurement of time made? Einstein had believed that any concept in physics earns its right to be used only if it can be connected clearly to experience.[16] He had been influenced by the writer Ernst Mach, who sought to rid physics of all traces of metaphysics. One of the notions that Mach had singled out was Newton's concept of "absolute time." Newton had argued that in addition to the apparent measures of time that we obtain with clocks and observations, there exists an exact and true time that flows constantly, uniformly, and independently of anything.[17] Nearly all physicists agreed with him. But Mach, in his 1883 book called *Mechanics*, had complained: "With just as little justice, also, may we speak of an 'absolute time'—*of a time independent of* change. This absolute time can be measured by comparison with no motion; it has therefore neither a practical nor scientific value; and no one is justified in saying that he knows aught about it. It is an idle metaphysical conception."[18] Besso gifted Mach's book to Einstein in 1897, and Einstein read it while he was in college, and again in Bern before 1905.[19] Now Einstein suddenly realized that there was something obscure, ambiguous, about the measurement of time when we consider, *specifically*, bodies in motion.[20] After his helpful conversation with Besso, he went home frustrated that he still had not solved the problem, but sensing that he had pinpointed the crux of the matter, the notion of time—the key to the solution.[21] We don't know exactly what he thought, but we can piece together some relevant notions in order to explain his key insight.

We know that sometimes, often actually, clocks disagree. We usually decide which ones are right by referring to the best and most accurate clocks. Consider clocks at the Royal Observatory at Greenwich in south-east England or at the National Institute of Standards and Technology in Maryland. They have precise clocks, but how well do they agree with each other?

We can well bring two clocks together and adjust one so that it marks the same time as the other. And that seems to work well enough. But how do we synchronize clocks that are far away from one another? We might, for example, synchronize the clocks together and then move one clock far away. But how do we know whether the clock that we moved remains synchronized with the stationary clock? What if moving it disturbs the rates of motion of its parts, even slightly? We might make the effect smaller by making the distance smaller, or the motion smoother. But there's still an effect, quite likely, however small.

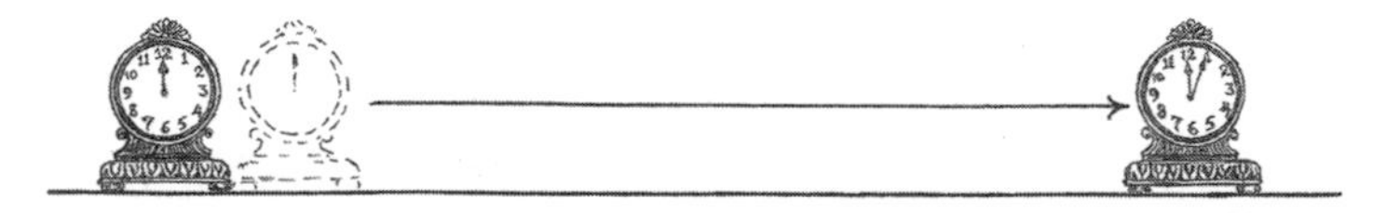

Figure 10.2. Two clocks are synchronized together, and then one of them is moved to a position a distance away.

One way to verify that the clocks are synchronized is to bring them back together. Then, either they do mark the same time, or they don't. If they mark the same time, then either they were indeed synchronized all along, or, they *became* synchronized only when we brought them back together. The problem of just moving one clock or both is that we know, in fact, that accelerated motion can and does affect the rate of clocks. (To accelerate a body is equivalent to hitting it with hammers, even extremely small hammers.) This was known for decades, as clockmakers tried to devise ever-improved chronometers for traveling.

So, to synchronize distant clocks, we should try to avoid moving them. Instead, we might connect such distant clocks with a long rigid beam, such that when we yank the beam the clocks start ticking. For example, given a beam connecting two clocks, both may be activated to start ticking by yanking the beam to one side. The problem is that we don't know whether the effect of yanking the beam is transmitted in the same way along both of its sides. On one side, the parts of the beam (molecules and atoms) are pushed together, on the other they are pulled, and we don't know if both distortions travel at equal speeds along the beam.

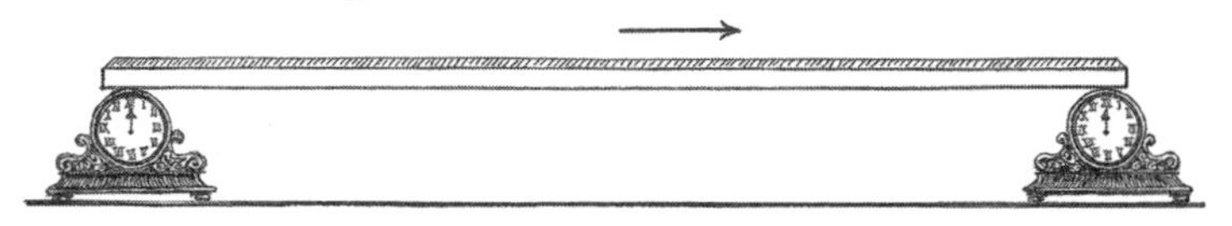

Figure 10.3. An attempt to synchronize clocks by yanking a beam between them.

So what about pushing down the beam at the center? The same problem occurs. If the force does not travel in the exact same way along both sides, then one side will activate the clock first, and the two won't really be synchronized. How can we check whether the "push" traveling along both sides of the beam reaches both extremities at once?

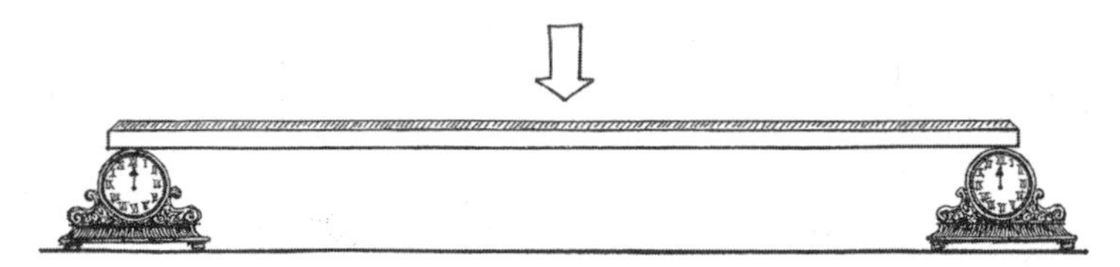

Figure 10.4. An attempt to synchronize clocks by pushing a beam at its center.

And there's another problem: what if the beam tilts as you push it down? Even if it tilts very slightly, the two clocks won't be exactly synchronized. The only way to ensure that the beam does not tilt is to confirm, as it moves down, that both of its extremities are at the same height

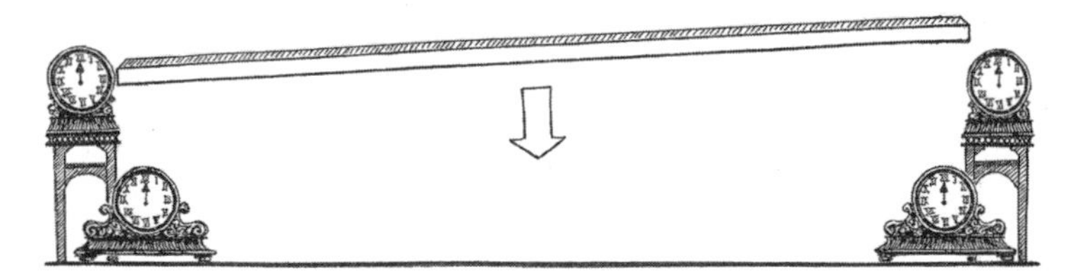

Figure 10.5. As the beam descends toward two clocks, to synchronize them, two more clocks are needed to check whether the beam tilts, that is, whether both of its extremities are located at the same height simultaneously.

at the same time. But for that we need at least two more clocks, and those clocks need to be synchronized. But we're trying to synchronize distant clocks in the first place!

Another plausible procedure is to synchronize clocks by sending a signal of some sort, such as a light ray, from one to the other. Light travels extremely quickly, so we send the light ray and expect that it synchronizes the clocks. But wait, the clocks are not quite synchronized precisely, because light takes a little bit of time to travel from one clock to the other; there's a very slight delay because light is not infinitely fast.

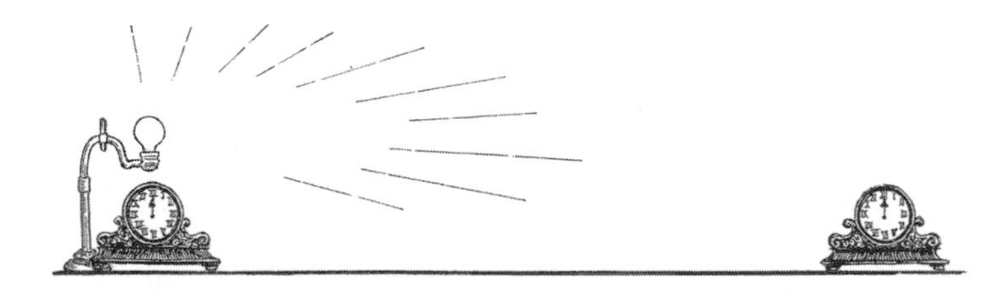

Figure 10.6. Light is emitted from one clock to synchronize another.

We might account for this delay by making the first clock start ticking a few moments after the light ray is sent outward from it, just long enough for light to have reached the other clock. Then we expect that the two clocks start at once. But to do so, we need to know, for a fact, how long it takes light to travel from one clock to another. So how do we measure the speed of light?

Back in 1676, a Danish astronomer, Ole Rømer (sometimes written Olaf Roemer), measured the speed of light and presented his results at the French Academy of Sciences in Paris.[22] Rømer measured the speed of light using one of Jupiter's moons. As the innermost moon, Io, moves behind Jupiter (its orbit takes about 42.5 hours), its moonlight is eclipsed, so from Earth we don't see it for a while. After a time *t*, Jupiter's moon becomes visible again.

But later in the year, when Earth had moved to the other side of the sun, Rømer again looked at the eclipse of Jupiter's moon and found that it did not reappear after the same time *t*. Instead it took a longer time.

He inferred that Jupiter's moon took longer to become visible because its moonlight had to travel a greater distance, across Earth's orbit, to reach Earth. By comparing the two time delays, Rømer found a value for the speed of light. Newton and other scientists were impressed.

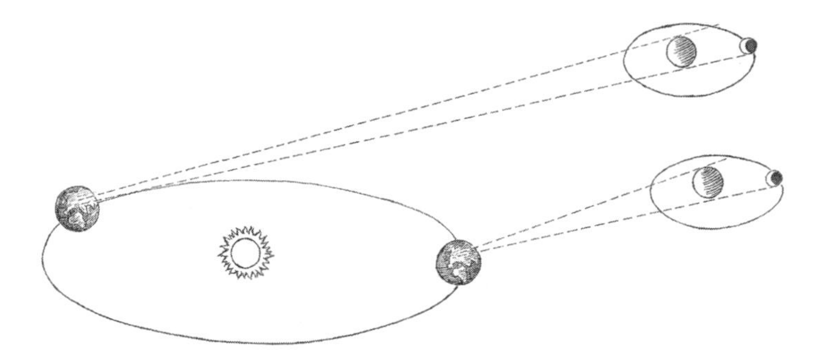

Figure 10.7. The eclipses of Jupiter's moon Io, as seen from Earth at two different times during the year.

But do we really therefore know the speed of light? Notice that the only thing we have directly measured are the time delays for the eclipses. But, what if there is any change of speed in Jupiter's moon? To make Rømer's procedure work, we have to assume that the speed of Jupiter's moon is constant. But it isn't. Also, Jupiter itself is moving, so what if the speed of Jupiter changes? The only way we can observationally know the speeds of Jupiter and its moons are by measuring the light that comes from them. To do that, we need to know the speed of all those light rays. Moreover, sunlight bounces off of Jupiter and its moons, but physicists did not know whether the motions of Jupiter and its moons, as they move away from the sun or toward it, affect the speed of the reflected light.

How do we know that light rays all have the same constant speed? Rømer didn't know from his observations, he just assumed it. Thus, consider the critical words of the French mathematician and physicist, Henri Poincaré, writing in 1898: "When an astronomer tells me that some stellar phenomenon . . . happened [at a certain time] . . . I seek his meaning, and to that end I shall ask him first how he knows it, that is, how he has measured the velocity of light. He has begun by *supposing* that light has a constant velocity, and in particular that its velocity is the same in all directions. That is a postulate without which no measurement of this velocity could be attempted. This postulate could never be verified directly by experiment."[23] Accordingly, Rømer's observations are insufficient to ascertain the speed of light, contrary to the claims of many textbooks that fail to explain this matter.

One alternative was to carry out some sort of terrestrial measurement of the speed of light rays. In 1849, the French physicist Armand Fizeau successfully carried out an experiment using a rotating gear wheel with 720 teeth. Fizeau placed the first mirror on the high belvedere of a house in Suresnes, a suburb west of Paris, and he placed the other mirror on a belvedere on Montmartre hill, north of the center of Paris. The distance between the two mirrors was 8.63 kilometers (5.4 miles). Using the first mirror, he sent a narrow beam of light toward the spinning gear wheel, at a position we may call *A*. The light beam crossed between a gap in the teeth of the spinning wheel at a time $t_1$, it traveled the 8.63 kilometers and there bounced against a mirror at *B*, making it go in the opposite direction, returning at a time $t_2$ back to the spinning wheel at *A*.

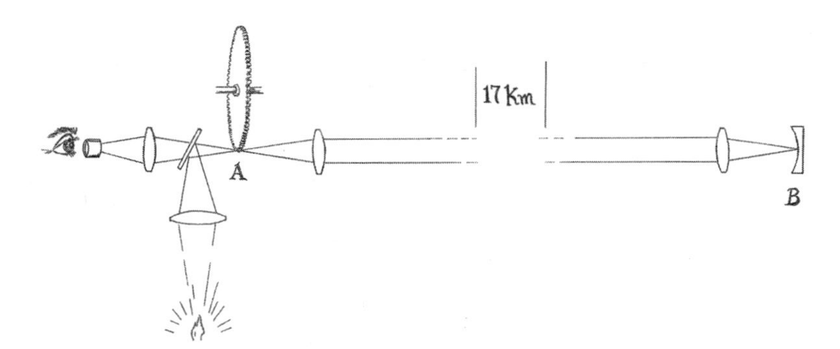


Figure 10.8. Fizeau's experiment of 1849. Light is emitted from a lamp at the bottom, a lens and a mirror redirect it to make it pass by a spinning toothed wheel, then the light continues out a to another lens and mirror where it is reflected, so that it returns to the spinning wheel.

By varying the speed of rotations of the wheel, Fizeau controlled whether or not the light beam, having crossed the wheel once, made it back through on its return. At first, for various speeds, light crossed a gap in the moving wheel and returned right through the same gap. But at

12.6 revolutions per second, light did not make it back through; that is, the returning ray was blocked by a tooth of the moving wheel. By taking into account the total distance light traveled, $2AB = 17.26$ kilometers (10.7 miles), and the total travel time, Fizeau calculated the speed of light.[24] We may write it as:

$$\textit{speed of light} = \frac{2AB}{t}$$

So here we have a terrestrial experiment, and a value for the speed of light. It would seem that finally, knowing the speed of light, we're ready to synchronize those clocks!

Notice, however, that we have not really measured the velocity of light, we have only measured the *round-trip average speed*. That is, we now know, pretty much, the time $t$ it took a light ray traveling from $A$ to $B$ and back to $A$, but we do *not* know the time it took to travel *from A to B alone.*

We might perhaps assume that the speed of light from $A$ to $B$ alone is the same as the round-trip speed. That sounds reasonable. But that is just an assumption, we have not proven *by experiment* whether the speeds in opposite directions are truly equal.

Someone might think: "That's weird! Why the heck would light take two different times to travel equal distances in opposite directions?" Mainly, because the distances might not really be equal. As Earth moves in a given direction, the mirror in Fizeau's experiment moves, for example, away from the approaching light ray. Once the light is reflected, it travels toward the source that approaches it as well, carried by Earth. Hence the distance light traversed to meet the mirror is longer than the distance back to the source. In that case, it takes longer for light to travel from $A$ to $B$ than from $B$ to $A$. It does not help that the source of light moves as well, carried by Earth, because that would not make light move any faster toward the receding mirror. (Physicists expected that the speed of light is *independent of the motion of its source.*[25]) So, contrary to many textbooks, Fizeau's experiment says nothing about the one-way velocity of light.

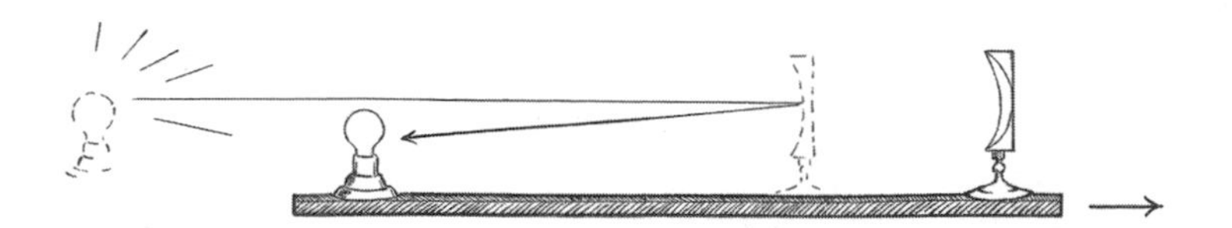

Figure 10.9. On a moving platform, such as Earth, light is emitted and then reflected.

Figure 10.10. An attempt to synchronize clocks with a flash of light.

Suppose we synchronize clocks by sending light from a lamp in opposite directions, to activate the clocks, to start them in synchrony. This procedure would work *if* light rays have the same speed in opposite directions. But does light take the same travel time in opposite directions? To know that, we have to measure the speeds of light rays and then compare them. In order to measure any such speed, we need two more clocks.

Figure 10.11. To measure the speed of light in any one direction, it is necessary to use synchronized clocks.

But to measure a speed, those clocks need to be synchronized. How? Let's place a light bulb at the center between them, and turn it on. But that only works if light takes the same speed in opposite directions. So how do we test *that*? We need two more clocks. And how do we synchronize those clocks?

There is a circularity here! To synchronize distant clocks you need to know a velocity. To know a velocity you need to have synchronized clocks. Thus, in spring 1905, Einstein realized, "there is an inseparable relation between time and signal velocity."[26] He wasn't the first to notice this circularity. For example, Poincaré also noticed it earlier and wrote about it in a paper of 1898 that was cited in a book that Einstein read before 1905.[27]

Like Poincaré, Einstein decided that the only way out is by "convention."[28] Einstein argued that the equality of the speeds of light in opposite directions is "a *stipulation* which I can make of my own free will in order to arrive at a definition of simultaneity."[29] He also decided that the speed of light is the same in opposite directions essentially as a matter of "*definition*."[30]

In March 1905 Einstein had been reading the *Treatise on Human Nature* (1739–1740) by the Scottish philosopher David Hume.[31] Hume argued that even basic physical notions, like the notion of cause and effect, involve certain assumptions.[32] In no case do we really know whether one thing is the cause of something else, we only know that the two occur in a regular sequence. That is, when we believe that *x* is the cause of *y*, it is mainly a matter of habit. While studying these ideas, Einstein concluded: "All concepts, even those which are closest to experience, are from the point of view of logic freely chosen conventions."[33]

Thus Einstein posited that the speed of all light rays is equal and constant—that the time it takes light to travel from *A* to *B* is the same as it takes to travel from *B* to *A*—as a fundamental *assumption*. By contrast, some books simply claim that the constancy of the speed of light is "an experimental fact."[34] But they give no evidence about how the speed of light rays traveling in opposite directions can be measured and compared. Contrary to the myth that the postulates of special relativity are experimental facts, a very few good books do note the essential difference between the velocity of a light ray and the average round-trip speed of light.[35]

Einstein also considered the question of simultaneity. To understand the synchrony of clocks, we need to have some way to verify whether any two events happen at once. The morning after his conversation with Besso, it occurred to Einstein, just as he was getting out of bed, that events that are simultaneous to one observer might not be simultaneous to another moving relative to the first.[36] To explain this notion, Einstein envisioned an imaginary experiment with a train.[37] Suppose that lightning strikes at two distant places on a railroad track. How can we decide whether these two events are simultaneous? Einstein argued that you should stand at the midpoint between the two events, holding an angled mirror to reflect light to you from both directions. So, suppose you see both light flashes at the same time. Then you say: the two lightning bolts were simultaneous events.

Figure 10.12. As the train moves along, lightning bolts strike the railroad tracks.

But consider now a train that, in the meantime, travels along the railroad tracks. Now, when lightning strikes, it burns marks on the train. So there can be no doubt regarding where on the train the lightning struck. And there are people sitting inside all the train cars, so we seek and find the one passenger who happened to be sitting exactly midway between the two burnt marks, and we ask: Did you see the lightning strokes?

"Yes."

"Did they happen at the same time?"

And the passenger replies: "No, the lightning on the front struck first, clearly."

We might argue: "The lightning strokes happened at the same time; the reason why you on the train saw one of them first is just because *you were moving* forward along its direction."

But then the passenger objects: "I wasn't moving at all, I was just sitting here."

We reply: No, you were moving, the entire train was moving.

The passenger might insist: "No it wasn't, I didn't feel it to be moving at all. I had a cup of hot coffee in my hand. And actually, what I saw was that all the trees and the guy outside were moving."

Maybe we smile and reply that that was just an illusion, really, that the trees weren't moving at all.

But then the passenger might continue: "How do you know? I say they were moving because *the entire planet is moving,* just like you say I was moving because of the train."

For Einstein, it didn't matter at all that the train is smaller than Earth. Both observers applied the same method to determine simultaneity. Yet they obtained different results. Einstein says both are right. Simultaneity is relative.

Einstein argued that the usual notion of simultaneity is a prejudice. We usually think that if events are simultaneous, then they are simultaneous for everyone. Einstein reasoned that that is just an assumption, that there is no evidence for it. For one observer, *A* happens before *B*; for another observer, *A* happens after *B*; and for another, they happen at the same time. Einstein said that they're all right, so long as they each applied the same procedure to determine simultaneity.

This relativity business is not entirely arbitrary; there should be actual visual evidence to correspond with it. Suppose that each door on either end of a train car is rigged to open automatically at the very instant when light hits it, and that the car is moving quickly when a lightbulb inside, at its center, gets turned on. We're outside, and suppose we see that the rear door opens first. We reason that the rear door opened first because it raced to meet the light, while the front door moved away. An important point is that this would be empirical data: *we saw* the rear door open first; it's not an opinion.

But what about someone inside the train car? According to the assumption that the speed of light is equal in opposite directions, a passenger expects that light should reach both doors at the same time, making the respective mechanisms open each door. We expect that a person standing in the car, midway between the two doors, should actually see both doors open at once.

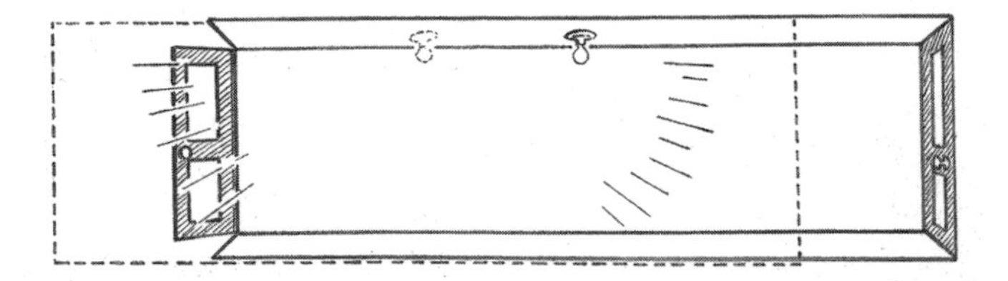

Figure 10.13. Light is emitted inside a moving train car. Standing outside, we see that the rear door opens first.

Figure 10.14. An observer inside the train car sees both doors open simultaneously.

But how can that happen, since we, standing outside, actually *saw* that the rear door opened first? Well, suppose that the train car is moving to the right, so that light takes less time to reach one door than the other, as it seems to us, standing outside. We say that the rear door opened first. But for the passenger inside the car, the only way to know whether a door opens is by seeing it. Light needs to travel from the door to the passenger. So, if the rear door opened first, light has to travel back to the observer at the center. But the observer is moving with the car. Both light rays will return to the middle observer at the same time. It doesn't matter whether one door opened first; he will literally *see* that both doors opened simultaneously. He will judge that the outside observers, us, see that the rear door opens first only because *we* are moving to the rear.[38]

So who's right, the person inside or outside? Both, says Einstein. That's the relativity of simultaneity. Events that are simultaneous for one observer are not necessarily simultaneous for another moving relative to the first.

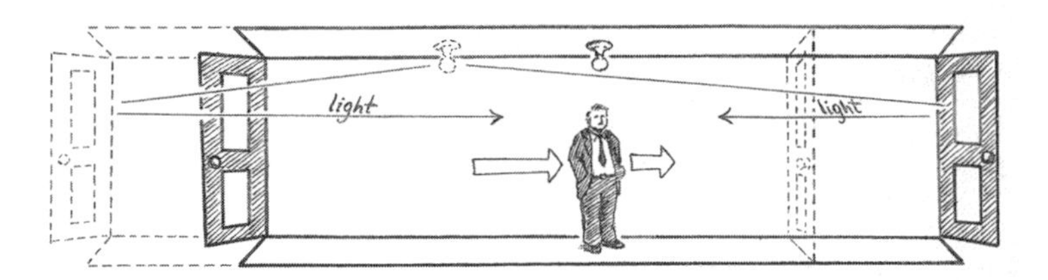


Figure 10.15. Light travels to each door and back.

The relativity of simultaneity has some bizarre consequences. Consider one example. We see that a ruler is twelve inches long. But what's the length of that ruler when it's moving? We might say that its length is the same. But unless we establish and carry out some procedure for actually measuring its length, the claim that it is still twelve inches long is just an assumption. So how do we measure a moving body? One way is to project a flash of light to cast its shadow on photographic film. Another way is to shoot a series of inkjets toward the ruler, such that they outline its shape on a backboard.

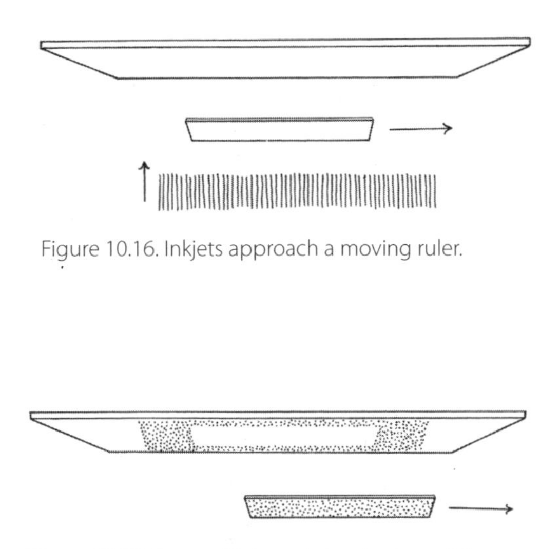

Figure 10.16. Inkjets approach a moving ruler.

Figure 10.17. Spots of ink mark the shadow of the moving ruler.

The result of this process, we expect, would be that the outline or shadow has the same length as the ruler. We then measure the length of that shadow and call it "the length of the ruler."

But suppose the ink jets to the right were shot first, and the rest in a sequence. Then the resulting shadow will be shorter.

Thus, the length of the shadow depends on whether all the ink jets are fired simultaneously. But how can we fire them simultaneously? If one procedure satisfies us, then that procedure might not satisfy an observer moving relative to us. Since we disagreed about the simultaneity of distant events, the moving observer would say that the inkjets were not emitted simultaneously, and will conclude that our measurement is wrong.

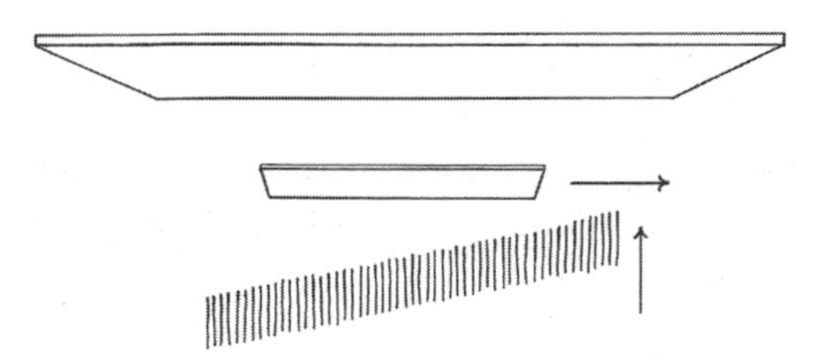

Figure 10.18. The inkjets are not emitted simultaneously.

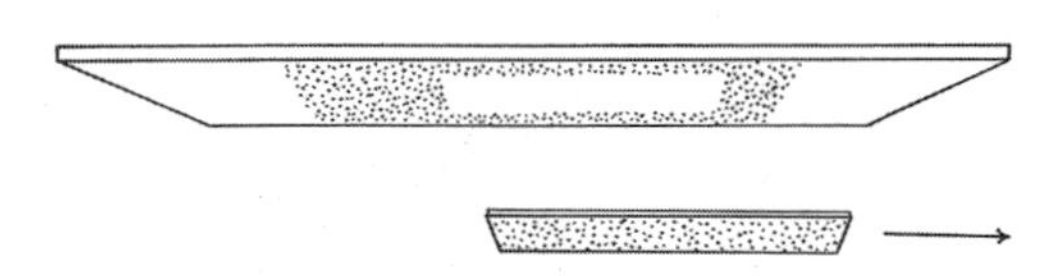

Figure 10.19. Spots of ink mark the shadow of the ruler, contracted.

Lengths depend on simultaneity; if we disagree about simultaneity, we'll disagree about the lengths of moving objects. Furthermore, if we disagree about lengths, we'll disagree about volumes. If we disagree about volumes, we'll disagree about density. Plus, if we disagree about simultaneity, we'll disagree about time intervals. When we disagree about time intervals, we'll disagree about accelerations. Moving observers disagree about *forces, energy, mass,* and so on. That's why Einstein formulated a so-called theory of relativity, to interrelate systematically how various physical quantities compare and vary among reference systems.

Einstein established that measurements should depend on the postulates of relativity and the constancy of the speed of light. To derive the equations of his theory, the so-called Pythagorean theorem is useful. Consider just one example, the relativity of durations, or time intervals. Suppose that a light ray is emitted from the ceiling of a train car straight to the floor, and let us call its vertical speed $c'$ (we call it "$c$ prime" because we have not yet assumed that it is equal to the speed of light outside the train). If $t'$ is the time interval it takes to go from the ceiling to the floor as determined by clocks inside the train car, then it covers a vertical distance $c't'$. But meanwhile, from the embankment outside, the train is seen to move forward at a speed $v$, measured by clocks on the ground, and from the instant when the light ray is emitted downward, the train covers a distance $vt$, as measured by those clocks on the ground. Then, to observers standing on the embankment, the ray did not just travel downward, it traveled diagonally forward and down, covering a distance $ct$, where $c$ is the speed of light as measured by clocks on the ground. By relating these displacements into the so-called Pythagorean theorem, we have:

$$(c't')^2 + (vt)^2 = (ct)^2$$

which can be rewritten as:

$$\frac{ct}{c't'} = \frac{1}{\sqrt{1 - v^2/c^2}}$$

If we now assume, with Einstein, that the speed of light is the same on both the train and the embankment, $c' = c$, we have:

$$t = \frac{t'}{\sqrt{1 - v^2/c^2}}$$

This equation states that the time interval marked by clocks inside the train *differs* from the time interval marked by clocks on the embankment.

Einstein's equations hardly implied any measurable anomalies in the behavior of bodies moving in everyday circumstances, such as trains and clocks, because most things move at speeds far slower than light. But the equations applied aptly to describe the motions of electrons. Their implications seemed most fascinating when applied to larger objects.

Returning to the start, relative to a spaceship traveling at an immensely high speed, planet Earth is contracted by a certain amount. Is that wrong? Not according to Einstein's theory, and neither does Earth *seem* to be contracted. Instead, Earth *does not have a single length of diameter*, nothing does. Objects all have various lengths in relation to various systems of reference. If we like the idea that the real length is the length relative to an observer at rest, then Einstein would disagree; we have a preference for a relative speed of zero, but that's just one speed, one perspective. It's just like the relativity of motion. What's the speed of the floor? Zero miles per hour, but only relative to you on your chair—relative to the moon, the floor beneath your feet moves very fast, and relative to the sun it moves at another speed, and relative to another star at yet another. Just as speed is relative, Einstein argued, the simultaneity of distant events is also relative, and lengths too, and so on.

And what about the guy in the spaceship? He turns on his flashlight which sends a beam of light forward. He moves relative to Earth at 160,000 miles per second. And the light moves relative to him at 186,282 mps. One might expect that the light ray moves relative to Earth at a speed of 346,282 mps. But Einstein would say that we're just assuming that velocities combine according to the simple addition rule:

$$c' = v + c$$

He argued that this is just an assumption, that there is *no evidence that this rule is exactly right*. Instead, he managed to show that given the relativity of simultaneity, we can derive a different rule for the composition of velocities, namely:

$$w' = \frac{v + w}{1 + vw/c^2}$$

If the speed of the spaceship is $v$, and the speed to be added is $c$, then the speed of the light ray relative to Earth is:

$$c' = \frac{v + c}{1 + vc/c^2}$$

And this becomes:

$$c' = \frac{v + c}{c/c + v/c}$$

$$c' = (v + c)\left(\frac{c}{v + c}\right)$$

$$c' = c$$

So the speed of the light ray relative to Earth is the same as relative to the spaceship, 186,282 miles per second. It looks like magic, but it isn't. We can grasp this seemingly impossible result once we remember that speed, simply put, is just distance divided by time; and since the people on Earth and the guy on the spaceship *disagree* about times and distances, they can therefore agree about the speed of light.

The point is that throughout time, scientists have had to distinguish between the properties that belong to objects and the properties that exist only as relations among objects. People used to think (and many still do) that objects have certain colors, intrinsically, say, that a given apple is really red. But it turns out, as we know thanks to Newton, that colors are not in objects, they're in the light. If we turn off the lamp, an apple ceases to be red. Colors also vary depending on the speed with which we move relative to each object. Similarly, we used to think that weight is something that is an intrinsic attribute of a body. But again, thanks to Newton, we know that weight is a relational property. Your favorite book would weigh much more if it were sitting on the surface of Jupiter. And its weight there is no less real than its weight here. Einstein argued that notions such as length and time are also relational properties. If we're going to state the length of a body, then we'd better specify the reference frame. And events that are simultaneous relative to you, strictly speaking, need not be simultaneous relative to other observers.

Einstein showed that despite such disagreements, one could still formulate a physics in which certain relations hold generally. The net result was that physics as a whole became reformulated and statements that we used to know as laws became just approximations, while new statements came to replace them. Space and time, which for ages had been imagined as absolute, like mythical gods indifferent and unaffected by human affairs, came to be construed instead as variable relative concepts. Although Einstein based his theory on convenient assumptions, the old habit of viewing physics as based on universal facts continued. To this day, many scientists tend to construe, in particular, the constancy of the speed of light as a brute experimental fact. While Einstein construed his special theory of relativity as a makeshift and preliminary construct, many of his followers did not. Einstein became religiously devoted to the spirit of scientific *inquiry*, but many scientists remained devoted to scientific doctrine, even if it changed.

# 11
## *The Cult of the Quiet Wife*

Here's an intriguing tale: having enjoyed decades of extraordinary fame, Albert Einstein never admitted that his acclaimed theory of relativity owed partly to the secret contributions of his modest wife. Not only had they lived together during his most creative year, they had studied physics together and when he won the Nobel Prize he gave the money to her. Was she his secret collaborator? It's a good story, but is it true?

Proponents of Einstein's wife have been arguing about this for years. It would be awful to discover that historians and physicists have systematically lied, based on some sexist bias, to deny credit where it is long overdue. If you trust authoritative historians you might simply disbelieve the story, dismiss it as a modern myth. Personally, I'd be glad to learn that Mileva Marić was Einstein's secret collaborator. I *want* her to be the secret coauthor. But we should set aside our speculative preferences and instead look at the evidence.

Better yet, we can trace how stories evolve. People sell reams of print by taking historical tidbits and stretching and sculpting them into provocative shapes. In 2003, television stations in the United States and other countries began to broadcast a documentary called *Einstein's Wife*.[1] It dramatized the life of Mileva Marić, highlighting the idea that she contributed to Einstein's scientific works. It was accompanied by a PBS website (since updated in response to concerns about historical accuracy) that featured an online poll that asked: "Was it really possible for Albert alone to produce all the phenomenal physics generated during 1905?" It continued: "Did Mileva Marić collaborate with Einstein? You

Decide! Take our online poll." Thus viewers were invited *to decide the past* by voting. In a couple of years, more than 75 percent of the people polled had replied that Marić did collaborate with Einstein.[2] How did so many people come to believe that?

The fuss began in 1987, when historians, led by John Stachel, began publishing comprehensive compilations of Einstein's works, manuscripts, and correspondence. Among the documents, they published old letters between Einstein and Marić. In a few of those letters, written around 1900, Einstein briefly used expressions such as "our research," "our paper," and once, "our work on the relative motion."[3] Historians of physics were fascinated as they analyzed such letters, but concluded that they are just too vague and insufficient to establish whether Marić contributed to Einstein's publications.

Still, plenty of non-specialists also pondered roles that Marić could have played. The lure to speculate was understandable. For example, Einstein's most intriguing comment, translated, reads: "How happy and proud will I be, when we both together have brought our work on the relative motion victoriously to its end!"[4] Non-specialists quickly concluded that this letter refers to the theory of relativity. Written by Einstein's own hand, could it be any clearer?

But wait. The letter was written in 1901, and Einstein had no concept of the theory that he later formulated which became known as relativity. At that time he still believed in the invisible ether and sought ways to detect its relative motion experimentally. This problem of "the relative motion" was a widespread concern; many physicists aimed to solve it. As a college student in 1899, Einstein began trying to design experiments to exhibit the relative motion of the ether. As he mentioned to Marić: "I also wrote to Professor Wien in Aachen about my paper on the relative motion of the luminiferous ether against ponderable matter."[5] Then in 1901, Einstein shared his speculations or aspirations with Marić. But by 1902, he had abandoned the idea of detecting the ether motion. He abandoned the concept of the ether and hypothesized instead that light behaves like bullets—its speed is affected by the speed of its source.[6] Later, in 1904, he discarded those conjectures too and hypothesized instead that the speed of light is independent of its source.[7]

He struggled to modify the leading theory, that of Hendrik Lorentz, to improve it.[8] But he failed again, and only in spring 1905 did he abruptly formulate a radically new theory that became known as special relativity, after ten years of reflection, including more than seven years of intensive struggles.[9] On the basis of abundant documentary evidence, we know that Einstein obsessively worked on physics before 1905 and for decades afterward. What about Marić?

She began college in the same year as Einstein, in 1896, and she too studied physics and mathematics to try to earn a teaching degree. They sometimes studied together, and in particular, read together works in physics that Einstein had read previously.[10] Although she was more than three years older than him, Einstein regarded her as "my student."[11] When they each had to write their diploma projects, Einstein chose his own topic (contrary to standard practice), and he also chose one for Marić.[12] One of her friends noted that they "devised their topics together, but Mr. Einstein relinquished the nicer one to Miss Marić."[13] But still, the final examinations were difficult, and Marić failed them. She retook the exams, but again did not pass. A major complication was that Marić became pregnant by Einstein in 1901. Another difficulty was that she had several arguments with her supervising professor.[14] She then ended her efforts to obtain the teaching degree. Letters also show that she chose to abandon her efforts to do a PhD thesis.[15]

Still unmarried, Marić gave birth to their daughter in 1902. They kept it a secret, and their daughter then disappears from the historical record. The build-up to those events seems to have changed Marić's focus from academic aspirations to familial concerns and occupations. At the time, she wrote to an intimate friend: "I believe that human happiness is more satisfying than any other success."[16]

Regardless, one writer, Dord Krstić, later claimed: "From the spring of 1898 until the fall of 1911, Mileva worked daily at the same table with Albert—quietly, modestly, and never in public view."[17] But this is a charitable speculation. Krstić never met Einstein or Marić, and he wrote nearly ninety years after the events in question. From mid-1900 until late 1902, they lived mostly in different cities, even in different countries. Plainly, the two could not work "daily at the same table" because they were not always in the same place. Moreover, there is no evidence that they regularly worked together on physics once they reunited in Bern. Regardless, Krstić claimed: "Almost simultaneously, Marie Curie opened the door into the world of radiophysics and radiochemistry and Mileva Einstein bravely began to explore the secrets of quantum and relativity—the fields that even today we call modern physics."[18] Here, as in the tales about Pythagoras, fact and speculation mix.

Consider another charitable speculation. One of the leading historians in researching the lives of Einstein and Marić is Robert Schulmann. He was interviewed for the program *Einstein's Wife*, in which he commented, "It is very conceivable that Mileva had input on the paper on capillarity. That of course has nothing to do with special relativity. But, I think it's fair enough to say that Mileva contributed—could have conceivably contributed, to that very first paper of his." The article on capillarity was Einstein's first scientific paper, completed in December 1900. Schulmann's comment might give the impression that Marić did collaborate on that scientific work. But notice the words. Schulmann, very properly, used the word "conceivable." Yes, we can imagine that Marić contributed to that paper, she well could have, but did she? Actually, we have Marić's own words on the matter. In a letter to her intimate friend, Helene Savić, she wrote: "This is not just an everyday paper, but a very significant one, it deals with the theory of liquids. We sent also a private copy to Boltzmann, and would like to know what he thinks about it; let's hope he is going to write to us."[19] The last words are suggestive, as if she were the coauthor indeed. But I quoted the passage without its initial sentences. Those sentences specify who actually wrote the paper: "Albert has written a paper in physics that will probably be published very soon in the physics *Annalen*. You can imagine how proud I am of my darling. This is not just an everyday paper . . ." Why then did Marić hope for the physicist's reply "to us"? Maybe she was Einstein's secret collaborator—or maybe she just penned the copy of the paper, or maybe she just mailed it,—we don't know. We do know that she proudly acknowledged Einstein's authorship.

In 1901, Marić again bragged to her friend: "Albert has written a magnificent study, which he has submitted as his dissertation. He will probably get his doctorate in a few months. I read it with great joy and real admiration for my little sweetheart who has such a good head on his shoulders. I'll send you a copy when it gets printed. It deals with the investigation of the molecular forces in gases using various phenomena. He is really a splendid fellow."[20]

Still, did Marić play any scientific role once she and Einstein lived together? It is well-known that in 1902 Einstein and two friends, Moritz Solovine and Conrad Habicht, started a discussion group that they jokingly called "the Olympia Academy." Their readings and discussions influenced Einstein's physics. Some writers claim that Marić, too, was an active participant. For example, in *Einstein's Wife*, the narrator claimed, "Maurice Solovine writes: Mileva would sit in the corner during our meetings listening attentively. She occasionally joined in. I found her reserved, but intelligent, and clearly more interested in physics than in housework." Where did the producers get this information? It echoes the novelized book *Einstein in Love* (2000), in which Dennis Overbye, who never met Einstein or Marić, wrote, "Marriage had made Mileva a de facto member of the Olympia Academy, and Solovine later recalled her sitting quietly in the corner during the meetings at their apartment, following the arguments but rarely contributing. He found her reserved but intelligent, and clearly more interested in physics than in housework."[21] Since Einstein and Marić lived together, one might readily *imagine* that Mileva now participated in the discussion group. While Solovine warmly appreciated Marić, he *actually* reported that once Einstein married her, "That event did not effect any change in our meetings. Mileva, intelligent and reserved, listened to us attentively, but never intervened in our discussions."[22] Compare this original passage to the derivative accounts. Solovine did not write that Marić "occasionally" or "rarely" contributed, nor even that she was "clearly more interested in physics than in housework." Instead, there is no documentary evidence that she was an active participant, and in none of their correspondence does Marić appear as a "member" of the academy, not even in her own letters to intimate friends.

Einstein had lively discussions with Solovine and Habicht, to the extent that neighbors complained. He also discussed his research extensively with his friend Michele Besso, whose help he acknowledged in his first paper on relativity. What about discussions with Marić? We find in the extant letters, that wherever Einstein raised a scientific argument, Marić did not reply to that, but focused instead on everyday personal topics. Philipp Frank, a friend who interviewed Einstein, noted that Marić "was taciturn and reticent," but that "Einstein in his zeal for his studies hardly noticed it." And, "When he [Einstein] wanted to tell her, as a fellow specialist, his ideas, which overflowed from him, her reaction was so scant and faint, that often he just did not know whether she was interested or not."[23]

Proponents of Marić have staked their case on alleged evidence from a presumed witness. Marić's unauthorized biographer, Desanka Trbuhović-Gjurić, who never met or corresponded with Marić, felt a great kinship toward her because they were both Serbian. In 1969, Trbuhović-Gjurić claimed that the Russian physicist Abram Joffe, whom she also never met, once pointed out (in his 1955 article titled "In Remembrance of Albert Einstein") that the 1905 papers were originally signed "Einstein-Marić."[24] In 1991, writer Evan Harris Walker came to a similar conclusion, translating Joffe's article about Einstein's 1905 papers in this way: "Their author was Einstein-Mariti."[25] Furthermore, in 1999, Michele Zackheim claimed that "Joffe, a Russian scientist, wrote in *Meetings with Physicists: My Reminiscences of Foreign Physicists*, that three original manuscripts, including the one describing the Special Theory of Relativity, were signed 'Einstein-Marity.'"[26] Then, in 2003, the claim that Joffe cited Marić's name on the 1905 manuscripts was broadcast in the documentary *Einstein's Wife*, and its website boasted, "there is at least one printed report in which Joffe declared that he personally saw the names of two authors on the 1905 papers: Einstein and Marity."[27] The list of people who echo such claims goes on and on.

Joffe was a reputable physicist who later knew Einstein and even met Marić at least once.[28] In his book *Meetings with Physicists*, Joffe claimed nothing about how the 1905 manuscripts were signed, and he did not even claim to have ever seen them. Thus the claim by Zackheim and others is just false. As for Joffe's brief article "In Remembrance of Albert Einstein," it was an obituary for Einstein published in the top Russian journal on physics. Literally translated, without changing the order of words, the relevant passage reads: "In the year 1905, in *Annals of Physics*, there appeared three articles, thereupon beginning three most important, relevant directions in the physics of the 20th century. Those were: the theory of Brownian motion, the photon theory of light and the theory of relativity. Their author—unknown until that time, a bureaucrat at the Patent Office in Bern, Einstein-Marity (Marity—the last name of his wife, which by Swiss custom is added to the last name of the husband)."[29] So Walker's translation: "Their author was 'Einstein-Mariti.'" is a gross misrepresentation. Likewise, other writers stretch and twist Joffe's words to make provocative claims. They read much between the lines. Joffe's plain words say that the author was just one person, a male employee at the Swiss patent office, onto whose name Joffe added the spouse's name.

Still, proponents of Marić have tried to make something out of the fact that Joffe happened to write "Marity." For example, Walker claimed that Joffe must have seen a manuscript bearing the name Marity, for otherwise he would not have known that alternative spelling because it "apparently is not found in any of the Einstein biographies."[30] But Walker was wrong; that name appears, for example, in a very popular biography of Einstein published in 1954.[31] Also, when Joffe first tried to visit Einstein at his home in Switzerland, he happened to meet Marić, who then used the name Einstein-Marity. (In Switzerland in the early 1900s, some spouses, male and female, did use joint names.[32]) It is odd that decades later Joffe once happened to refer to Einstein by the compound name, but that small oddity does not mean that Joffe ascribed authorship to Marić.

Joffe did not claim that Marić wrote or collaborated in any scientific papers. He did not claim that her name was on the 1905 manuscripts, nor that he ever saw any such manuscripts. In multiple places throughout his career, Joffe acknowledged Einstein for having authored the famous works of 1905. Yet the producers of *Einstein's Wife* and the companion website pictured a fragment of a page, purportedly by Joffe, which reads that the articles were "signed Einstein-Marity." That was a mistake; the page pictured was really from a popular science book from 1962 by a Russian writer who did not claim to have ever seen the original manuscripts nor to have known anyone who had.[33] The claims regarding Joffe dissolve into nothing.

Another apparent witness was the first son of Marić and Einstein. Hans Albert Einstein was interviewed by a few authors and historians who asked him about his mother, yet he did not claim that she was Einstein's secret coworker. But then in 1962, for two days, Peter Michelmore, who had never met Einstein or Marić, interviewed Hans Albert Einstein to gather information for a brief biography of Einstein. In the resulting booklet, Michelmore wrote that once Marić fell in love with Einstein, by their final year of college, "Her personal ambition had faded." Yet Michelmore briefly noted that later, while Einstein struggled to solve puzzles of relative motion in electrodynamics, "Mileva helped him solve certain mathematical problems, but nobody could assist with the creative work, the flow of fresh ideas." Did Einstein's son say that? We do not know. We do not know for certain what parts of what Michelmore published were voiced by Hans Albert. The author admitted that Hans Albert did not see or proofread the manuscript for the book: "he answered all my questions, and waited while I wrote down the answers. He did not ask to check my notes, or edit my book. He trusted me. It was the sort of naïveté his father had. Thank God for all naïve people, and I use the word in its noblest sense."[34] Unfortunately, when interviewers' accounts remain unchecked, inaccuracies increase.

Alongside some verifiable statements, Michelmore's book also includes incorrect information. For example, he mentioned that while Einstein studied in Zurich, he befriended "Maurice Solovine, a Frenchman taking the physics course".[35] In fact, Moritz Solovine was Romanian, born and educated in Romania, until he moved, not to Zurich, but to Bern, where he met and befriended Einstein in 1902, almost two years *after* Einstein had graduated in Zurich. Such errors diminish the credibility of an author's words. Consequently, John Stachel, editor of the *Collected Papers of Albert Einstein*, inquired whether Michelmore's family happened to possess Michelmore's notes from the interview with Hans Albert. No such luck; lacking such notes, we don't know precisely what Hans Albert told Michelmore.

Faced with such ambiguities, each historian must decide what to do with hearsay in a historical reconstruction. In my book on the history of relativity, I chose to incorporate Michelmore's words about Marić. But I hope that readers will realize that the sentence in question is not a photograph of actual events. It is but a passing claim that appears in a popular biography written by an author who only interviewed a son of the individuals in question, a biography which was not proofread by those individuals, nor even by the interviewee. It was written and published almost sixty years after the event in question. Moreover, Hans Albert could not possibly have witnessed such an event, since he was only a one-year-old baby in spring 1905. Hence, if he did actually speak such words in 1962, he was merely voicing a conjecture or echoing words voiced by someone else. The point is to distinguish between late indirect claims and evidence from early sources.

Seldom do we try to articulate, systematically, the extent to which different sources warrant different degrees of credibility. Therefore, it seems useful to illustrate such differences. Historians sometimes disagree on how much weight to attribute to any one document, but I can at least sketch my own outlook in the accompanying table, which describes some of the different kinds of information that may exist pertaining to the genesis of a scientific work. To distinguish them, I have ranked them in order of proximity to the historical event, the instance of scientific creativity. The greater the number of an item, the less credibility I tend to ascribe to it as a likely source of precise information about that moment in time. The list is not exhaustive; my aim is only to distinguish among some different kinds of information. The line following item 5 sets a boundary between evidence generated during the production of the scientific work and various kinds of hindsight and conjecture.

In this hierarchy, the biography written by Michelmore falls on level 18. In contradistinction, a letter by Einstein to his friend Conrad Habicht, written in May 1905, while he was drafting the paper on relativity, counts as evidence of level 4; because he alluded to his work without describing it in detail (which would raise it to level 3). That letter, which historians cite often, is a precious though narrow window to the creative moment. Thus there are many different kinds of information to which we ascribe various degrees of reliability.

**Table 11.1.** A scale of likely reliability for various kinds of sources, from most to least credible

| | |
|---|---|
| 1. | Original notes and drafts of the scientist's labors and ruminations |
| 2. | Contemporary private diaries of the scientist, peers, or friends |
| 3. | Contemporary documents such as letters to friends |
| 4. | Contemporary accounts of statements among scientists and peers |
| 5. | Manuscripts, the original scientific work |
| 6. | Early retrospective accounts by the scientist |
| 7. | Early interviews of the scientist, proofread by the scientist |
| 8. | Later retrospective accounts by the scientist |
| 9. | Later interviews of the scientist, proofread by the scientist |
| 10. | Systematic interviews by historians, psychologists, or other specialists |
| 11. | Informal interviews of the scientist |
| 12. | Recollections that exist only in an indirect form, such as a transcribed lecture |
| 13. | Retrospective accounts that exist only in a doubly indirect form |
| 14. | Late recollections by an intimate acquaintance |
| 15. | Biography based on interviews, approved by the scientist and interviewees |
| 16. | Account based on multiple interviews but not proofread by the interviewees |
| 17. | Account of interviews with a close relative or peer, proofread by the interviewee |
| 18. | Material based partly on interviews from a relative, peer, or acquaintance |
| 19. | Rough translations of biographies or sources |
| 20. | Hearsay: late indirect accounts of what someone allegedly told someone else |

As another example, in 1922, Einstein delivered a lecture in Kyoto, Japan, titled "How I Created the Theory of Relativity." He delivered it in German without having written it down, and as he spoke, it was translated into Japanese. The translator kept notes that were soon published in Japanese. In the hierarchy listed in the table, I would rank this Japanese rendition as being on level 13. It is "doubly indirect" in the sense that Einstein did not write it, and that we only have the version in Japanese. It is not a very late document in Einstein's life, so we may imagine that forgetfulness perhaps did not distort it very much. But still, the transcript was not proofread by Einstein.

This effort to distinguish among various levels of reliability helps to add perspective to a document. It is especially sobering to try to carry out this sort of exercise to evaluate the credibility of ancient sources in the history of science and mathematics. For example, what reliability might we ascribe to Plutarch's comments on Pythagoras, written almost a thousand years after Pythagoras had died? It compels us to write in ways that acknowledge uncertainties rather than to pretend that the extant evidence faithfully echoes the distant past. But at least for more recent historical events we can say some very definite things.

Fortunately, several documents do shed light on Mileva Marić around 1905. For example, Krstić provided this translation of a letter from Marić to her friend Helene Savić, written soon after the 1905 papers were published: "My husband spends all of his free time at home, often playing with the boy; but . . . I would like to remark that this, together with his official job, is not the only work he does—he is writing a great number of scientific papers."[36] As usual in her letters to her intimate friend, Marić made no claim of working on science herself, not since she left college. Now notice the ellipsis, the three dots in the quotation above. What did Krstić omit? An uncut and proper translation of the original letter was published later by a grandson of Helene Savić. It reads: "My husband often spends his leisure time at home playing with the little boy, but to give him his due, I must note that it is not his only occupation aside from his official activities; the papers he has written are already mounting quite high."[37] So we see that Krstić deliberately omitted a phrase in which Marić herself further acknowledged Einstein's labors, "to give him his due." In 1909, when Einstein was receiving much recognition from physicists, Marić similarly wrote to her friend "I am very happy for his success, because he really does deserve it."[38]

Yet such happiness was diminishing. At the end of the documentary *Einstein's Wife*, Milan Popović, the grandson of Helene Savić, against a gentle backdrop of moving music, paraphrased the sad contents of a letter from Marić to Savić:

> *"She said she is like a shell,*
> *and Einstein is a pearl in the shell,*
> *and when the pearl is finished,*
> *then the pearl don't need more the shell."*

When I first heard Popović say these lines, I was transfixed and moved. Then later I looked up the original letter. Marić wrote the actual letter in the winter of 1909–1910. Literally translated, it reads: "with such fame there is not much time left over for the wife at all. I read a certain impishness between the lines, when you wrote that I must be jealous of the science. But what can one do? the one gets the pearl, the other the box."[39] Compare her words to Popović's interpretation. It is too awkward, at least for me, to twist the actual quotation into Popović's account, where Einstein is the pearl and Marić is the creative shell. Instead, it simply seems that Marić complained that Einstein, being both the pearl and the shell, unfairly gave his best part to science while Marić received only his shell.

Einstein and Marić eventually divorced. It's not quite true that he then gave her the money award from his Nobel Prize of 1921. He proposed to invest it so that Marić and their sons could benefit from the earnings.[40] Through all their difficult and sometimes friendly interactions over financial support, the many letters on the matter give no evidence that there was any intellectual debt involved.[41]

Any document can include errors, omissions, inaccuracies, or even lies. Likewise, information of all kinds might include truthful claims. The important point is that each step away that a document is separated from the period it purportedly describes, introduces additional layers of potential inaccuracies that can arise in the translations, rewording, additions, and so forth. A letter written by a participant in the events in question, even decades later, can be very informative. But we should still be careful with its contents. A still later account by one who was not present at such events involves greater uncertainties. If we cannot dissipate such uncertainties, we should at least acknowledge them. We should cultivate a fair skepticism, especially against outstanding stories that resonate with what we would personally like to believe. Too often, writers enamored with a sensational conjecture tend to misread evidence. They seek not to test a conjecture but to confirm it. But what makes a good story does not necessarily make good history.

Nevertheless, enough evidence discloses important roles for Marić. In the words of historian Gerald Holton, "Ironically, the exaggeration of Mileva's scientific role, far beyond what she herself ever claimed or could be proved, only detracts both from her real and significant place in history, and from the tragic unfulfillment of her early hopes and promise. For she was one of the pioneers in the movement to bring women into science, even if she did not reap its benefits. At great personal sacrifice, as it later turned out, she seems to have been essential to Albert during the onerous years of his most creative early period, not only as anchor of his emotional life, but also as a sympathetic companion with whom he could sound out his highly unconventional ideas."[42]

For those of us who prefer the voice of firsthand witnesses rather than historians, consider the agreeable recollections of Solovine. He remarked that Einstein's productivity benefited from his good life at home in the company of Mileva: "I am convinced that her influence was beneficial, in the environment and affection and in allowing him to work peacefully. I cannot forget that the memoir On the Electrodynamics of Moving Bodies, which established his reputation, was published in 1905, where a perfect harmony reigned between them."[43]

# 12
## *Einstein and the Clock Towers of Bern*

NEAR the center of the old city of Bern, there stands a massive medieval tower that, in the 1300s, served as a prison for women who had illicit relations with clergymen. In 1405, a great fire burned it severely. The structure was rebuilt and converted into a bell tower bearing a great astronomical clock. On two faces, golden clock-hands tipped by suns point to hours, while another dial shows the phases of the moon. Five ancient gods—Mercury, Saturn, Jupiter, Mars, and Venus—illustrate five days of the week as well as the five planets of Ptolemy's heavens. And near a corner, beneath a grinning jester, sits a bearded Chronos, the god of time, holding an hourglass.

This medieval clock tower straddles the very street where Einstein lived in early May of 1905, when he first thought of the relativity of time. Every day as he walked to work, Einstein walked by that tower or even under its arches. This fact gave rise to a myth comparable to that of Galileo and the Leaning Tower of Pisa—a myth about Einstein and the clock towers of Bern. Whereas the growth of the tale about Galileo is somewhat difficult to ascertain, because it began centuries ago, the story about Einstein and the towers is a recent development, so we can track its growth quite thoroughly.

In 1905, in his first paper on relativity, Einstein briefly illustrated a definition of the notion of time by alluding to the arrival of a train and the pointers of a clock. Since he was then an employee at the Swiss patent office, one might imagine that his job led him to think about clocks and trains in relation to physics and that he thus came to solve problems that other physicists failed to solve. But actually, Einstein's allusion to the

Figure 12.1. The very old Zytglogge clock tower on Kramgasse street in Bern, Switzerland, where Einstein lived in early 1905.

generic, commonplace technologies of clocks and a train were not unique to him. Other physicists who did not work at patent offices also analyzed notions of motion and time using examples involving trains and clocks.[1] But this is not widely known, so it might seem that Einstein's abstract relativity stemmed from his practical job.

In 1993, Alan Lightman, in a bestselling book, imagined discussions between Einstein and his friend Michele Besso against a backdrop of clock towers that loom in Einstein's dreams.[2] Lightman's short narrative was intended to be a thoughtful yet fictitious account. Still, historical claims also arose.

By 1995, a story had emerged: "Einstein is said to have been prompted to the Theory of Relativity as he watched the receding clock tower while travelling by tram to his job at the Swiss patent office in Berne."[3] In 1999, Steven Pinker, professor of psychology at Harvard University, published an article in the *New York Times* in which he briefly stated how Einstein formulated relativity: "from imagining himself riding on a beam of light and looking back at a seemingly frozen clock tower, he developed the theory of special relativity."[4] Here the valid historical account given by Einstein, about imagining catching up to a light beam, appears arbitrarily mixed with the image of a clock tower. (Interestingly, notions of "frozen" time show up in Lightman's short fictional story).

At the same time, also at Harvard University, historian Peter Galison carried out research to argue that Einstein's relativity stemmed from the intersection of technology, physics, and philosophy, focusing on an apparently neglected chronometric dimension. In 2000, Galison wrote: "This summer I was standing in a northern European train station, absentmindedly staring at the turn-of-the-century clocks that lined the platform. They all read the same to the minute. Curious. Good clocks. But then I noticed that, as far as I could see, even the staccato motion of their second hands was in synchrony. These clocks were not simply running well, I thought, these clocks are coordinated. Einstein must have seen such coordinated clocks. . . . Every day he must have seen the great clock towers that presided over Bern with their coordinated clocks." Galison further claimed: "Pointing up at a Bern clock tower—one of the famous synchronized clocks in Bern—and then to a clock tower in nearby Muri (not yet linked to the Bern mother clock), Einstein laid out his synchronization of clocks."[5] For this claim, he cited only one piece of evidence, by Josef Sauter, which I'll discuss below.

At roughly the same time, a writer of a biography of Einstein, Dennis Overbye, commented: "It would be pretty to think" that maybe Einstein's breakthrough happened one day as he walked with his friend Michele Besso under that great clock tower.[6]

In a book of 2001, historian Arthur I. Miller argued that Einstein had been influenced by discussions about clocks: "We must not forget that at the Patent Office Einstein . . . spoke often with friends from the Federal Postal and Telegraph Administration about issues in wireless telegraphy and synchronization of clocks." Miller characterized this as an important fact to remember, but it was just a plausible speculation; there is no evidence that Einstein often discussed clocks at the patent office. Regardless, Miller further claimed that the "practical problems of wireless telegraphy" provided a "key input into Einstein's thinking into relativity in 1905."[7] Again, there is no evidence for such a claim.

Two years later, in a book published in 2003, Peter Galison repeatedly claimed that, in 1905, while Besso and Einstein stood on a hill northeast of downtown Bern, Einstein excitedly gestured toward the clock towers in Bern and Muri as he explained his realization that time should be defined by exchanging signals.[8] This anecdote stemmed from two sources: a biography by Albrecht Fölsing and a recollection by Josef Sauter. Writing in 1993, Fölsing had never met Einstein or Besso, yet he wrote that Einstein explained to Besso, and later to Sauter, his procedure for synchronizing clocks—by pointing to a bell tower in Bern and another in Muri.[9] Since those towers are not visible from downtown Bern, Galison claimed that "Besso and Einstein must have been standing on the hill shown to the northeast of downtown Bern."[10] That, however, would be impossible; the map in Galison's book is upside down; the hill is really south of downtown Bern. Galison's and Fölsing's only source was the account by Sauter, a coworker at the patent office. Sauter voiced his recollection in 1955, when he was eighty-four years old. He explained how Einstein once explained to him his new definition of synchrony: "to pin down the ideas, he told me, let's suppose that one of the clocks is atop a tower at Bern and the other on a tower at Muri."[11] Sauter did not claim that Einstein pointed to any actual clocks on towers, nor that he had his sudden creative insight by thinking about clock towers, nor even that he told Besso anything about clock towers. Thus the story about Einstein and Besso discussing the coordination of clock towers while standing on a hill dissolves in the air—or it should, but instead it becomes a myth.

In 2003, hundreds of thousands of readers had the opportunity to read a book review in the *New York Times* that claimed: "In May 1905, on a hill from which he and his friend Michele Besso could see both the electrically synchronized clocks of Bern and the as yet uncoordinated clock tower of suburban Muri, Einstein realized in a flash that . . ."[12] The reviewer was trying to summarize Galison's book, but actually, Galison did not claim that Einstein had his great idea while he and Besso looked at clock towers, only that Einstein explained it to him in that way soon afterward (which is fiction as well). Aspects of the story spread into other books, acquiring new details.

Years ago, one day in Cambridge, Massachusetts, on Church Street, I was at a store that repairs Swiss clocks and watches. On the wall, nicely framed, was a review article on Galison's book. Right then a man said to his companion: "Hey honey, did you read this one? How they got it all from Swiss clocks." She smiled: "Oh yeah. . . . That was a good one." Maybe we could draw a line between experts and laypersons, but what I want to highlight is the continuity between the biographers, historians, fiction writers, reviewers, physicists, and laypersons at a store. We're all connected by the common reflex to slightly misread, to decorate impressions with speculations. It is a common and pervasive habit. This is the same pattern that we have seen in the evolution of other historical myths: a plausible conjecture ("he may have . . ." or "he must have . . ."), voiced by an authoritative source (a prominent professor of history, a famous physicist), becomes misconstrued as an actual happening.

Aside from readers' urge to conjecture, the tale shines because of its plausibility. The overall circumstance is striking: Einstein formulated the relativity of time while living in the clock-making capital of the world, Switzerland. He worked at the Swiss patent office, a clearinghouse for chronometric technologies, as Galison argued. Maybe patents on timing devices actually influenced Einstein? Unfortunately, there is no evidence that Einstein in fact was influenced by any such technologies or that he evaluated any patent applications for chronometric devices.[13]

Still, was Einstein influenced by the patent office? Researching this question, the best sign of an apparent connection that I found was a single sentence in an early book on Einstein, written by a man who followed him around in the mid-1910s. In that book, published in 1921, Alexander Moszkowski briefly remarked about Einstein: "He recognizes a definite connection between the knowledge acquired at the patent office and the theoretical results which, at that same time, emerged as examples of the acuteness of his thinking."[14] Moreover, there's another bit

**Table 12.1** Evolution of the myth about Einstein and the clock towers

| | | |
|---|---|---|
| | Albert Einstein worked at the Swiss patent office. From 1904–1905, he walked by the Zytglogge clock tower on his way to work. | |
| 1955 | Josef Sauter, a coworker, recalled words from 1905 | "To pin down the ideas, he told me, let's suppose that one of the clocks is atop a tower at Bern and the other on a tower at Muri." |
| 1974 | Max Flückiger, historian | It is interesting that Sauter mentioned the bell towers. Einstein sometimes visited his friends and colleagues in Bern and Muri. |
| 1993 | Albrecht Fölsing, biographer | "He was observed gesticulating to friends and colleagues as he pointed to one of Bern's bell towers and then to one in the neighboring village of Muri. Michele Besso was the first person and Josef Sauter the second to whom he explained in this manner that the synchronization of spatially separated clocks . . ." |
| 1993 | Alan Lightman, science writer, writes a fictional account | While theorizing about the nature of time with his friend Besso, Einstein dreamed of clock towers. |
| 1999 | Steven Pinker, psychologist | "From imagining himself riding on a beam of light and looking back at a seemingly frozen clock tower, he developed the theory of special relativity." |
| 2000 | Dennis Overbye, science writer | "It would be pretty to think" that maybe Einstein's breakthrough happened as he and Besso walked under the great clock tower. |
| 2001 | Arthur I. Miller, historian | Einstein often spoke with coworkers about the synchronization of clocks. |
| 2003 | Peter Galison, historian | Standing on a hill northeast of downtown Bern, Einstein excitedly gestured to clock towers of Bern and Muri as he explained to Besso that time should be defined by exchanging signals. |
| 2003 | William Everdell, history teacher and author | "In May 1905, on a hill from which he and his friend Michele Besso could see both the electrically synchronized clocks of Bern and the as yet uncoordinated clock tower of suburban Muri, Einstein realized in a flash . . ." |
| 2005 | Michio Kaku, physicist | Einstein "imagined what would happen if his street car raced away from that clock tower at the speed of light." |
| 2007 | Walter Isaacson, biographer | The synchronized clocks of Bern were not synchronized with the steeple clock visible in the neighboring village of Muri. |
| 2008 | Hans Ohanian, physicist | Einstein "reviewed patent applications for electromagnetic devices used for the operation of citywide networks of synchronized clocks." He pointed at a clock tower in Bern and to another in Muri to exemplify to Besso his crucial idea about synchronizing clocks. |

of evidence: decades later, Einstein himself commented, "the work on the final formulation of technical patents was a true blessing, and also provided important inspiration for physical conceptions."[15] Yet both of those statements are vague. What was the connection? Did it allude to Einstein's relativity or to his other papers over seven years? Does it have to do with timing technologies? Did it help his decisive creative step to relativity? We have no evidence of that.

Perhaps we should ask: What evidence is there of *any way* in which Einstein's job helped him as a physicist? We find that Einstein learned to write more clearly, thanks to the stern critical teachings of the director at the patent office, Friedrich Haller. As reported by one of Einstein's friends, Haller warned Einstein: "If you do not write clearly, I will throw you out."[16] And Josef Sauter quoted Einstein on this matter: "This man [Haller] taught me how to express myself correctly; he was stricter than my father." Also, Einstein's stepson-in-law Rudolf Kayser reported that: "He [Haller] taught them [the examiners] to think sharply and logically, and to select every word in its most exact sense."[17] This is a less dramatic connection between Einstein's patent work and his physics, but it is warranted by the evidence.

Regardless, consider now the ways in which the myth about the clock tower has spread. In his number-one-bestselling 2007 biography of Einstein, Walter Isaacson duly mentions the "synchronized clocks of Bern and the unsynchronized steeple clock visible in the neighboring village of Muni."[18] But that too was a mistake; the clock towers actually marked a common time. By now, some writers, even historians, have misconstrued Galison's conjectures as historical facts, as they now claim that Einstein actually often evaluated patent applications for synchronizing the clocks of Bern.[19]

Galison's account has engendered similar stories with additional imaginary details. In a book in French, for example, theoretical physicist Thibault Damour writes:

> Soon the two friends were climbing the hill of Gurten at a happy pace, from where one has magnificent view of Bern. Let's try to imagine their lively dialogue. . . .

> A. E.: "Wait . . . Yes, that's it! I think I've got it . . . Look at the tower of the Clock, down there in the center of Bern. If one had binoculars, one could read there *the time*. But that would not be *our time*. It would be necessary to subtract the time taken by the light to come from the clock to us. I sense that this will modify the notion of *time* for an observer in motion. Thank you Michele! I am sure that now it will work! Tonight, I will figure out what follows in detail."[20]

Galison's account has also entered books on literature.[21] Moreover, some physicists believe the story. For example, Hans Ohanian claimed that Einstein reviewed patent applications for the operation of synchronized city clocks (fiction), and that he explained this creative breakthrough to Besso by pointing at the clock towers (fiction again).[22] Another physicist, Max Jammer, echoes a version of the story, casting it as conjecture, but as a likely one: "In fact, Einstein may have already encountered the problem of time synchronization on his daily walk to the patent office when passing near the famous clock tower on the Kramgasse, where he lived, and seeing the distant clock on the church in Muri, a nearby suburb of Bern."[23] This apparently plausible scenario is actually impossible, because the church bell tower of Muri simply cannot be seen at all when walking from Einstein's apartment in Kramgasse to the old patent office building. Regardless, Jammer carries out a kind of inversion to add plausibility to the story. Instead of beginning with the scant evidence, Sauter's account, Jammer first posits that Einstein "may have" been influenced by the clock towers of Bern and Muri, and afterward he quotes Sauter's words. Thus we encounter a common device in some myths: a story arises from *misreading* a genuine source, and later, that source is used as *evidence* for the myth, by presenting the story first.

I've focused on the development of this tale mainly in the English language, but note that Galison's book has already been translated into German, Spanish, French, Portuguese, Italian, Greek, and perhaps others of which I am unaware.

Other writers have presented different versions of the myth by which Einstein's relativity work was inspired by the clocks. In 2000, Walter Mih presented one such version as a "legend": "Streetcars routinely pass through the archway below the astronomical clock. Legend says that Einstein developed the Special Theory of Relativity when he was riding in a streetcar and watching a clock."[24] And the brief claim echoed by psychologist Steven Pinker in 1999 has engendered its own strain of myths. That tale simply mixed Einstein's light wave thought-experiment with a clock tower. Again, Pinker was not the first to state it, but I associate it with him because some writers cite him as their source. Pinker's claim has spread quickly, passing into books on law, the environment, and education. One such book claims that Einstein's "first insights into the theory of relativity, for example, occurred when he imagined himself riding on a beam of light looking back at a frozen clock tower."[25]

Even physicists add fanciful details. Michio Kaku is a professor of theoretical physics at the City University of New York. He is also a bestselling author of books that popularize science and a host of television and radio shows. Here is Kaku's account of Einstein and the clock tower:

> Einstein was depressed, his thoughts were still churning in his mind when he returned home that night. In particular, he remembered riding in a street car in Bern and looking back at the famous clock tower that dominated the city. He then imagined what would happen if his street car raced away from that clock tower at the speed of light. He quickly realized that the clock would appear stopped, since light could not catch up to the street car, but his own clock in the street car would beat normally. . . . He had finally tapped into "God's thoughts."[26]

Here, the story is portrayed as Einstein's own recollection. Yet, every sentence is complete fiction, except for one simple phrase: "his thoughts were still churning in his mind when he returned home."

Kaku's elaborate story has been picked up by various writers. It was echoed by political commentator George Will.[27] It has been repeated in various books on physics, economics, scientific literacy, and creative thinking.[28] Also, this story about the tram and the tower has promptly entered into tourist books of Switzerland. For example, one book from 2006 states: "It's said the clock tower helped Albert Einstein. . . . The great scientist surmised, while traveling on a tram away from the tower, that if the tram were going at the speed of light, the clock tower would remain on the same time, while his own watch would continue to tick—proving time was relative."[29] The story has also entered into educational books.[30] I much expect, someday soon, to come across a schoolbook on physics that will duly mention the supposedly chronometric roots of Einstein's relativity. It might be a brief human interest sidebar. Maybe it will have a nice picture of an old clock tower, like the drawings of Galileo's legendary experiments from the Leaning Tower of Pisa. And fine, maybe it will help students to become more interested in the history of physics and its relation to society, which would be good. Already, I find an advanced textbook on quantum chemistry that tells a story about how Einstein was influenced by a tram and a clock. It stops short of referring to any tower: "Einstein recalls that there was a clock at a tram stop in Bern. Whenever his tram moved away from a stop, the modest patent clerk asked himself what would the clock show, if the tram had the velocity of light. While other passengers probably read their newspapers, Einstein had questions which led humanity on new pathways."[31]

I read the growing echoes of the myths about towers and trams with a mixture of amusement and discomfort. It is revealing to see how many writers don't bother to check the truth of a story before they repeat it. I guess that they just trust their sources. And it is entertaining to see how they bury facts beneath layers of thoughtful but needless conjectures. It will be interesting to see what forms these tall tales will take, if they reach schoolbooks, standardized tests, and children's story books. But I also look forward to a time when writers bypass this cloud of fiction to get back to the task of finding out what really happened in the past.

# 13

## *The Secret of Einstein's Creativity?*

Many theories have been proposed to explain Albert Einstein's creativity. For example, Peter Galison argued that Einstein's relativity arose from the fruitful intersection of three fields: science, philosophy, and technology. Although Galison wrote more about the technological side, he acknowledged that science and philosophy were very important too. In physics, technology, and philosophy, people were concerned with notions of time. There's something pleasant in the idea that the intersection of these three fields generated Einstein's conceptual breakthrough. But one reviewer of Galison's book fairly complained: "The inevitable question is then to decide what weight to attach to these different factors, and Galison refuses to address the matter."[1]

And what about art? In 2001, Arthur I. Miller argued at length that special relativity arose from the intersection of *four* fields: science, philosophy, aesthetics, and technology.[2] Was the influence of each field equally great, about 25 percent, upon Einstein's creativity? The problem is that then we are chopping up creativity equally among certain academic disciplines. Is that fair? What about economics? Why don't we include it and assign each field a 20 percent influence? But perhaps the influences were not equal. The historical question—how do we ascertain the *actual* roots of someone's creativity, rather than just the plausible imaginable roots?

Einstein lived for fifty years after 1905, and he became ridiculously famous. Friends, strangers, coworkers, relatives, reporters, writers, biographers, psychologists, and historians all asked him about the roots of his creativity. "How did you do it?" How did he come to think of the relativity of time? Despite plenty of inquiries, interviews, letters, and casual questions, Einstein never mentioned any influence from art or from any timing technologies at the patent office. And neither did any of his coworkers. Instead, Einstein, his friends, and his peers pointed to several other factors and influences.

For example, Michele Besso was Einstein's close friend since 1897 and coworker at the patent office since 1904. Besso had trained as a mechanical engineer and was experienced with electrical technologies. Einstein acknowledged that Besso helped him to clarify the thoughts that resulted in his special relativity, with critical discussions and valuable suggestions. Hence Besso was very well positioned to judge whether patents or chronometric innovations influenced Einstein's path to relativity. Yet in none of their extensive correspondence, written over decades, did either of them mention any such thing. In a letter written in 1947, when Besso was seventy-four years old, he actually asked Einstein how he had come to think of clocks and measuring rods in relativity. It was an opportunity to state or insinuate any influence from timing technologies, yet Besso did not. Instead, he asked whether perhaps Einstein's early reading of a book by Ernst Mach, following Besso's suggestion, had been at the root of Einstein's thoughts about clocks and measuring rods.[3] Mach? No, replied Einstein. Now he had a perfect opportunity to point to clocks and clock towers. But again he did not allude to any technological issues, nor even to art. He acknowledged a great influence of Mach on his intellectual development in general, but he noted that his reading of the philosopher David Hume, whom he discussed with Solovine and Habicht, had been of greater importance in thinking toward relativity.[4]

Accordingly, John Stachel, founding editor of *The Collected Papers of Albert Einstein*, has argued that Hume's notion of time, in particular, may have influenced Einstein.[5] But Stachel acknowledged that this was essentially a conjecture since Einstein did not specify Hume's views on time, just acknowledged the influence of Hume's critical outlook in general.

Over the years, writers have proposed many other hypotheses to try to explain Einstein's path to relativity. Besides Hume, Mach, patents, Mileva Marić, and aesthetics, writers variously have conjectured that Einstein was crucially influenced by the physics of H. A. Lorentz, experiments on light, the writings of Poincaré, and even by reflections on God.[6] Thus Stachel has noted that there is still no consensus on the history of special relativity. More broadly, philosophers have variously attributed notions of relativity to many earlier authorities, including even the alluring Pythagoras.[7] It seems as if the roots of relativity are not a matter of history, but of personal preference.

Several years ago, some historians met at a centenary conference on Einstein. One speaker was giving his new account of how Einstein reached special relativity—what, allegedly, were Einstein's series of steps. Two or three times during his talk, he punctuated his statements by loudly and pointedly saying "*John Stachel*," as if to emphasize that Professor Stachel should agree with some particular point he was making, but didn't. And Stachel remained quiet, he was sitting behind me, and he scribbled something on a sheet of paper and handed it to me. It said: "What song *did* the syrens sing."[8] There's no consensus. Trying to figure out precisely the thoughts of one guy back on a spring day in 1905 seems to be unsolvable—like specifying which songs were sung by mythical mermaids. But some writers are quite sure that their favorite way in which Einstein *may have* reached special relativity was in fact the way it happened.

Figure 13.1. What song *did* the syrens sing?

One possible reason why historians have not converged on a shared explanation is that some accounts are too reflective of the historians themselves rather than the subject. Also, many writers tend to neglect participant testimony, such as what Einstein himself said about how he made his conceptual breakthrough. A common reflex is to say that a participant's answer, whatever it be, would likely be irrelevant. Some authors lie, or they forget their seminal thoughts, or they are quite unaware of their essential motivations. Such dismissive expectations are useful as a historical strategy because they shove aside the authors' presumed authority to explain themselves, such that the historian or commentator instead can seize that right with all the benefits it entails.

Regardless, suppose we care enough to listen to the dead. What if we could ask Einstein, today, why was it *you* who formulated the theory of relativity? What would he answer? Actually, we have a document that responds directly to that question. Einstein once explained to another physicist, James Franck:

> When I asked myself, how it came to be that I in particular found the theory of relativity it seemed to me to lie in this: The normal adult person does not think about space-time problems, he has already done all that there is to think about that in early childhood. By contrast, I was so slow in my development that I first began to wonder about space and time when I was already grown up and hence naturally I penetrated deeper into the problem than a child.[9]

Here we have a documentary trace pertaining explicitly to the actual roots of Einstein's creativity. We might place it under the banner of developmental psychology.

There may, however, be doubts over Einstein's quotation. Was he serious when he said this? Did he really say this? It is not a statement from Einstein's hand but a transcription by Franck in a letter he sent to Einstein's biographer Carl Seelig in 1952. If Einstein did say this, when did he do so? We don't have the date, so even if he said it, in the 1940s for instance, that does not mean that it really described the situation in 1905, when he was formulating his theory of relativity. There are also questions about the content: Did Einstein really develop intellectually slowly? Did he really think in some childish way about notions of time in 1905?

To analyze this quotation historically, let's trace backward, against chronology, to find its apparent roots in the distant past. Part of the exciting thing about history is the research process by which one piece of evidence leads to other earlier bits, so I want to convey that process of gradually unveiling earlier layers of documentary facts.

Regarding the legitimacy of the quotation: Einstein approved it. We know so because he made a detailed list of corrections to Seelig's biography of him after it was published, and he kept this quotation.

Given that Einstein said this late in his life, can we trust that it really happened? Did he, for one, "develop so slowly"? Luckily, there are sources that attest to this. For example, when Einstein was a boy he began to talk so late that his parents were worried that he was developmentally challenged. According to his sister, for the young boy: "Normal childhood development proceeded slowly, and he had such difficulty with language that those around him feared he would never learn to speak."[10] Eventually he did speak, but at the age of five, according to his parents, he was still phlegmatic and distracted.[11] He spoke, but with a strange habit: "Every sentence he uttered, no matter how routine, he repeated to himself softly, moving his lips. This odd habit persisted until his seventh year."[12] We also know that at the age of eight or nine he was shy and unsociable.[13] Only much later, ten or twelve years later, did he become social and talkative. But in school, Einstein "was considered backward by his teachers. . . . his teachers reported to his father that he was mentally slow, unsociable."[14] Thus, several pieces of evidence substantiate the claim that Einstein developed slowly as a child.

What about Einstein's claim that he analyzed notions of time like a child? It is reminiscent of Jean Piaget's work on child psychology. In 1946, that Swiss psychologist published his book on *The Development of the Notion of Time in Children*. So it might seem appropriate that if Einstein's words are from the 1940s, then he might just be echoing ideas of the time, as what Piaget called "genetic epistemology" was then pretty popular. It turns out, however, that the influence was the other way around.

Back in 1928, Einstein was invited to preside over the International Congress on Philosophy and Psychology held in Davos, Switzerland. At the time, Piaget was chair of psychology, sociology, and history of science at the University of Neuchâtel. At the conference, Einstein asked Piaget's group to investigate the following questions: "The subjective intuition of time, is it primitive or derived, and intermixed, or not, to that of speed? Such questions, do they present a concrete meaning in the analysis of the genesis of notions in a child, or well the construction of temporal notions: is it finished before being translated to the plane of language and conscious reflection?"[15] It was only after that suggestion, according to Piaget, that his group began to research the development of notions of time in children, resulting in Piaget's book of 1946.

Returning to Einstein's explanation of his creativity, he claimed that there was a kind of developmental delay, a childish thinking about basic problems, that led him to relativity. Is it true that he thought about problems like a child in 1905, when he was twenty-six years old? Moritz Solovine, a close friend of Einstein's and one of the three members of their informal discussion group from 1902 until 1905, recalled that at that time: "In the examination of fundamental notions, Einstein employed with predilection the genetic method. He used it to clarify them by what he had been able to observe in children."[16] Solovine reported this in 1956. At first blush it sounds anachronistic, as if Solovine were projecting into 1905 the kind of work that arose only later: the psychological studies on children led by Piaget on genetic epistemology in the 1930s.

That is not, however, what Solovine meant by "the genetic method." Because by 1900 when Piaget was not even five years old, the "genetic method" already was construed to investigate the formation of ideas and habits, and to replace static metaphysical presuppositions with scientific developmental statements.

This line of research was pioneered in the 1890s by American psychologist James Mark Baldwin. He performed systematic experiments on children, including his own daughter, advocating what he called "the genetic method." Baldwin was a proponent of the experimental psychology coming from Germany, from the laboratory of Wilhelm Wundt, where he had studied. Baldwin was a professor of psychology at Princeton, a founding member of the American Psychological Association, and, in 1897, its sixth president. For Baldwin, genetic psychology was the new field of studying the mind—not as a static soul or substance with fixed structure and attributes, but instead as a growing, evolving entity. Baldwin argued that to begin to understand elaborate conceptions, one should begin by analyzing the relatively simple activities, behaviors, and movements of children; how they respond to their environment. Baldwin argued: "The study of children is often the only means of testing the truth of our mental analyses."[17]

Did Einstein read any writings by Baldwin? We don't know; I have no evidence that he did. Yet at least there really was such a thing as "the genetic method" in 1905 when Einstein was formulating his theory of relativity.

Next, do we have any evidence that Einstein, on or before 1905, cared about making observations about how notions in children develop? Yes, but very little. In early 1902, Einstein's fiancée Mileva Marić gave birth to their daughter. When the baby, Lieserl, was just one month old in February 1902, Einstein wrote to Mileva: "Can she already turn her eyes well toward anything? Now you can make observations. I'd like to make a Lieserl myself sometime, it must be interesting! She can certainly cry already, but won't know how to laugh until much later. Therein lies a profound truth."[18] There is also a letter from Mileva Marić that refers to how children develop physical concepts. Her bright and thoughtful letter, from 1897, told him: "I do not believe that the structure of the human brain is to be blamed for the fact that man cannot grasp infinity; he certainly would if only in his young days, when he is learning to perceive, the little man wouldn't be so cruelly confined to the Earth, or even to a nest between four walls, but instead be allowed to walk out a little into the universe. Man is very capable of imagining infinite happiness, and he should be able to grasp the infinity of space."[19]

Even if Einstein and Marić were unfamiliar with Baldwin's writings and observations on infants and children, was there anything similar in other writers' works? There are several significant leads. For one, prior to 1905 Einstein read with enthusiasm the works of Ernst Mach. Mach was the well-known Austrian physicist who actually had a strong interest in psychology. His interest in the development of general notions in our minds is one of the things Einstein admired. And Einstein later praised Mach for having been a scientist who "looked into the world with the curious eyes of a child."[20] (Friends and physicists made similar comments about Einstein.)

In 1886 Mach published his book, *The Analysis of Sensations*, which Einstein read before 1905. Through systematic research in physics, physiology, and history, Mach tried to resolve the traditional apparent conflict between the physical and the psychological, arguing that the fundamental elements of physics are sensations. Mach argued that "the aim of all scientific research" is the "adaptation of thoughts to facts" and that *that process* can be essentially observed in children.[21] This resonates with Solovine's claim: "In the examination of fundamental notions, Einstein employed with predilection the genetic method. He used it to clarify them [such fundamental notions] by what he had been able to observe in children."

Mach traced the historical development of physical theories and also tried to understand the gradual development of abstract concepts and intuitions. For that, he considered how people develop such notions, especially at early ages. Mach even gave examples from what he had observed in his son. For example, he wrote: "Almost every new fact necessitates a new adaptation, which finds its very expression in the operation known as *judgment*. This process is easily followed in children. A child, on its first visit from the town to the country, strays, for instance, into a large meadow, looks about, and says wonderingly: 'We are in a ball. The world is a blue ball.' [note: This case is not fictitious, but was observed in my three-year old child.] Here we have two judgments. What is the process accompanying their formation?"[22] Mach discussed how concepts are gradually formed in the mind of a child, how such concepts are distinct from perceptions or images, and how they are independent of words. Those are the kinds of distinctions that Einstein later echoed.

Einstein also read works by the German physician and physicist, Hermann von Helmholtz, who had cultivated interest in the growth of concepts in children. Significantly, Helmholtz argued that a person's knowledge of the principles of geometry and mechanics stems not from any transcendental source, but from ordinary everyday experiences.[23] Helmholtz repeatedly expressed his interest in the "stages of development of our mental life," and in the formation of basic concepts or intuitions. To that end, he studied, for example, experiments in physiological optics. In

1878, at the University of Berlin, Helmholtz gave a speech titled "On the Facts of Perception," which later circulated widely. He argued:

> The new-born human child is extremely unskilled at seeing; it requires several days before it learns to judge the direction of the visual image towards which it must turn its head so as to reach the mother's breast. To be sure, young animals are much more independent of individual experience. What, however, is this instinct which directs it? Is direct inheritance of the parents' circles of representation possible? Does it concern only pleasure or displeasure, or a motoric drive, which attach themselves to certain aggregates of sensation? About all that we know as good as nothing. Plainly recognizable residues of these phenomena still occur in humans. In this area, clean and critically conducted observations are highly desirable.[24]

The first and last lines of this quotation sound much like the words from Einstein's letter from 1902: "Can she already turn her eyes well toward anything? Now you can make observations." We don't know if Helmholtz influenced Einstein, but the similarity is noteworthy.

So we find a common interest in childhood learning in Baldwin, Mach, and Helmholtz. How did they get their interest? There are some leads to follow, but consider just one. Baldwin noted that he became interested in the formation of the mind by studying how children learn partly owing to Darwin's theory of evolution. Thanks to the success of Darwin's works, notions of evolution became increasingly applied in the social sciences.

In his book of 1872, *The Expression of the Emotions in Man and Animals*, Darwin analyzed the evolution of notions in children. He analyzed how emotions are manifested in animals, humans, and in children to figure out how emotions echo evolutionary history. Darwin collected data on how infants act and react and he made and reported systematic observations of his own children. He observed when infants' vocal sounds become expressive, when infants begin to blush, how they react when startled, what sounds and visual stimuli make them blink, and more. And one noteworthy point is this: that Darwin emphasized that while babies cry since birth, they learn to smile gradually, and to laugh only much later, after three or four months in the case of his own children.[25] That very point was raised by Einstein in his letter of 1902. Again, it is surprising to see that it had actually shown up in the fledging literature on developmental psychology.

Furthermore, before 1905, Einstein was very familiar with the writings of the philosopher Arthur Schopenhauer, especially his work of 1851, on maxims of life. There, Schopenhauer argued that children are not merely occupied by individual objects, but also that in their early years children are intensely occupied in analyzing their sensations, "and without any clear consciousness of what it is doing, the child is always silently occupied in grasping the nature of life itself—in arriving at its fundamental character and general outline by means of separate scenes and experiences; or, to use Spinoza's phraseology, the child is learning to see the things and persons about it *sub specie aeternitatis*—as particular manifestations of universal law."[26]

Thus Schopenhauer too was interested in how children learn. Was Einstein concerned with universal laws as a child? If you believe him, yes, because he told his stepson-in-law that "since earliest childhood he had always shown an attraction toward the universal."[27] Schopenhauer continued:

> The younger we are, then, the more does every individual object represent for us the whole class to which it belongs; but as the years increase, this becomes less and less the case. That is the reason why youthful impressions are so different from those of old age. And that is also why the slight knowledge and experience gained in childhood and youth afterwards come to stand as the permanent rubric, or heading, for all the knowledge acquired in later life—those early forms of knowledge passing into categories, as it were, under which the results of subsequent experience are classified; though a clear consciousness of what is being done, does not always attend upon the process.[28]

Einstein agreed with Schopenhauer (and Mach and Helmholtz) that *intuition* was the kind of knowledge that develops, especially at early ages, by direct personal experience and reflection, not from education. In the process of scientific creativity, as Einstein emphasized to Moszkowski, "the really valuable factor is *intuition*!"[29] According to Peter Michelmore, who interviewed Einstein's son, Einstein's "imagination conjured up the various approaches. His intuition told him when he was on the right track. It was that way with Relativity."[30]

Another piece of evidence is that throughout his life, Einstein emphasized the essentially artificial character of the fundamental elements of physical theories. He argued that creativity consists of the "free play" of concepts. Consider the famous thought-experiment in which Einstein, at the age of sixteen, imagined that he chased light. That experiment marked the beginning of a ten-year odyssey that resulted in relativity. Einstein described it as a "childish thought-experiment."[31] This license to playfully manipulate, modify, and rearrange concepts was encouraged by his childlike approach to physics.

And there's more. Looking back, Einstein recalled that his Olympia Academy "delighted with a child-like joy in all that was clear and reasonable."[32] And throughout the years, he maintained a playful interest in how children learn.[33] But let's sum up:

- Einstein stated that *the distinctive factor* that led him to relativity was childish thinking.
- From about 1902 through 1905, his favorite method of analyzing fundamental notions was to ponder how notions develop in children.
- To some extent, Einstein was interested in observations of children.
- Some of Einstein's favorite readings highlighted the formation of concepts through ordinary experiences, especially in childhood.
- In the process of scientific creativity, Einstein often emphasized the importance of *intuition*, as developed especially in childhood.

So what am I saying? That Einstein's breakthrough to relativity came essentially from child psychology? No. What I've shown is that Einstein's words to James Franck make historical sense. There was actually a significant growing tradition of researchers, even physicists, who seriously turned to the analysis of children's behaviors to understand the formation of fundamental notions. My point is not that the most important factor in Einstein's creativity was developmental psychology. There is evidence, however, that this played a larger role than, say, new timing technologies, art, religion, or his wife.

While childish thinking, on the basis of the documentary evidence, played an important role in Einstein's creativity, there were other factors for which there is even more evidence. I will not discuss them here, but we should at least mention them. In particular, Einstein often acknowledged the foremost influence of H. A. Lorentz's works in physics. He also reflected on various experiments, and his critical outlook was influenced by the writings of Hume, Mach, and Poincaré.

The reason why we can construct convincing narratives about the importance of one or another "key" component in someone's creativity is because there is often an abundance of evidence, such that one freely grabs pieces to cite and emphasize. Ultimately, the goal is to pull together *all* the various kinds of evidence but also to weigh the relative influences of the various factors. What we find depends on what we look for, but the evidence itself can lead us to places we did not anticipate. The value of really understanding someone's creative paths is that such paths might teach us some productive ways to think.

Darwin's constant productivity, for example, greatly impressed Alfred Russel Wallace, who independently conceived the notion of evolution by natural selection. In a review of the book on *The Expression of Emotions*, Wallace praised Darwin for never having lost the full force of "the restless curiosity of the child."[34]

Likewise, owing to his childish thinking, Einstein believed that education should not begin from books or generalizations, but from particular experiences. He wanted children's education to be based on concrete experiences that would lead to abstract concepts. As he told Moszkowski:

> The first beginnings should not be taught in the schoolroom at all, but in open Nature. A boy should be shown how a meadow is measured and compared with another. His attention must be directed to the height of a tower, to the length of his shadow at various times, to the corresponding altitude of the Sun; by

this means he will grasp the mathematical relationships much more rapidly, more surely, and with greater zeal, than if words and chalk-marks are used to instill into him the conceptions of dimensions, of angles, or perchance of some trigonometrical function. What is the actual origin of such branches of science? They are derived from practice, as, for example, when Thales first measured the height of the pyramids with the help of a short rod, which he set up at the ultimate point of the pyramids shadow. Place a stick on the boy's hand and lead him on to make experiments with it by way of a game, and if he is not quite devoid of sense, he will discover the thing for himself. It will please him.[35]

# 14

## *Eugenics and the Myth of Equality*

WHEN you go to the supermarket, you can see that the tomatoes are round, smooth, very red, and not nearly as bitter as you might imagine. Can we make them better? Yes, and we already have: those tomatoes have been engineered systematically to be redder, sweeter, and to last longer. Tomatoes have been improved to withstand various germs, bugs, and disease. Similarly, breeders create dogs with tailored characteristics such as modified bodies, hair, and teeth. Dogs descended from wolves, we hear, yet dogs are easier to get along with. You can have one in your house without constantly worrying that it will kill you. Not only are most dogs smaller and therefore less dangerous than wolves, their *behavior* is also different. They're trustful, playful, and attentive. How did they get those behaviors? We might imagine that maybe dogs were bred from an ancient species of wolves that were naturally friendly to humans. We might imagine that although breeders modified the *physical* traits of that species, its *behavioral* traits remained utterly unchanged. But maybe not. Maybe breeders modified the behavioral traits too. After all, some breeds of dogs have different behaviors, don't they? If we can do that with wolves, and since we can make better tomatoes, then what about humans? Can we improve human behaviors by selective breeding?

Throughout history, some people have believed that not much can be done to change human nature, our innate tendencies. Many teachers accept that children are born with different talents, which can be identified and cultivated but not essentially changed. For example, Pythagoras allegedly examined his prospective disciples thoroughly to find their innate and hereditary talents in order to decide what each should study: "he inquired about their relation to their parents and kinsfolk.... He considered their frame's natural indications physionomically, rating them as visible exponents of the invisible tendencies of the soul."[1] Allegedly, Pythagoras had proclaimed: "*You cannot make a Mercury of every log*," that is, "Not every mind will answer equally well to be trained into a scholar."[2] It is interesting to see how widely such ancient notions still propagate, for example, my mother has told me an amusing Spanish saying, Latinized: "Lo que natura non da, Salamanca non presta"—what nature gives not, the University of Salamanca cannot lend either.

However, the notion of evolution increasingly challenged the idea that human nature is immutable. According to Darwin, humans were not always as they are—a grotesquely ridiculous idea to many people in Victorian England. Some of Darwin's peers became convinced of evolution, but others were deeply disturbed by it. Captain FitzRoy, for one, regretted that Darwin's work was a byproduct of the *Beagle* voyage. He denounced it as incompatible with a literal reading of the Bible. Owing also to other misfortunes, FitzRoy became increasingly depressed and disturbed, and in 1865, he committed suicide: he cut his throat.[3] Originally, FitzRoy had asked Darwin for company during the *Beagle* voyage partly because he feared that he might have inherited the insanity of an uncle who had killed himself. Is insanity inherited?

If humans evolved from animals, then they evolved from animals that did not behave the way we do. Maybe environments plus selective mating gradually produced the human behaviors that now seem natural to us. But if so, should humans control their own evolution? This question was investigated by Darwin's cousin, Francis Galton.

In the 1840s, Galton studied mathematics at Cambridge University. He wanted to earn a degree with honors, but his studies were so intensive and exhausting that he suffered a nervous breakdown, and settled for merely passing. Afterward, he inherited monies from his father that enabled him to freely pursue his interests. He loved to quantify things, and he wanted to apply mathematics for the good of mankind.

Anthropologists had tried to establish correlations between people's bodies and their behaviors. They measured bodies, profiles, and bumps on people's heads, yet they failed to find the biological bases of behavior. Galton too became increasingly fascinated by the quantification of human traits. Traveling in Africa, he discreetly measured the shapes of women's bodies. He later attempted to numerically analyze fingerprints as he became an early advocate for using fingerprints to identify criminals.

Owing partly to his interest in Darwin's theory of evolution, Galton became curious about whether mental abilities are inherited. There was a view that great men tend to have stupid sons. Or were geniuses related? So, Galton studied biographical encyclopedias to count family relations among prominent statesmen, jurists, military leaders, scientists, and artists. He found that a surprisingly high number of them were related. He was especially impressed by the recurrence of scientific and artistic achievements among blood relatives, because in such fields, nepotism and social forces were not as strong as in politics and social institutions. He concluded that heredity affected not only physical features, but also talents.

Galton acknowledged that social advantages, of course, help individuals gain opportunities and recognition but he claimed that such advantages were insufficient to explain the extent of success among blood relatives. He published his conclusions in his book *Hereditary Genius* in 1869.[4] Darwin, who believed that men differed not mainly in inborn mental talents but in dedication and efforts, became convinced by studying Galton's work that genius tends to be inherited.[5]

According to Galton, education was not the cause of intelligence. At the time, education for the middle and lower classes in Great Britain was not as good as education in America. Yet the Americans, he argued, did not produce better masterpieces of literature, philosophy, and art. He therefore concluded that some talents were inherited. Yet historian Daniel Kevles has noted that Americans at the time did not fuss as much with canonically high art because they were busy building their nation. Moreover, the biographical encyclopedias used as data by Galton were not a fair measure of the population's talents; Kevles explained that Galton confused individuals' reputation as an indication of their natural abilities.[6]

Regardless, Galton was aware that the lower classes in Great Britain were reproducing at a greater rate than the more privileged. While he and his wife were unable to have children, immigrants and poor people seemed to flood the cities. Thus Galton feared that Britain's talents would diminish over time, while incompetence, frailties, and poverty multiplied. He disagreed with Thomas Malthus's call for reproductive restraint, because only moral and capable individuals would exercise such restraint, diminishing only their own numbers. Civilization seemed to have an indisputable effect: it deterred the operation of natural strife among humans. Thus, natural selection seemed to fail in the case of humans. The "unfit" were propagating.

Breeders, it was well known, were remarkably successful at controlling the traits of dogs, horses, and pigeons. Accordingly, Galton expected that human qualities too, even mental talents, could be improved by selective breeding. He hoped to find mathematical laws of heredity, to find the laws of nature that would save mankind.

Galton painstakingly developed statistical methods to analyze heredity. His studies of the distributions of traits in populations among succeeding generations led him to formulate mathematical methods (the so-called coefficients of regression and correlation) that decades later became widely used—not just in genetics, but also in medicine, economics, sociology, anthropology, and more.[7]

In 1883, Galton applied the name *eugenics*, meaning "in good birth," to the study of planned breeding to improve human inheritance.[8] Eugenics is an aspect of the history of biology which biology teachers generally avoid. Most science teachers, if they discuss history of science at all, deal mainly with its heroic and positive aspects. They avoid "wasting time" on erroneous old science, stuff that was later denounced as non-science. Among such embarrassing failures, eugenics stands out. It was a greater shame than the worst aspects of astrology and alchemy.

Yet eugenics was fueled by the noble hope to cure society of its ills. After all, if we can make better tomatoes, why not make better humans? Galton urged that the British government should measure people's abilities and rank them accordingly. Couples having higher ranks would then be encouraged to have plenty of children, while couples with lower ranks would be discouraged from having as many children. Moreover, Galton hoped that the lowest-ranked individuals would be segregated from society to prevent them from having any children. Galton did not believe that people are naturally equal.

Moreover, rising indices of crime, poverty, and disease seemed to suggest that the British peoples were degenerating. One of Galton's followers, the statistician Karl Pearson, argued: "How is the next generation of Englishmen to be mentally and physically equal to the past generation which has provided us with the great Victorian statesmen, writers, and men of science?"[9] The eugenicists aimed to make the science of inheritance a moving force in politics and social customs. Their urge to measure and quantify human traits led to the collection of disturbing evidence that humans are not equal in any traits—not in height, or skin color, or in their performance on written exams. It seemed that different races scored differently in physical and mental skills. Against that urge to quantify and proclaim inequity, there stood, across the Atlantic, the ideology of equality. The Declaration of Independence of the United States of America asserted: "We hold these truths to be self-evident, that all men are created equal."[10] Simultaneously, the nation of immigrants itself did not entirely believe in equality, as various rights were denied to women and to the descendants of slaves.

Interest in eugenics grew as biologists came to accept "Mendel's laws." I use quotation marks because contrary to what is stated in many books on genetics, the Austrian monk Gregor Mendel actually did not discover such laws; that is, he did not claim to have found laws of heredity valid for all species. Instead, Mendel had struggled to find whether there was a general law for the development of plant hybrids.[11] In the early 1900s, Mendel's results on hybrids were fairly construed as evidence that certain physical traits are transmitted as pairs of units. Biologists spoke of "laws of heredity."

Accordingly, Charles Davenport, a biologist in the United States, showed that certain human traits, such as eye color, are transmitted in accord with such laws. Consequently, Davenport came to believe that important human traits were also transmitted in their entirety, even if sometimes they were not manifest. It seemed to him that traits such as alcoholism or stupidity could not be cured by simply treating the ailing individual, because that individual would pass that trait onto his children. While effective techniques for healthy breeding were used in horses, for example, it seemed outrageous that they were not used also in humans. Hence Davenport began to advocate eugenics.

In 1904, thanks to funding from the Carnegie Institution, Davenport established a center for the study of human inheritance and evolution, located in Cold Spring Harbor, New York. The center began to widely solicit replies to questionnaires on the physical and mental traits of individuals and their families. Davenport analyzed such family histories to find numerical patterns in the incidence of: stump-fingers, polydactyly (having more than five fingers per hand), albinism, hemophilia, insanity, alcoholism, criminality, and especially "feeblemindedness." He tried to fit many of these traits into the simple mathematical patterns of inheritance laws.

Davenport recognized that social environment was an important factor in determining how a person develops. Yet he also argued that some inborn traits cannot be curbed, no matter how good the environment be—just as some "bad seeds" cannot grow well even in good soil.[12] Davenport attributed social deviance to a lack of self-restraint, the result of a supposedly defective nervous system caused by hypothetical bad genes. Yet his evidence was anecdotal, lacking systematic measurements.[13]

Thus, American eugenicists claimed that just one or two pairs of genes determined bad traits: feeblemindedness, violent temper, epilepsy, criminality, manic depression, and even poverty. Eugenics seemed to promise techniques to fight epidemics and enact social reforms. Eugenicists sought to reduce degeneracy and racial mixtures by somehow restoring humans to a former condition. Instead of bothering with historical research, some freely imagined a glorious age long gone: "Thus in primal days was the blood of the race kept high and pure, like mountain streams. One may not admire the harsh conditions of the savage life of our German forefathers in their Teuton forests; but one must admit the high purity of their blood, their high average sanity, soundness, and strength. They were a well-born, well-weeded race."[14] Moreover, American eugenicists conjectured that certain "races" had inborn troublesome tendencies.

Supposedly Italians were violent, Jews were prone to thieving, and so forth. Davenport therefore called for laws to oppose the influx of "inferior blood" into the Nordic population. He argued: "The biological basis for such laws is doubtless an appreciation of the fact that negroes and other races carry traits that do not go well with our social organization."[15]

Politicians advanced restrictive immigration policies, along with sterilization laws supported by President Theodore Roosevelt. Alarmed by decreasing birth rates, Roosevelt also led a crusade against "race suicide." By 1907, more than four hundred prisoners had been sterilized in the state of Indiana, and then Indiana approved a sterilization law for "degenerates." By 1916, the American Eugenics Society explained that just one troublesome family, the "Jukes," had cost the state of New York more than $2,000,000 in criminal and institutional expenses, whereas segregation of the original Jukes couple from society would have cost only $25,000; further, their sterilization would have cost a mere $150 instead.[16] By 1917, sixteen states had sterilization laws.

Eugenic concerns also impelled intelligence testing. Building on the contributions of various psychologists, in 1916 Lewis Terman published *The Measurement of Intelligence*, in which he argued that no amount of instruction could work to qualify certain people as voters and leaders in society. Terman called for intelligence tests to determine whether, as was apparent to him, workers and servants who were "Indians, Mexicans, and negroes" had distinct mental traits or limitations.[17] Intelligence tests became a means for segregating students according to their supposed inborn abilities. Terman advocated IQ tests, ascribing a single number to a person's intelligence.[18] He and others also advocated that immigrants be tested.

Eugenics enthusiasts increasingly denounced the American declaration of equality as a myth. A book by Alfred Schultz, *Race or Mongrel*, complained: "The principle that 'all men are created equal' is still considered the chief pillar of strength of the United States. . . . Only one objection can be raised against it, that it does not contain one iota of truth."[19] Likewise, the president of the American Museum of Natural History proclaimed: "The true spirit of American democracy that *all men are born with equal rights and duties* has been confused with the political sophistry that *all men are born with equal character and ability to govern themselves.*"[20]

In 1921, Vice President Calvin Coolidge complained that the United States was seen as a "dumping ground" for unwanted foreigners. He noted that biological laws showed that while "Nordics" propagate successfully, certain "other races," when mixed, lead to deterioration, and therefore, that "Quality of mind and body suggests that observance of ethnic law is as great a necessity to a nation as immigration law."[21]

In 1924, the state of Virginia approved its "Racial Purity Law," which outlawed interracial marriage. Several other states copied that law. Meanwhile, Harry Laughlin carried out IQ tests on immigrants. He found that Italians and Africans scored less than Nordics, and so he inferred that they lacked equal mental abilities. Laughlin's subsequent testimony about his conclusions before the federal government helped to pass the Immigration Restriction Act of 1924, signed by Coolidge, who had become president. The act established quotas to restrict the influx of people from certain countries and ethnicities.[22]

Furthermore, the public increasingly embraced eugenic ideals. Several state fairs came to feature eugenic exhibits and "Fitter Families" contests. The governor of Kansas, for example, awarded trophies and medals to the healthiest exemplar families. Medals bragged: "Yea, I have a Goodly Heritage." In Philadelphia, the Eugenics Society displayed exhibits on the slow birthrate of "high grade" people compared to the alarmingly fast birthrates of "deficient" and "abnormal" people. The exhibit asked, "How long are we Americans to be so careful for the pedigree of our pigs and chickens and cattle—and then leave the *ancestry of our children* to chance or to 'blind' sentiment?"[23]

In 1926, Leta Hollingworth, a professor of education at Columbia University, authored a study of "gifted children." She ascribed students' abilities to their inborn predispositions. She also claimed that, in the United States, children of Africans and Italians had a lower than average intelligence. She also claimed: "Modern biology has shown that human beings cannot improve the qualities of their species, nor permanently reduce its miseries, by education, philanthropy, surgery, or legislation. Such attempts are palliative merely and leave a worse condition for the next generation to face. A philanthropy that succeeds in relieving the chronic pauperism of a thousand individuals of this generation, bequeaths at least two thousand paupers to be relieved by generations immediately following, for it has enabled a thousand organisms of pauper quality to live and breed." Like stupidity, poverty was cast as an unfortunate inherited trait. To solve social problems, Hollingworth too advocated eugenics: "It would ultimately reduce misery if the stupid, the criminal, and other mentally, physically, and morally deficient would refrain from reproduction."[24] Other eugenicists argued further that procreation should be legislated, not left to choice.

Not everyone was pleased with eugenic laws. Numerous biologists and lawyers criticized especially the punitive sterilization laws. Hence, a sterilization case reached the Supreme Court in 1926. The state of Virginia had ordered that Carrie Buck be sterilized because she, her mother, and her infant daughter were allegedly feebleminded. Without having met Carrie Buck, Harry Laughlin declared in an expert deposition that her feeblemindedness was indeed hereditary and that she belonged to the "shiftless, ignorant, and worthless class of anti-social whites of the South."[25] The court ruled that just as the best citizens sometimes sacrifice themselves for the public welfare, so too the leeches of society should sacrifice to protect the state from being flooded with incompetence. Justice Oliver Wendell Holmes declared: "It is better for all the world, if instead of waiting to execute degenerate offspring for crime or to let them starve for their imbecility, society can prevent those who are manifestly unfit from continuing their kind. . . . Three generations of imbeciles are enough."[26]

It was a great victory for eugenicists, so they kept pushing their social programs. By 1928, nearly four hundred universities in the United States offered courses on eugenics.[27] By 1929, twenty-four states had sterilization laws, and by January 1935, more than 21,500 individuals had been involuntarily sterilized by law.[28]

Still, a growing coalition of scientists and critics denounced eugenics as nonsense. They complained that it reeked of prejudices disguised as science. Reform activists argued that the apparent problems of racial degeneracy were really just problems of social disorder. Even for diseases that were indeed inherited, it became clear that eugenic policies could hardly cure them.

For example, Charles Davenport had shown that Huntington's chorea was inherited, so it seemed that this disease could be wiped out by sterilizing every person who exhibited it. But that would rid only one generation of all offspring who carried two genes for the disease. Others who carried only one such gene, which did not manifest itself, would continue to transmit the disease. Society would still need to sterilize people in the next generation, and so on. Geneticist Reginald Punnett calculated how many generations it would take to reduce the frequency of a supposedly simple trait such as "feeblemindedness." Punnett found that to diminish its incidence from 1 in 100 persons, for example, to 1 in 10,000, it would actually take *90* generations of sterilizations.[29] So, more than two thousand years of eugenics would still fail to wipe out feeblemindedness. Moreover, there was a greater problem: that there was no reliable evidence that traits such as "idiocy," violence, and criminality really depended on single pairs of genes. Instead, such behavioral traits were hardly definable genetically, and would seem more likely to depend on countless many genes.

Nevertheless, eugenics gained popularity in other countries. In Germany, in particular, it met a receptive audience. Albert Einstein, like many people, became fascinated by genetic inheritance. Having first loved his wife Mileva Marić as an equal, he came to despise her as "a physically and morally inferior person."[30] She had a congenital hip displacement and suffered depression and nervous breakdowns, which he attributed to her genes. Her sister was insane and catatonic. Likewise, Einstein and Marić's second son, Eduard, became mentally and emotionally unstable, and Einstein ascribed that to "the severe hereditary flaw" in Marić's bloodline.[31] Einstein privately approved of the ancient Spartans' practice of abandoning their weakest children to die, to strengthen society. In 1917, he wrote to an intimate friend, "To keep something alive that is not viable beyond the years of fertility is undermining civilized humanity. . . . So it would be urgently necessary that physicians conducted a kind of inquisition for us," to sterilize "without leniency in order to sanitize the future."[32]

Those were the grotesque prejudices of a private individual. Yet more awful opinions were more publicly voiced. Adolf Hitler advocated the natural superiority of the German peoples. For him, *inequality* among races was a permanent and unchangeable aspect of the natural order. Hitler argued that "anyone who wants to cure this era that is inwardly sick and rotten, must first summon the courage to expose the causes of this disease." He believed that racial mixing was poisoning the most valuable and natural German resource—the pure Aryan blood—and claimed that only the healthy German people, beautiful and spiritually superior, could produce the highest works of culture and creativity. Above all, he despised Jews as intrinsically evil and degenerate and as guilty for Germany's woes. He blamed them and the Marxists for supposedly undermining the natural superiority of the German blood by their "theory of the *equality of men*."[33]

Hitler proposed that people who were physically degenerate and mentally sick should be prevented from procreating for at least six hundred years to help cleanse society of its ills. He claimed that it was the sacred racial mission of the Germans to protect "the most valuable stocks of racially primal elements" and to raise them to a dominant position. Hitler ranted that the State "has to take care that only the healthy beget children," acting as "the guardian of a thousand years' future, in the face of which the wish and the egoism of the individual appears as nothing and has to submit. It must put the most modern medical means at the service of this knowledge. It must declare unfit for propagation everybody who is visibly ill and has inherited a disease, and must carry this out in practice."[34]

Some German theorists saw human societies as living organisms, such that individuals with defects or ailments appeared as dangerous imperfections. To keep society healthy as a whole, the "racial scientist" Adolf Jost had advocated "the right to death" in his book of 1895. Hitler similarly argued, "*If the power to fight for one's own health is no longer present, the right to life in this world of struggle ends.*"[35]

For Hitler, German nationality was founded on race: "*the purity of blood.*" His National Socialist Party was impressed by the American laws for sterilization and against immigration and interracial marriage. The

Nazis developed the Race Purification Program through which they sought a sort of racial hygiene to cure the state of expensive and "unproductive lives." For Hitler, the aim of the state was to promote "*a community of physically and mentally equal living beings*."[36]

In 1933, the Nazi government approved the Law for the Prevention of Hereditary Diseases in Future Generations. They announced that 400,000 Germans would be promptly sterilized beginning on January 1, 1934.[37] It became essentially criminal to suffer from mental retardation, schizophrenia, epilepsy, blindness, deformations, alcoholism, and other conditions. Physicians were required to report any such individual to Hereditary Courts, which in turn would judge whether any such individual should be segregated or sterilized to be deterred from procreating. The goal was to sterilize "lives not worth living." The deputy leader of the Nazi Party, Rudolf Hess, proclaimed that "National Socialism is nothing but applied biology."[38]

The Nazis sterilized thousands of Germans in just months. In 1934, an American eugenicist in Virginia complained, "The Germans are beating us at our own game."[39] By 1937, roughly 225,000 people had been sterilized in Germany, about ten times more than in the United States. Furthermore, the Nazis enacted a medical program to kill disabled individuals. Hitler was seen as "The Physician" of the German peoples, one who would implement coercive corrective procedures to cure the health of the social organism. One of his followers exclaimed: "Our characteristics are deeply rooted in our race. Therefore, we must cherish them like a holy shrine, which we will—and must—keep pure. We have the deepest trust in our Physician and will follow his instructions in blind faith, because we know that he will lead our people to a great future. Hail to our German people and *der Fürher*!"[40]

Alongside Nazi eugenics, there emerged a movement for "German physics." Philipp Lenard, Johannes Stark, and a few other German physicists claimed that science developed by Aryan men was superior to other science. They claimed that dead greats such as Galileo, Kepler, and Newton were Aryans, and praised the "pragmatic" approach of Rutherford, as opposed to the "dogmatic" Einstein.[41] They ridiculed Einstein's relativity as "Jewish physics," as theoretical conjectures and lies. They burned books and prevented Jews from teaching at universities. Einstein received death threats. Many Jewish professors including Einstein, James Franck, and Lise Meitner fled Germany.

Under Nazi rule, eugenics became such a powerful movement that some commentators traced its origins to purported influences centuries older than Galton's mathematical schemes. For example, one writer even associated it with Pythagoras.[42] According to Iamblichus, Pythagoras examined his pupils to ascertain their innate talents. And allegedly, some Pythagoreans had believed that men should not be allowed to act freely, lest they become degenerates, but should instead be ruled even regarding birth control: "This is the most powerful and manifest cause of the vice and depravity of the greater part of mankind, for the generality undertake procreation on impulse, like beasts."[43] Other writers ascribed the eugenic drive to Plato, since he discriminated between citizens whom God had made different: "Some of you have the power to command, and these he has composed of gold, wherefore also they have the greatest honour; others of silver, to be auxiliaries; others again who are to be husbandmen and craftsmen he has made of brass and iron; and the species will generally be preserved in children."[44]

Centuries later, in the colonial United States, racism infected the minds of even the most prominent leaders. When Benjamin Franklin, for example, wrote his essay on population growth (first published anonymously), he included a final passage that was deleted from various subsequent editions:

> the Number of purely white People in the World is proportionably very small. All Africa is black or tawny; Asia chiefly tawny; America (exclusive of the new Comers) wholly so. And in Europe, the Spaniards, Italians, French, Russians, and Swedes, are generally of what we call a swarthy Complexion; as are the Germans also, the Saxons only excepted, who, with the English, make the principal Body of White People on the Face of the Earth. I could wish their Numbers were increased. And while we are, as I may call it, *Scouring* our Planet, by clearing America of Woods, and so making this Side of our Globe

> reflect a brighter Light to the Eyes of Inhabitants in Mars or Venus, why should we, in the Sight of Superior Beings, darken its People? Why increase the Sons of Africa, by Planting them in America, where we have so fair an Opportunity, by excluding all Blacks and Tawneys, of increasing the lovely White and Red? But perhaps I am partial to the Complexion of my Country, for such Kind of Partiality is natural to Mankind.[45]

By the 1940s there was no shortage of great dead men who apparently endorsed or celebrated eugenic practices.

Hitler fired ambitious military campaigns to capture more "living space" for his so-called master race. But after devastating battles, his Slavic enemies along with the British and American armies defeated the Nazi army. Before World War II ended, though, the Nazi medical program had compulsorily sterilized hundreds of thousands of people. Moreover, the Nazis killed millions of Jews, Roma, and persons of various ethnic minorities.

Meanwhile, eugenics had become an important part of reforms in American education. Among biology textbooks used in high schools in the United States in the 1940s, nearly 90 percent of them included sections advocating eugenics.[46] But when people learned of the horrors of the Nazi racial programs, widespread support for eugenics collapsed.

Overall, eugenics was fueled more by speculation than science. It had the yearning fervor of a young cult. In 1904, Galton acknowledged: "I see no impossibility in eugenics becoming a religious dogma among mankind." Yet he hoped that it would be carefully based on scientific quantification. Galton feared that "Overzeal leading to hasty action would do harm, by holding out expectations of a near golden age, which will certainly be falsified and cause the science to be discredited."[47] This is indeed what happened.

In the 1960s, the notion that people are all naturally equal cast a shadow of shame on the efforts of scientists who insisted on quantifying differences in inborn abilities among individuals or races. One eugenicist complained that serious research was obstructed because "This nonsense about 'equality' has affected nearly everyone!"[48]

The drive to make a science of improving human heredity depended on the ancient aspiration to understand phenomena in terms of numbers. Yet its results clashed with the idea of equality. In 1976, more than a thousand members of the Genetics Society of America endorsed a statement declaring that "We deplore racism and discrimination . . . because they are contrary to our respect for each human individual. Whether or not there are significant genetic inequalities in no way alters our ideal of political equality, nor justifies racism or discrimination in any form."[49]

Eugenics did not end with the Nazis. The word itself lost currency, especially in schoolbooks. But policy makers continued to struggle with its principles, and geneticists continued to search for inheritable causes of people's ailments, behaviors, and intelligence. Nowadays, geneticists have found that many hundreds of maladies are caused by "single-gene" disorders, even though most of those maladies are not those that the eugenicists targeted. Still, some geneticists today continue to search for genetic causes of many behaviors: alcoholism, criminality, schizophrenia, depression, violence, hyperactivity, even shyness.

Regardless, the old ideals were framed in speculations and common fictions. Despite its currency, the old notion of race is a myth. It was based mainly on skin colors, whereas the degrees of genetic differences among human groups do not mirror the apparent "races" that people and anthropologists had established on superficial impressions and skin colors. For example, Africans with dark skin are genetically much more similar to Europeans with light skins than they are to Australian aborigines with dark skins.[50]

Large-scale social programs to restrict or control breeding lost public favor, while new technologies for genetic screening have given greater reproductive control to individuals. Unfortunately, most people only want to have "normal" babies, the very kind that eugenicists wanted. Einstein's second son, for example, became melancholic and schizophrenic and was interned in an insane asylum. His father attributed such ailments to heredity, and he wrote to his son, "The deterioration of the human race is surely a bad thing, one of the worst possible things," and he added, insensitive: "forgive me for your existence."[51] Likewise, Einstein bitterly opposed the marriage of his first son to a woman whom he disdained as genetically inferior: she was older, shorter, driven, and complicated, much like Marić, so Einstein rudely tried to prevent the supposedly "risky," "wretchedness," or "disaster" that Hans Albert have children with her. His prejudiced objections failed.[52]

Now we know that certain genetic anomalies are not entirely disadvantageous. Consider one example. Some American eugenicists had opposed the influx of "inferior blood" from southern and eastern Europe. Accordingly, it turns out that the blood of many people of eastern Mediterranean descent carries an anomaly that makes them susceptible to being poisoned by certain chemicals that don't affect most people. In particular, they become poisoned by eating beans, fava beans. Thus the ancient Pythagorean prescription, abstain from beans, makes good sense, at least for some people from the very regions where Pythagoras lived. Other chemical compounds, other than fava beans, might also be toxic to people with that blood condition. This heritable blood disorder, geneticists have found, is caused by a "genetic defect" that, however, has a positive consequence: it gives blood a resistance against malaria, especially the most deadly form of that disease.[53] And so, biologists have conjectured that thousands of years ago this genetic mutation arose in a region affected by malaria, and it propagated by natural selection. Is that *inferior* blood? It might also have other advantages.

And what about behavior? Was the aim of eugenics merely a myth? Is it impossible to predict or predetermine behaviors merely on the basis of selective breeding?

Eventually, geneticists did make certain discoveries that once could have seemed impossible. For example, in 1958, the French geneticist Jérôme Lejeune discovered that people with Down's syndrome have forty-seven chromosomes instead of forty-six. He then realized that this anomaly was the underlying cause of Down's. Here finally was an instance in which the distinctive future behavior and intelligence of certain infants could actually be *predicted* genetically. This finding was so remarkable that a fellow geneticist described Lejeune's chromosomal photograph as "just about as astonishing as a photograph of the back of the moon."[54] At that time, a Russian space probe had actually photographed the far side of the Moon, which humans had never seen—an ancient mystery. Moreover, researchers soon found that one kind of Down's syndrome (namely, Robertsonian translocation) was actually transmitted by inheritance, thus explaining its recurrence in families. Here finally was an example of a behavioral anomaly that is *inherited*.

What about the question posed at the start of this chapter—is there a way to improve human behaviors by selective breeding, to prevent, for example, aggression and crimes? There is no evidence for it, but consider again the case of dogs. One might doubt whether dogs' behaviors were designed by humans by selective breeding. Yet evolutionary theory requires that species' behaviors, like their bodies, have changed over time. Accordingly, scientists have carried out experiments to see whether the behaviors of certain animals can really be modified by selective breeding.

Beginning in 1959, Dmitri Belyaev in Novosibirsk, Russia, began to selectively breed wild silver foxes to try to make them tame. Each caged fox reacted with more or less fear when presented with a human hand. Those that exhibited less fear and aggression were segregated to breed. None were trained, and contact with humans was strictly limited, to ensure that any behavioral changes resulted only from intrinsic factors.

Surprisingly, after only ten generations, the selected foxes exhibited very tame behaviors, similar to those of domestic puppies—including eagerness for human contact, sniffing and licking humans, and wagging their tails.

Meanwhile, the more violent foxes were bred together, resulting in a strain of particularly aggressive foxes.[55] After Belyaev's death in 1985, Lyudmila Trut continued the experiments. In 1999, she reported that having bred forty-five thousand foxes over forty years, they had produced a variety of fox as tame as a dog.[56] The friendly behaviors of these foxes became evident before they were even one month old. And there was another stunning result: the more passive, friendly foxes were *physically* distinct. Some parts of their fur lacked pigmentation creating, in particular, a star-shaped pattern on their face; their ears became floppy, their tails curled, and after fewer than twenty generations, some of the increasingly tame foxes were born with shorter tails, shorter legs, overbites and underbites; the skulls tended to be smaller, their snouts shorter and wider. On the whole, there was an increased incidence of doglike characteristics. Therefore, we can practically identify the genetic strain of such a fox not only by the animal's behavior, but by its physique; we can *predict* its overall behaviors from its physical traits. There it is, in one species: a connection between behavior and physical traits.

Belyaev and his colleagues also began to domesticate other mammals: river otters, minks, and wild rats. Like the foxes, the Siberian gray rats were segregated to breed according to their tolerance toward humans. After only sixty generations, two distinct varieties had arisen: one that readily let itself be handled freely, and another that behaved viciously, screeching and lunging against cage bars, ferocious against being handled. In the words of an animal behavior expert at the University of St. Andrews in Scotland: "Imagine the most evil supervillain and the nicest, sweetest cartoon animal, and that's what these two strains of rat are like."[57]

Eugenics and dysgenics already have had behavioral effects in some animal species. Geneticists are now trying to identify differences in the DNA between the two strains of rats. Such differences might exist in other mammals as well. And history shows that our attempts to assert discontinuity between humans and the rest of nature have often been wrong. As shown in dogs, foxes, rats, it seems clear that eugenics was not merely a myth.

Behavioral genetics creeps closer to disturbing revelations, but fortunately, the prospect of mandatory eugenic policies among humans remains in disrepute. Traditional mathematical notions of beauty foster a mystical appreciation for equalities, which seem eternal and unchanging. But a renewed appreciation for differences should encourage us toward an ethics that kindly values fluid inequalities of all individuals.

# Epilogue

STORIES of innate genius still generate interest in science, but they also disguise. Pythagoras, Newton, and Einstein are sometimes portrayed as nearly divine. But that hides answers to questions: What did they really do? And how did they do it? Because they were gifted, well born?

At the start of this book, I said that portrayals of science often take the shapes of myths. I was referring to some questionable aspects of old religious cults, such as the tendency to deify charismatic leaders, to celebrate their achievements as miracles, to shroud knowledge in esoteric language, to neglect genuine understanding, and to echo traditional stories rather than historical findings. We began by tracing how astronomy became infused by Pythagorean myths, and finally, we traced how the would-be mathematical science of eugenics decayed into murderous sects driven by myths and millennial visions. Having analyzed a series of myths and historical episodes, one might be expected to synthesize some common pattern by which such myths develop. Let me therefore say a few comments along these lines.

Writers and researchers who wish to tell the past go through a process of selection. Following their needs or curiosities, they carry out a limited search for source materials. Driven by personal or practical motivations, and by the interests they expect from a prospective audience, they search and select whichever elements seem worthwhile, plausible, and compelling. That limited search stops when writers or researchers become satisfied that they have enough material, that it is reasonably reliable, and that they have something worth saying. While composing their account, they not only omit, as they must, material that is beyond the present scope of interest, they also often add phrases and notions that are absent from the original sources. To compose their narrative, they imagine scenes, and consequently, imaginary details from those scenes become woven into the historical excerpts. That imaginative process is not arbitrary, it responds to certain notions.

In the myths that we have considered, we see a common pattern of compression. Key elements in a story are increasingly pushed together. Consider a few examples. First we read that Darwin fancied that variations in finches arose from some sort of evolution. Then writers imagine that Darwin entertained such thoughts while he was on his voyage of discovery and saw strange finches on the Galápagos Islands. Then they also assume that, being a naturalist, Darwin "would have" systematically measured the beaks of the finches, studied their eating habits, and recorded their geographical distributions; and then writers infer that Darwin "must have" concluded that finches had evolved, and that this was the seed that led to his theory of evolution. Likewise, first we hear that Newton was inspired by seeing an apple fall in his garden. Next, the apple falls at his feet, or on his head, or on his nose. Or it hits him hard on the head. And rather than being merely an interesting event that accompanied a series of thoughts, it becomes construed as the trigger, *the cause*.

The small event leads to great consequences. I've met many people who believe that the apple story conveys the fact that "Newton discovered gravity," as if gravity, which is just a Latin expression for heaviness, had been unknown for thousands of years. The point of the story, instead, is that Newton wondered whether gravity, as we know it on Earth, extends beyond the atmosphere, into outer space. But the three elements, Newton, apple, gravity, become compressed into a mythical, grand discovery. Likewise, many people now think that Franklin flew a kite in a thunderstorm and that lightning struck the kite, whereby "he discovered electricity." Similarly, Einstein lived in Switzerland, a country famous for its clocks. There is a large clock tower on a street where he lived, and in 1905 he formulated the relativity of time. Wouldn't it be interesting if the clock tower triggered his ideas? Then writers make the connection. But there were many clock towers all throughout Europe.

Likewise, Giordano Bruno believed in the theory of Copernicus, and he was killed by the Inquisition; wouldn't it be dramatic if Bruno was killed for believing in Copernicus? These stories develop by a kind of *compression*, their elements are brought together.

Once a writer publishes an account, it begins to compete against other similar stories in the market. If readers like it, then it propagates. Or, if the story is carried by a popular book, which itself does not focus on that topic, the story spreads. This process of public selection is determined partly by the public's preconceptions of what makes a story sound good, rewarding, dramatic, inspirational, and true. One reason why some stories sound true is that they echo dramatic notions that readers previously hold. For example, a story might echo the true notion that small things can sometimes have extraordinary effects. The plain apple led to the greatest discovery, the system of the universe. A child's toy, a kite, led a man to understand a most powerful force in nature: lightning. A telling moment shows up in the movie *The Lord of the Rings: The Fellowship of the Ring*. One of the characters, Boromir, picks up a thin, silver chain from which there hangs a golden ring, the One ring, the ring of power. He stares at the ring, entranced, and says: "It is a strange fate that we should suffer so much fear and doubt over so small a thing . . . such a little thing."

As with small objects, another element that makes a story appealing is that it includes ordinary people achieving extraordinary things. The third-class patent clerk formulates a theory that revolutionizes physics. The quiet, modest wife allegedly turns out to be, secretly, a brilliant mathematician and covert mother of the theories of modern physics.

Another factor that makes stories sound true is the inclusion of definite settings and actions. At *that particular place*, so-and-so was doing such-and-such when this happened. Allegedly, Einstein was examining patents for the synchronization of clocks, or Newton was reading under an apple tree, or Galileo was dropping objects from the Leaning Tower of Pisa, and so forth.

What makes such stories even more compelling is when they come from an authoritative source: a famous scientist, a reputable historian, and so on. Readers assume that authoritative experts are less prone to invent the past. However, my impression is that there is a certain danger in authority. Once someone has written extensively about, say, Galileo, they sometimes tend to develop a kind of empathy, a sense of how that person *would have* behaved, what he *must have* thought. It is almost as if the investment of many thousands of hours of work had given the researcher a special power, an ability to divine the past, as if human actions were consistent, as if social and intellectual contexts involve conveniently few factors. The trouble with authority is that it often deters the first process of selection that I mentioned above. A writer or teacher needs to say something about Galileo, so they consult an authority on the matter and they trust that authority at their word. Authority can thus work to stop the already limited search that someone undertakes in order to ascertain the past.

The solution is to trust evidence instead of experts. If someone claims something, even if it is Galileo writing about Aristotle, or Newton writing about Galileo, or even the latest, best biographer writing about Einstein, we should abstain from simply believing what they say, *unless* they cite the specific evidence to which they refer.

Once a popular writer has told a story, it propagates immensely. It soon reaches science textbooks and school materials. It also spreads to juvenile literature and children's books. Later, when researchers manage to critique and correct a myth, it eventually loses ground in schoolbooks. For example, after the myth of Galileo and the Leaning Tower was clearly exposed in the 1930s, it still took many years for the tale to become relatively rare in textbooks. Still, hundreds of thousands of people learn it today, because it retains a strong presence in children's books.

We might have assumed that myths develop mainly by a process of loss of information, that stories are copied and echoed without full accuracy, and therefore some details are lost. But a greater factor is the addition of information. Writers, teachers, and historians sometimes embellish stories with key words and elements that resonate with their expectations, interests, and concerns. It is a process mostly of unconscious invention.

For example, there is a well-known story that Karl Marx wanted to dedicate his famous book, *Das Kapital,* to Charles Darwin. The story began in 1931, when Russian scholars of Marx's works claimed that in a letter of 1880, Darwin had turned down Marx's request to dedicate a book to him.[1] This claim led writers to misrepresent a letter from 1873 by writing that Darwin thanked Marx for wanting to dedicate a book to him, but cordially declined.[2] Actually, the letter in question was from 1880 and does not state Marx's name; it was merely addressed to "Dear Sir."

Even so, in the 1960s and 1970s, scholars continued to claim that Darwin had mailed this letter to Marx.[3] To account for the discrepancy in time (*Das Kapital* had been published in 1867, years before the letter was written or even thought to be written), one writer conjectured that Marx wanted Darwin to read a French translation of some passages by Marx.[4] Other writers supposed that the letter referred to an English translation of *Das Kapital*.[5] Still others speculated that Marx hoped to dedicate the *second* volume of his work to Darwin.[6] They disregarded the fact that Friedrich Engels had reported that Marx intended to dedicate the second and third volumes to his wife (as pointed out by Margaret Fay).

In 1974, Lewis S. Feuer established that Darwin's letter was actually directed to Edward Aveling, who planned to publish a book about Darwin and free-thinking in opposition to Christianity.[7] Darwin did not want to get involved in such matters. He denied being an atheist; he called himself an agnostic. Darwin's papers in the Robin Darwin Archive in Cambridge, England, include a letter from Edward Aveling, dated 1880, in which Aveling requested permission to dedicate his new book to Darwin.[8] In the end, Darwin's 1880 reply became mixed up with Marx's correspondence mainly because Marx's daughter, Eleanor, was Aveling's common-law wife. She committed suicide in 1898, and for a short time Aveling alone held both his own letters and those of Marx. So the claim that Marx wanted to dedicate his book to Darwin was just a misapprehension, one that is still sometimes repeated.

Nevertheless, Darwin and Marx did briefly correspond. In 1873, Marx mailed a copy of his book, *Das Kapital*, with a handwritten inscription to Darwin. In turn, Darwin wrote a letter to Marx, in which he thanked him for the "great book." The autographed book is in German, and it still exists in the collection of Darwin's books. Out of 822 pages, only the first 105 have had the edges cut apart, as if someone meant to read them, but Darwin barely knew German, and he made none of the marginal annotations that he usually made in books.[9] Also, Marx was not entirely a fan of Darwin; over the years, he occasionally disdained his expressions, and he overtly disliked Darwin's reliance on the work of Thomas Malthus.

The Darwin and Marx story grew from a series of conjectures. We have a compulsion to speculate, to fill in the blanks. It is the urge that moves us to infer meaning in a gesture and to draw constellations in the stars. How much less appealing it would be to look at several disconnected points of light in the night sky, not thinking about the belt and sword of Orion. With his telescope, Galileo tried to draw all the stars that he could see in Orion, but there were just too many of them, he could not replace the simple mythical figure of Orion with a comprehensive accounting of the usually invisible stars. Likewise, too often we prefer to pick and choose the pieces that fit whatever story we would like to tell; it is an urge from which I could hardly escape in this book.

Wherever selective excerpts can be used as evidence, it is difficult to undermine myths. For example, although I studied history of science in graduate school, even taking courses in history of biology, I think that I was not taught that the story of Darwin's finches was a myth. Several years later, I read a brief mention that it was a myth, but even then, for years thereafter, I continued to echo variations of the myth, based partly on selected excerpts such as the relevant passage from Darwin's second edition (1845) of his *Voyage of the Beagle*. I wrongly taught versions of this myth to my own college students. Only after I finally read the historical articles by Frank J. Sulloway did I begin to grasp the extent of the myth, and I then confirmed it in further detail by turning to primary sources. But if such corrective stories about myths are not clearly told and retold, the myths grow again, like tree branches in various directions.

Some readers who, before reading this book, already knew some stories about Einstein's first wife might well have read other books or articles that briefly dismiss such stories as myths. And hence such readers might imagine that it's just an old story, that it was debunked years ago, that it's not worth any more paper and ink. But these stories defeat facts. Having studied their evolution, I suspect that they will never go away. Details, even single words, will charm and change and grow in fertile imaginations. My point is not that occasional pruning is necessary; obviously it is. My point is that the evolution of such myths should become part of the stories themselves; that we should work to track and enjoy the history of our mistakes. I believe that in studying that history, we learn to think more clearly; we find recurring patterns that illuminate this powerful urge to ever-so-slightly misread and misrepresent.

Much of what Einstein said became widely echoed and distorted. He complained that his fate was similar to that of King Midas: "Like the man in the fairy tale who turned everything into gold, what he touched, so with me everything turns into newspaper hype."[10] After his death, new stories about him continue to be generated, even based on nothing.

In January 2009, I was watching television, and the History Channel was showing a program titled *Nostradamus: 2012*, which discussed and provided prophecies for ever-present new audiences who might believe whatever sounds occult and agreeable. The program capitalized on the entertaining urge to read between the lines and to project the imminent future onto the past. One of the speakers on TV said: "Einstein was into alchemy, believe it or not, his wife said that what he would do every night before he went to sleep was he would read ancient books on alchemy."[11] So easily said, so easily injected into people's thoughts. Many thousands of people across the world got to hear that statement and likely some of them believed it. By contrast, how many fewer will make it to the back of a bookstore to maybe find a book that might deny that claim? Aside from the immediate reaction that this statement is just not true, let's appreciate it for its beauty. It plainly asserts that "Einstein was into alchemy," as if this claim were authoritative and evident. It connects two bright dots, like stars in a constellation: Einstein and alchemy. It matches what hopeful viewers might want to believe: that successful scientists were privy to occult secrets that we too might access in our nightly informal readings. And it allegedly purports a credible source: his wife. But which wife, first or second? And to whom did she say this? When? Where? Just as the claim is bolstered by the alleged reference to the credible witness, so too the content is boosted by the qualifier: Einstein didn't just read books on alchemy, allegedly he read *ancient* books on alchemy.

Writers trying to fairly capture the past navigate a difficult path. Some toil and crawl slowly in a crammed labyrinth of antiquarian details, while others rush to lose their way in a haze of conjectures. It reminds me of a moving song by the Cuban singer Silvio Rodríguez, a fable about brothers. One brother began a journey, wanting to get far, but he was so careful that he closely looked at every step he took, so much that he became enslaved by caution, his neck curved down, he became old and failed to get far because of his nearsightedness. Another brother also yearned to get far, so he began his journey by focusing his eyes on the horizon, but then he kept stumbling over stones and holes at his feet, so he too became old without getting far. A third brother became cross-eyed.

In researching to authenticate stories, I increasingly saw the extent to which many books blend speculations with evidence. There are patterns, for example, that when writers claim that *he must have* (done this or known that), or that *it cannot be doubted that,* such expressions usually belie, to the contrary, gratuitous conjectures. Words such as *doubtless, probably, evidently, to be sure, certainly, always,* are often used as patches precisely where writers really don't know the certainty of what they want to claim. So for this book I decided to not use such words and similar guesswork: *he may have,* nor common symptoms of imprecision, such as: *scientists in the nineteenth century* and *the modern age.* I am not saying that nobody should ever use such expressions, but the present book involved the experiment of not using such expressions at all, aside from quotations.

Myths connect us to famous individuals by placing such individuals in situations that we can visualize and that convey some idea that we find appealing. Most heroic myths are not malicious. Some fictions satisfy and empower: hearsay legends inspire us even when misconstrued as history. There's an urge to use intriguing historical figures—like Pythagoras, Galileo, and Einstein's wife—as characters in a morality play, to edit the past, to try to make it teach us what we're eager to learn. We project our concerns onto what we read. There remains, though, a need for genuine accounts that struggle to answer the questions: What happened? What can we fairly say about the past? By engaging in this struggle, we contribute to the evolution of stories: the gradual replacement of speculative myths with science and history. After having carefully studied how stories evolve in the recent past, when we are rich with documentary sources, we might again ponder the likelihood of much older stories and ancient accounts. Having traced the emergence of the story that Darwin was inspired by finches' beaks, or that J. J. Thomson discovered the electron, or that Einstein allegedly was inspired to relativity by evaluating patents for city clocks, we might return to analyze stories about Pythagoras. The more recent stories echo ancient forms; tales of heroes and unlikely feats.

On one hand, we may lose faith in some stories about feats of ancient genius, so it might seem that history loses some of its magic. On the other hand, there is no shortage of wonderful and astonishing stories that are true. For ages, alchemists failed to find the mythical Philosophers' Stone, but eventually chemists *did* find substances that emit rays, cure cancer, and are far more valuable than gold. It seemed impossible that elements might evolve, and that we might create gold. It seemed impossible that species too might evolve. It seemed impossible that we might breed animals to have innate friendly behaviors. It seemed impossible that there might be changes in the heavens. It seemed ridiculous that there might exist other worlds. Moreover, stories about how apparent impossibilities were overcome inspire us because they often involve rather ordinary people: Darwin was an average graduate from college, Coulomb was a retired engineer, Einstein was a third-class patent clerk. Whether they were bright, ordinary, or sometimes unpleasant, their successes inspire us to recall the proverb: What one fool can do, another can.

# Notes

## Preface

1. Albert Einstein to Max Brod, 22 February 1949, Einstein Papers Project, item 34–066, California Institute of Technology, Pasadena, Calif.

2. Walter Isaacson, *Einstein: His Life and Universe* (New York: Simon and Schuster, 2007), 1–7.

3. Jürgen Neffe, *Einstein, A Biography*, trans. Shelley Frisch (New York: Farrar, Straus and Giroux, 2007), 95.

4. Sweeping currents of meaning and debates precede my sentences and threaten to swallow up my point in the hollow suspicion that I am likely trying to criticize the sciences. How can we escape the seeping context of the "the two cultures" and "the science wars" and all that makes us suspect that someone hurls rocks either from within science or against it? We can try.

5. Bill Bryson, *A Short History of Nearly Everything* (New York: Random House/ Broadway Books, 2003), acknowledgments.

6. Tony Rothman, *Everything's Relative: And Other Fables from Science and Technology* (Hoboken, N.J.: John Wiley and Sons, 2003). John Waller, *Einstein's Luck: The Truth Behind Some of the Greatest Scientific Discoveries* (Oxford: Oxford University Press, 2002).

7. Rothman, *Everything's Relative*, xv.

8. Ronald N. Numbers, ed., *Galileo Goes to Jail and Other Myths about Science and Religion* (Cambridge, Mass.: Harvard University Press, 2009), 7.

9. "Amanda" (Becki Newton), in *Ugly Betty*, ABC television network, 2008.

10. Rothman, *Everything's Relative*, x, xi.

11. Karl Popper, "Science: Conjectures and Refutations," (lecture, Peterhouse, Cambridge, 1953); printed in Karl Popper, *Conjectures and Refutations: The Growth of Scientific Knowledge* (New York: Basic Books, 1962), 50.

## Chapter 1. Galileo and the Leaning Tower of Pisa

1. R. A. Gregory, *Discovery, or The Spirit and Service of Science* (London: Macmillan and Co., 1916), 2.

2. Ivor B. Hart, *Makers of Science; Mathematics, Physics, Astronomy*, with an introduction by Charles Singer (London: Oxford University Press, 1923), 105.

3. J. J. Fahie, "The Scientific Works of Galileo," in *Studies in the History and Method of Science*, ed. Charles Singer, vol. 2 (Oxford: Clarendon Press, 1921), 216.

4. James Stewart, Lothar Redlin, Saleem Watson, *College Algebra*, 5th ed. (Belmont, California: Cengage Learning, 2008), 293; Richard Panchyk and Buzz Aldrin, *Galileo for Kids: His Life and Ideas, 25 Activities*, rev. ed. (Chicago: Chicago Review Press, 2005), 33; Leon Lederman, with Dick Teresi, *The God Particle* (1993; repr., New York: Houghton Mifflin, 2006), 73–74; Wendy MacDonald and Paolo Rui, *Galileo's Leaning Tower Experiment: A Science Adventure* (Watertown, Mass.: Charlesbridge Publishing, 2009); Chris Oxlade, *Gravity* (Chicago: Heinemann-Raintree Library, 2006), 28; Rachel Hilliam, *Galileo Galilei: Father of Modern Science* (New York: The Rosen Publishing Group, 2005), 101; Ellen Kottler, Victoria Brookhart Costa, *Secrets to Success for Science Teachers* (Thousand Oaks, Calif., Corwin Press, 2009), 50–51; Gillian Clements, *The Picture History of Great Inventors*, rev. ed. (London: Frances Lincoln Ltd, 2005), 21; Stillman Drake, *Essays on Galileo and the History and Philosophy of Science*, ed. Noel M. Swerdlow and Trevor Harvey Levere, vol. 1 (Toronto: University of Toronto Press, 1999), 34–35; Gary F. Moring, *The Complete Idiot's Guide to Understanding Einstein*, 2nd ed. (New York: Alpha Books, 2004), 36; Kerri O'Donnell, *Galileo: Man of Science* (New York: Rosen Classroom, 2002), 7; Gerry Bailey, Karen Foster, Leighton Noyes, *Galileo's Telescope* (New York: Crabtree Publishing Company, 2009), 13; Stephen P. Maran, *Astronomy for Dummies*, 2nd ed. (Hoboken, N. J.: Wiley, 2005), 159; David Hawkins, *How to Get Your Husband's Attention*, rev. ed. (Eugene, Ore.: Harvest House Publishers, 2008), 187.

5. Lane Cooper, *Aristotle, Galileo, and the Tower of Pisa* (Ithaca: Cornell University Press, 1935); Stillman Drake, *Galileo at Work: His Scientific Biography* (Chicago: University of Chicago Press, 1978; New York: Courier Dover Publications, 2003), 415. For a fair review of the literature, plus valuable arguments, see Michael Segre, "Galileo, Viviani and the Tower of Pisa," *Studies in History and Philosophy of Science* 20, no. 4 (1989): 435–54.

6. Vincenzio Viviani, *Racconto Istorico della Vita del Sig. Galileo Galilei Nobil Fiorentino* (1657), Biblioteca Nazionale di Firenze, Mss. Gal.; first published in Salvino Salvini, ed., *Fasti Consolari dell'Accademia Fiorentina* (Firenze, 1717), 397–431, reprinted in Galileo, *Le Opere di Galileo Galilei Nobile Fiorentino*, vol. 1 (Firenze: Gaetano Tartini e Santi Franchi, 1718), lxvi, trans. Martínez.

7. Drake, *Galileo at Work*, 20–21.

8. Benedetto Varchi, *Questione sull'Alchimia* (manuscript, 1544); first published in Varchi, *Qvestione svll' Alchimia*, ed. Domenico Moreni (Firenze: Magheri, 1827), 34, trans. Martínez; I've written "evidence" for "la prova," but "test" is accurate too. Varchi briefly referred to Reverend Francesco Beato, professor of metaphysics at Pisa, and Luca Ghini, a physician and botanist at Bologna, as people who, among others, explained that Aristotle was mistaken about the speeds of falling bodies (34).

Likewise, Giovanni Bellaso asked about the "reason why, when letting fall from high to low two balls, one of iron and the other of wood, just as soon onto the ground falls the wooden one as the iron one." G. B. Bellaso, *Il Vero Modo di Scrivere in Cifra con Facilità, Prestezza e Sicurez* (Brescia: Giacomo Britannico, 1564), trans. Martínez.

9. Giuseppe Moletti, *On Artillery* (1576), manuscript at the Pinelli collection of the Bilioteca Ambrosiana de Milano, Ms. S. 100 sup.; also transcribed in Biblioteca Nazionale Centrale de Firenzi, Ms. Gal. 329; excerpt in Raffaelo Caverni, *Storia del Metodo Sperimentale in Italia*, vol. 4 (Firenze, 1891–1900; Bologna: Forni, 1979), 271–74.

10. Experiment by Donald R. Miklich and Thomas B. Settle, described (with photographs) in Thomas B. Settle, "Galileo and Early Experimentation," in *Springs of Scientific Creativity: Essays on Founders of Modern Science*, ed. Rutherford Aris, Howard Ted Davis, Roger H. Stuewer (Minneapolis: University of Minnesota Press, 1983), 12–17. To simultaneously release balls of different sizes and weights, physicist John Taylor used a hinged platform, at the University of Colorado, Boulder. Photographs show that three steel balls were indeed dropped simultaneously, from the Gamov Tower, and near the bottom, having fallen approximately 100 feet, the largest ball (5 inches diameter, 16 pounds) was 1 inch ahead of the medium ball (4 inches diameter, 8 pounds) and about 1 foot ahead of the small ball (1 inch diameter, 2 ounces). The effect of air resistance depends on the size of the object; it is smaller for the largest ball. See Allan Franklin, *Can That Be Right? Essays on Experiment, Evidence, and Science*, Boston Studies in the Philosophy of Science, vol. 199 (Boston: Kluwer Academic Publishers, 1999), 7–8, 11.

11. Galileo Galilei, *De Motu Antiquiora* [1590s?], manuscript at the Biblioteca Nazionale Centrale de Firenze, Ms. Gal. 71; in Galilei, *Le Opere di Galileo Galilei, Edizione Nazionale*, ed. Antonio Favaro, vol. 1 (Firenze: Barbèra, 1890), trans. Martínez.

12. Ibid.

13. Simon Stevin, "Res Motas Impedimentis suis non esse Proportionales," in *Liber Primus de Staticae Elementis. Statices Liber Secundus qui est de Inveniendo Gravitatis Centro. De Staticae Principiis Liber Tertius de Staticae Praxi. Liber Quartus Staticae de Hydrostatices Elementis* (Leyde, 1605), 151, trans. Martínez.

14. Iacobi Mazonii, *In Universam Platonis, et Aristotelis Philosophiam Præludia, sive De Comparatione Platonis, & Aristotelis* (Venetiis: Ioannem Geuerilium, 1597).

15. Galileo Galilei to Paolo Sarpi, 16 October 1604, in Galilei, *Le Opere di Galileo Galilei, Edizione Nazionale*, ed. Antonio Favaro, vol. 10 (Firenze: Barbèra, 1900), 115–16.

16. Giorgio Coresio, *Operetta intorno al Galleggiare de Corpi Solidi* (Firenze: Bartolommeo Sermartelli, 1612), translation from Cooper, *Aristotle, Galileo*, 29.

17. Vincenzo Renieri to Galileo Galilei, 13 March 1641, translation from Cooper, *Aristotle, Galileo*, 31. Renieri continued: "What was noted by me in such experiments

was this: it struck me that, the motion of the wooden balls being accelerated down to a certain mark, they began then not to descend perpendicularly but obliquely in the same manner as we see drops of water do as they fall from roofs, the which, coming near the earth, swerve aside, and here their motion begins to be less rapid. I have thought about this a little, and shall give your Excellency my notion of it."

18. For discussion of Viviani's writings, including claims that Viviani embellished history for the purpose of reaching an audience that was more interested in demonstrations rather than abstractions, see Michael Segre, "Viviani's Life of Galileo," *Isis* 80, no. 2 (1989): 206–31.

19. Antonio Favaro argued that Viviani deliberately lied about Galileo's date of birth, in Favaro, "Sul giorno della nascita di Galileo," *Memorie del Reale Istituto Veneto di Scienze, Lettere ed Arti* 22 (1887), 703–11; whereas Michael Segre has argued that Viviani may have just been mistaken, in Segre, "Viviani's Life of Galileo."

20. Antonio Favaro, ed., *Galileo Galilei*, 3rd ed. (1922; repr., Milano: Soc. An. Editr. Bietti, 1939), 17, trans. Martínez. See also Antonio Favaro, "Sulla Veridicita del 'Racconto Istorico della Vita di Galileo,' dettato da Vincenzio Viviani," *Archivo Storico Italiano*, tome 73, vol. 1, disp. 2 (Firenze, 1916), 1–24.

21. Cooper, *Aristotle, Galileo*, 21–31; Gregory, *Discovery*, 2; Francis Jameson Rowbotham, *Story-Lives of Great Scientists* (Wells, England: Gardner, Darton, and Company, 1918), 28–29; Fahie, "Scientific Works," 216.

22. Rowbotham, *Story-Lives*; Harold Moore, *A Textbook of Intermediate Physics* (New York: E. P. Dutton, 1923), 52; Harry Austryn Wolfson, *Crescas' Critique of Aristotle; Problems of Aristotle's Physics in Jewish and Arabic Philosophy* (Cambridge, Mass.: Harvard University Press, 1929), 127.

23. Cooper, *Aristotle, Galileo*, 22.

24. Rowbotham, *Story-Lives*, 27–29.

25. Lederman, *God Particle*, 73–74.

26. Cooper, *Aristotle, Galileo*, 17.

## Chapter 2. Galileo's Pythagorean Heresy

1. Aristotle, *De Caelo* [ca. 350 BCE], in *Aristotle: On the Heavens I–II*, trans. Stuart Leggatt (Warminster: Aris and Phillips, 1995), bk. 2, pt. 13.

2. Many writers claim that the problem was that the ancient astronomers had not detected parallax, the apparent shifting, for example, of nearer and distant trees and mountains as one drives by a landscape; but this kind of parallax cannot be what Aristotle and others referred to because he conceived the fixed stars as being all on one sphere, so there would be no background stars. Instead, the simpler effect involves only the apparent *separations* between stars all on one surface.

3. Archimedes, *Psammites (The Sand-Reckoner)* [ca. 220 BCE], quoted in Thomas Heath, *Aristarchus of Samos* (Oxford: Clarendon Press, 1913), 40–41.

4. Pliny the Elder, *Historia Naturalis* (*Natural History*) [ca. 77 CE], trans. H. Rackham (Cambridge, Mass.: Harvard University Press, 1949–54), bk. 2, sec. 19. According to Pliny's Pythagoras, the moon was 15,750 miles away from Earth; the actual distance is more than 233,000 miles from the surface of Earth to that of the moon.

5. Pliny the Elder, *Natural History*, bk. 2, sec. 20; here, by the ordering of the heavenly bodies, Pliny seems to assume that Pythagoras envisioned Earth at the center.

6. Pliny the Elder, *Natural History*, bk. 28, sec. 7.

7. Ptolemy, *Syntaxis Mathematica* [ca. 150 CE], published as Ptolemy, *Ptolemy's Almagest*, trans. G. Toomer (New York: Springer-Verlag, 1984), bk. I, secs. 5–7, pp. 41–45.

8. Ptolemy, *Almagest*, pp. 36, 141.

9. Geminos, *Eisagōgē eis ta Phainomena* [ca. 1st c. BCE], in James Evans and J. Lennart Berggren, *Geminos's Introduction to the Phenomena: A Translation and Study of a Hellenistic Survey of Astronomy* (Princeton: Princeton University Press, 2006), 117–19.

10. Owen Gingerich, *The Book Nobody Read: Chasing the Revolutions of Nicolaus Copernicus* (London: William Heinemann, 2004), 56–58.

11. An example of this myth: "Ptolemaic astronomers needed to add more and more epicycles to the system to keep it working. For a long time, they required only twenty-seven epicycles, but by Kepler's day, they needed nearly seventy—far too complicated. . . . Could a good and loving God, a rational God, all-wise and all-knowing, have created the epicyclic nightmare that the Ptolemaic system had become?" James A. Connor, *Kepler's Witch* (New York: HarperSanFrancisco, 2004), 65–66.

12. An equinox is when the sun is observed directly above Earth's equator, it is a day when the sun spends equal time above the horizon as below it. This happens twice per year. The word *equinox* might seem to suggest that the night is then as long as the day, but actually, during the equinox, daylight lasts longer than night.

13. Nicolai Copernici, *De Revolutionibus Orbium Cœlestium, Libri VI* (Norimbergae: Ioh. Petreium, 1543), f. iiij reverso. Copernicus listed Philolaus, Hicetas of Syracuse, Heraclides Ponticus, and Ecphantus (who actually was not a Pythagorean). Some of the works consulted by Copernicus, such as the *Placita* by "Plutarch" and Lysis's Letter, were not genuine. For discussion, see Bronislaw Bilinski, *Il Pitagorismo di Niccolò Copernico* (Wroclaw: Accademia Polaca delle Scienze, Biblioteca e Centro di Studi a Roma, Conferenze nr. 69, 1977), 111.

14. Plato, *Timaeus* [ca. 360 BCE], in *The Dialogues of Plato*, trans. Benjamin Jowett, vol. 3 (Oxford: Clarendon Press, 1892), 453.

15. The illustrations in this section of the chapter are based partly on the geometric schemes by W. D. Stahlman in Giorgio de Santillana, *The Crime of Galileo* (Chicago: University of Chicago, 1955), 30–31.

16. Blaise Pascal, *Pensées* [ca. 1650], in Pascal, *Pascal's Pensées*, introduction by

T. S. Eliot (New York: E. Dutton 1958), sec. 3: "Of the Necessity of the Wager," item 206, p. 61.

17. Copernici, *De Revolutionibus*, f. ii verso. See also Edward Rosen, "Was Copernicus a Pythagorean?" *Isis* 53 (1962), 504–8.

18. Copernici, *De Revolutionibus*, f. cij verso.

19. Martin Luther, *Sämtliche Schriften*, ed. Johann Georg Walch, vol. 22: *Colloquia oder Tischveden* (Halle: J. J. Gebauer, 1743), 2260.

20. Gingerich, *Book Nobody Read*, 136.

21. Copernici, *De Revolutionibus*, f. iv verso, f. iij reverso.

22. [Andreas Osiander], "Ad Lectorem de Hypothesibus huius Operis," in Copernici, *De Revolutionibus*, f. ij, verso. Johannes Petreius, the printer in Nuremberg, added the words *Orbium Cœlestium* to the book's title, which was simply *De Revolutionibus*.

23. John Calvin, "Sermon on 1 Corinthians 10 and 11, verses 19 to 24" (preached in 1556, edited in 1558); in Ioannis Calvini, *Opera Quae Supersunt Omnia*, ed. G. Baum et al., vol. 49 (Brunsvigae: C. A. Schwetschke, 1863–1900), 677, trans. Martínez.

24. Gingerich, *Book Nobody Read*, 23.

25. Leonard Digges, *A Prognostication Everlastinge of Righte Good Effecte*, corrected and augmented by Thomas Digges (London: Tomas Marsh, 1576), addition. Previously, Pliny the Elder (in *Natural History*) had reported that the astronomer Hipparchus had observed what seemed to be a new star (ca. 134 BCE), a claim that astronomers doubted.

26. In 1901, city officials in Prague opened Tycho Brahe's marble tomb to confirm the identity of the skeletal remains. Physicians found that Brahe's skull indeed had a defect in the upper end of the nasal cavity, rimmed by a bright green stain of copper. Dr. H. Matiegka, *Bericht über die Untersuchung der Gebeine Tycho Brahe's* (Prague: Bohemian Society of Science, 1901).

27. Digges, *Prognostication*, addition.

28. Victor E. Thoren, *The Lord of Uraniborg: A Biography of Tycho Brahe* (Cambridge: Cambridge University Press, 1990), 250–58.

29. Diego de Zuñiga, *Didaci a Stunica Salamanticensis Eremitaie Augustiniani in Job Commentaria* (Toleti: Ioannes Rodericus, 1584), 205–6.

30. Francisco Vallés, *De iis quae Physice in Libris Sacris Scripta Sunt* (Turin: Nicolai Bevilaquae, 1587), chap. 51.

31. Edward Rosen, "The Dissolution of the Solid Celestial Spheres," *Journal of the History of Ideas* 46 (1985): 25.

32. Johannes Kepler, "Observationes" [1601], in Willebrordus Snellii, *Coeli et siderum in eo errantium Observationes Itassiacae* (Lugduni Batavorum [Leiden]: Justum Colsterum, 1618), 83–84, trans. Martínez.

33. Bent Kaempe, Claus Thykier, and N. A. Petersen, "The Cause of Death of

Tycho Brahe in 1601," *Proceedings of the 31st International Meeting of The International Association of Forensic Toxicologists, TIAFT, Leipzig, August 1993*, ed. R. Klaus Mueller (Leipzig: Molina Press, 1994), 309–15. A recent best-selling book argues that Brahe was murdered by Kepler, who wanted to steal Brahe's astronomical data. See Joshua Gilder and Anne-Lee Gilder, *Heavenly Intrigue: Johannes Kepler, Tycho Brahe, and the Murder behind One of History's Greatest Scientific Discoveries* (New York: Doubleday, 1994). Their argument: that Kepler had the means, motivation, and opportunity. But their claims do not convince me, partly because Brahe leisurely said goodbye to his relatives before ingesting the second and deadly dose of mercury, so he expected to die. By a process of elimination, the authors claim that evidence points to Kepler; but I see no evidence pointing to anyone, and Kepler stands out as decent and kind-hearted.

34. Proclus, *A Commentary on the First Book of Euclid's Elements* [ca. 460 CE], trans. Glenn R. Morrow (Princeton: Princeton University Press, 1970), 53. The claim seems doubtful because Proclus wrote more than a thousand years after Pythagoras died and there are no early accounts.

35. Johannes Kepler to Michael Maestlin, June 1598, quoted in Max Caspar, *Kepler*, trans. C. Doris Hellman (London: Abelard-Schuman, 1959), 69.

36. Iamblichus, *On the Pythagorean Life* [ca. 300 CE], trans. and ed. Gillian Clark (Liverpool: Liverpool University Press, 1989), secs. 64–66, pp. 27–28; Johannes Kepler, *Harmonices Mundi, Libri V* (Linci, Austria: Godofredi Tompachii, 1619), translated as *The Harmony of the World*, trans. E. J. Aiton, A. M. Duncan, J. V. Field (Philadelphia: American Philosophical Society, 1997), bk. 2, p. 130.

37. Max Caspar, *Johannes Kepler*, 4th ed. (Stuttgart: Verlag für Geschichte der Naturwissenschaften und der Technik, 1995), 109.

38. Paul Henri Michel, *The Cosmology of Giordano Bruno*, trans. R. E. W. Madison (1962; repr., Paris: Hermann, 1973), 214–15.

39. "Just as no natural body at all is perfectly round and therefore has a simple center, so too among the sensible and physical motions that we observe in natural bodies there is notne that does not differ a lot from the truly circular and regular motion around a center." Bruno, *La Cena de le Ceneri* (*The Ash Wednesday Supper*), [1584], in Giovanni Gentile, ed., *Opere Italiane*, vol. 1 (Firenze: Gius. Laterza, 1907), pt. 3, p. 73, trans. Martínez.

40. Plutarch, *Placita Philosophorum, Lib.* 2 [2nd c. CE], chap. 13; it was not written by Plutarch, so historians attribute it to "Pseudo-Plutarch," and it is based on a work by Aetius (ca. 50 BCE).

41. For example, Dava Sobel, *Galileo's Daughter* (New York: Penguin Books, 2000), 4, 171. For examples of the recurring claim that Bruno was the first martyr for science, along with a refutation of this myth, see Jole Shackelford, "That Giordano Bruno Was the First Martyr of Modern Science," in *Galileo Goes to Jail and Other*

*Myths about Science and Religion,* ed. Ronald N. Numbers (Cambridge, Mass.: Harvard University Press, 2009), 58–67.

42. Noel Swerdlow, "Galileo's Discoveries with the Telescope and Their Evidence for the Copernican Theory," in *The Cambridge Companion to Galileo,* ed. Peter Machamer (Cambridge: Cambridge University Press, 1998), 245.

43. Galilei to Giuliano de Medici, January 1611, quoted in Mario D'Addio, *The Galileo Case: Trial/Science/Truth,* trans. Brian Williams (Rome: Nova Millennium Romae, 2004), 29.

44. "Plutarch" [Aetius], *Placita, Lib.* 2, chap. 30.

45. Ioh. Keppleri, *Somnium, seu Opus Posthumum De Astronomia Lunari, divulgatum à Ludovico Kepplero filio* [ca. 1609] (Sagani [Silesia]: Authoris, 1634). Kepler had read ancient works by Lucian and "Plutarch" that discussed the idea of exploring the moon.

46. Johannes Kepler, *Dissertatio cum Nuncio Sidereo* [1610], in Kepler, *Kepler's Conversation with Galileo's Sidereal Messenger,* trans. E. Rosen (New York: Johnson Reprint Corp., 1965), 27–28.

47. Pliny the Elder, *Natural History,* bk. 2, sec. 6. Pliny claimed that Pythagoras made this discovery at around the Forty-Second Olympiad, which is impossible because that Olympiad began in 612 BCE and Pythagoras was not even born until some twenty years later.

48. Galilei to Johannes Kepler, 19 August 1610, in Galilei, *Le Opere di Galileo Galilei,* ed. Antonio Favaro, vol. 10 (Florence: G. Barbèra, 1900), 421–23, trans. Martínez.

49. Galileo Galilei, *Istoria e Dimostrazioni intorno alle Macchie Solari e Loro Accidenti* (Rome, 1613), in Galilei, *Opere,* 5:190; also in Pietro Redondi, *Galileo Heretic,* trans. Raymond Rosenthal (1983; repr. Princeton: Princeton University Press, 1989), 37.

50. Paolo Antonio Foscarini, *A Letter to Fr. Sebastiano Fantone, General of the Order, Concerning the Opinion of the Pythagoreans and Copernicus About the Mobility of the Earth and the New Stability of the Sun and the New Pythagorean System of the World* [6 January 1615] (Naples: Lazaro Scoriggio, 1615); in Richard J. Blackwell, *Galileo, Bellarmine, and the Bible* (Notre Dame: University of Notre Dame Press, 1991), 218–21.

51. Blackwell, *Galileo, Bellarmine,* 226, 223, 234–35.

52. For an extensive discussion of this myth, see Dennis R. Danielson, "That Copernicanism Demoted Humans from the Center of the Cosmos," in Numbers, *Galileo Goes to Jail,* 50–58.

53. Ernan McMullin, "The Church's Ban on Copernicanism, 1616," in McMullin, ed., *The Church and Galileo* (Notre Dame: University of Notre Dame Press, 2005), 165–66.

54. Bellarmino to Foscarini, 12 April 1615, in Blackwell, *Galileo, Bellarmine,* 265–67.

55. Paolo Sfondarti, Bishop of Albano, "Decree of the Index" [5 March 1616] (Rome: Press of the Apostolic Palace, 1616), in Maurice A. Finocchiaro, *The Galileo Affair: A Documentary History* (Berkeley: University of California Press, 1989), 148–50.

56. Petrus Lombardus et al., "Consultants' Report on Copernianism" [24 February 1616], in Finocchiaro, *Galileo Affair*, 146.

57. Sfondarti, "Decree of the Index," in Finocchiaro, *Galileo Affair*, 149.

58. Kepler, quoted in James R. Voelkel, *Johannes Kepler and the New Astronomy* (Oxford: Oxford University Press, 1999), 77.

59. Connor, *Kepler's Witch*, 242, 287, 320–21.

60. Kepler, *Harmony*, bk. 3, p. 127.

61. Kepler, *Harmony*, bk. 1, p. 12: "Therefore that in the secrets of the Pythagoreans on this basis the five figures were distributed not among the elements, as Aristotle believed, but among the planets themselves is very strongly confirmed by the fact that Proclus tells us that the aim of geometry is to tell how the heaven has received appropriate figures for definite parts of itself."

62. Kepler, *Harmony*, bk. 4, p. 284.

63. Connor, *Kepler's Witch*, 266.

64. Galileo Galilei, *The Assayer* [1623], in *Discoveries and Opinions of Galileo*, trans. Stillman Drake (New York: Anchor Books/Random House, 1957), 237–38.

65. Pope Urban VIII, quoted by Francesco Niccolini to Lord Balì Cioli, 11 September 1632, in Finocchiaro, *Galileo Affair*, 229.

66. As quoted by Galilei to Elia Diodati, 15 January 1633, in Finocchiaro, *Galileo Affair*, 225. See also D'Addio, *Galileo Case*, 115.

67. To give just one example, Finocchiaro characterized Pythagoras simply as one of the earliest thinkers to advance the idea that Earth circles the sun. See Finocchiaro, *Galileo Affair*, 7, 15.

68. Diogenes Laertius, *The Lives and Opinions of Eminent Philosophers* [ca. 225 CE], trans. C. D. Yonge, bk. 8, *Life of Pythagoras* (London: Henry G. Bohn, 1853), sec. 12.

69. Ancient poem quoted in Iamblichus, *Pythagorean Way*, chap. 2, p. 35.

70. Iamblichus, *Pythagorean Way*, chap. 2, p. 35.

71. Hippolytus [traditionally misattributed to Origen], *Philosophumena* (*The Refutation of All Heresies*) [ca. 225 CE], trans. J. Macmahon, ed. Alexander Roberts and James Donaldson (Edinburgh: T&T Clark, 1867), bk. 6, chaps. 23, 24, 47. Hippolytus (bk. 4, chap. 13) denounced the "alliance between heresy and the Pythagorean philosophy," he criticized the "enormous and endless heresies" of those such as Colarbasus who attempted to explain religion by measures and numbers, and who deceived unsophisticated individuals with vain prophecies and calculation. He convicted Valentinus (bk. 6, chap. 24) of plagiarizing arithmetical philosophy, that he

"may therefore justly be reckoned a Pythagorean and Platonist, not a Christian." He complained (bk. 6, chap. 47) that Marcus too and his followers practiced "portions of astrological discovery, and the arithmetical art of the Pythagoreans" and therefore were not disciples of Christ. He also dismissed Monoïmus (bk. 8, chap. 8) as having copied Pythagoras. Furthermore, Hippolytus (bk. 9, chap. 9) denounced the heretical system of Elchasai as being derived from Pythagoras, and rejected especially the claim that Christ had been born repeatedly and would continue to be born as His soul transferred from body to body.

72. [Ambrosius Aurelius Theodosius] Macrobius, *Commentarii in Somnium Scipionis* [ca. 430? CE], in *Commentary on the Dream of Scipio*, trans. William Harris Stahl (New York: Columbia University Press, 1952), 134: "Pythagoras also thinks that the infernal regions [or the empire of death] of Dis [Pluto] begin with the Milky Way, and extend downwards because souls falling away from it seem to have withdrawn from the heavens. He says that the reason why milk is the first nourishement offered to the newborn infant is that the first movement of souls slipping into earthly bodies is from the Milky Way." See also Porphyry, "De Antro Nympharum," *Select Works of Porphyry*, trans. Thomas Taylor (London: T. Rodd and J. Moyes, 1823), 193: "According to Pythagoras, also, the *people of dreams*, are the souls which are said to be collected in the galaxy, this circle being so called from the milk with which souls are nourished when they fall into generation." Also, Hippolytus complained that Pythagoras claimed that the world is eternal, originated from the un-begotten monad, that the stars are fragments of the Sun, and that from the stars come the souls of animals, which are then buried into bodies, until later death separates them from bodies, whence their souls become immortal. Human souls could pass between animals and plants, and souls who philosophized would eventually ascend to a kindred star, but if a soul did not escape the passions it could become mortal. See Hippolytus, *Refutation*, bk. 4.

73. Claudius Ælianus, *Varia Historia* [ca. 220 CE], in *Claudius Ælianus, His Various History*, trans. Thomas Stanley (London: Thomas Dring, 1665), bk. 4, chap. 17; Iamblichus, *Pythagorean Way*, chap. 28.

74. For discussion on whether the *Metamorphoses* were "dangerously pagan," see Ursula D. Hunt, *Le Sommaire en Prose des Métamorphoses d'Ovide dans le manuscrit Burney 511 au Musée Britannique de Londres* (Paris: Presses Universitaires de France, 1925), xiii; Alan Cameron, *Greek Mythography in the Roman World* (Oxford: Oxford University Press, 2004), 24.

75. Diogenes Laertius, *Life of Pythagoras*, see sections 4, 15, 19, 20. The ancient belief that the sun is a god was briefly noted by Copernicus, e.g., "Trimegistus uisibilem Deum," that for Hermes Trismegistus the sun was a visible god. See Copernici, *De Revolutionibus*, f. cij verso.

76. Porphyry, *Life of Pythagoras* [ca. 300? CE], trans. Kenneth Sylvan Guthrie

(Alpine, N.J.: Platonist Press, 1919), sec. 28. Porphyry claimed that Pythagoras had a golden thigh, evidence that he was divine, related to Apollo. He insisted that Pythagoras predicted earthquakes and stopped violent winds, hail, and storms over rivers and seas. The heretical nature of Porphyry's portrayal of Pythagoras was well conveyed by Reverend Waddington, who argued that by comparison to Porphyry's books *Against the Christians*, "that which being more insidious, may have been more pernicious was his 'Life of Pythagoras.' Early in the third century, one Philostratus, a rhetorician at Rome, had composed a fabulous account of Apollonius of Tyana, a celebrated philosopher and magician; and so wrought out the supposed extraordinary incidents of his life, as to establish a close resemblance between them and the miracles of Christ. Porphyry imitated this example; and he represented the peaceful Pythagoras as having worked by his own power many stupendous prodigies—and having, moreover, imparted the same power to his principal disciples, Empedocles, Epimenides, and others. Such is the sort of weapon, which as it proceeds from the imagination, and addresses the imagination, and eludes the grasp of reason, has proved at all times the most dangerous to Christianity." George Waddington, *A History of the Church from the Earliest Ages to the Reformation*, 2nd ed. (London: Baldwin and Cradock, 1835), 103.

77. Porphyry, "On the Philosophy Derived from Oracles" [ca. 270 CE], quoted in Eusebius, *Praeparatio Evangelica* [ca. 314 CE], as translated in Amos Berry Hulen, *Porphyry's Work Against the Christians: An Interpretation*, Yale Studies in Religion, no. 1 (Scottsdale, Penn.: Mennonite Press, 1933), 16.

78. Porphyry, *Adversus Christianos* [ca. 290 CE]; Porphyry appreciated the historical Jesus but attacked Christians and contradictions in the Gospels; he criticized some of the alleged deeds and sayings of Jesus. He ridiculed the claim that Jesus was God incarnate and he denied the resurrection of Jesus and of select humans on Judgment Day. He denied the doctrine that the world has a beginning and an end. He criticized Christianity for its emphasis on faith, unreason, and its appeal to the gullible poor and uneducated rather than to the educated and to philosophers. He denied that Jesus was the sole path to salvation. Porphyry believed instead in the existence of many gods and demons, and he asserted the eternity of the universe. He also stated that human souls transmigrate into various bodies as they travel for nine thousand years, descending from the moon, spending time on each planet, and finally going to the sun. See Hulen, *Porphyry's Work*; R. Joseph Hoffman, *Porphyry's Against the Christians: The Literary Remains* (Amherst, N. Y.: Prometheus Books, 1994); Jeffrey W. Hargis, *Against the Christians: The Rise of Early Anti-Christian Polemic* (New York: Peter Lang, 1999).

79. Pope Leo X, Papal Bull: *Exsurge Domine* [15 June 1520], in Hans. J. Hiller, *The Reformation in Its Own Words* (London: SCM Press, 1964), 80.

80. Augustine of Hippo, *De Civitate Dei Contra Paganos* [ca. 412–427 CE], ed.

R. W. Dyson (Cambridge: Cambridge University Press, 1998); *City of God Against the Pagans*, bk. 7 [ca. 417 CE], chap. 35, p. 310. Augustine attributes this claim about Pythagoras to Marcus Terentius Varro (ca. 40 BCE), also known as Varro Reatinus. Likewise, Cicero had commented on divination that Pythagoras "added a great weight of authority to this belief—and indeed he himself wished to acquire the skill of an augur." Marcus Tullius Cicero, "On Divination" [ca. 44 BCE], in *Treatises of M. T. Cicero*, trans. C. D. Yonge (London: Henry G. Bohn, 1853), bk. 1, sec. 3. Plutarch also claimed that Pythagoras engaged in false divination: Plutarch, "A Discourse Concerning Socrates's Daemon," in *Plutarch's Miscellanies and Essays. Comprising All His Works Under the Title of "Morals,"* ed. William W. Goodwin, vol. 2, 6th ed. (Boston: Little, Brown, and Company, 1898), sec. 9. Also, Hippolytus (*Refutation*, bk. 9, chap. 9) denounced the followers of Elchasai for their incantations and for pretending "to be endued with a power of foretelling futurity, using as a starting-point, obviously, the measures and numbers of the aforesaid Pythagorean art. These also devote themselves to the tenets of mathematicians, and astrologers, and magicians, as if they were true. And they resort to these, so as to confuse silly people, thus led to suppose that the heretics participate in a doctrine of power."

81. Iamblichus, *Pythagorean Way*, chap. 19.

82. Hermias the Philosopher, *Irrisio Gentilium Philosophorum* [ca. 250–550 CE?], in Demetrii Cydonii, *Oratio de Contemnenda Morte* (Basle: Ralph Seiler, 1533); also in *The Writings of the Early Christians of the Second Century*, trans. J. Giles (London: John Russell Smith, 1857), 193. See also R. P. C. Hanson, *Hermias. Satire des Philosophes Païens*; Sources Chrétiennes 388 (Paris: Cerf, 1993).

83. Niceforo Callistos, *Ecclesiasticae Historiae III* [ca. 1300–1330?], in J.-P. Migne, ed., *Patrologiae Graecae* (Paris: Migne, 1857–1866), 145; Gianfrancesco Pico della Mirandola, *De Rerum Praenotione* [*On the Foreknowledge of Things*] (1507), in Joannis Francisci Pici Mirandulae, *Opera Omnia* (Basel: 1519), 664–74. See also Maria Dzielska, *Apollonius of Tyana in Legend and History*, trans. Piotr Pienkowski (Rome: L'Erma di Bretschneider, 1986).

84. Caesar Longinus, *Trinvm Magicvm, sive Secretorum Magicorvm Opvs* (Frankfurt: Conradi Eifridi, 1630), 45, 373, 385–91; Henning Grosse, ed., *Magica de Spectris et Apparitionibus Spiritu: de Vaticiniis, Divinationibus, &c* (N.p.: Franciscum Hackium, 1656), 186–87. Pliny the Elder had claimed that Pythagoras studied magic, in *Natural History*, bks. 24, 25, 30, secs. IC, V, II, respectively.

85. Anonymous authors, manuscript [late 1500s]: Ms. Marshall 15 (5266), University of Oxford, Bodleian Library, f. f66v; Christofo de Cattan, *La Géomance dv Seigneur Christofe de Cattan, gentilhomme Geneuoys. Liure non moins plaisant & recreatif. Auec la roüe de Pythagoras*, rev. and trans. Gabriel du Preau (Paris: G. Gilles, 1558); *The Geomancie of Maister Christopher Cattan, Gentleman. A booke no lesse pleasant and recreatiue, then of a wittie inuention, to knowe all thinges, past, present, and to come.*

*Whereunto is annexed the Wheele of Pythagoras*, trans. (from French) Francis Sparry (London: John Wolfe, 1591); Robert Fludd [Roberto Flud], *Utriusque Cosmi Maioris scilicet et Minoris Metaphysica, Physica atqve Technica Historia: in Duo Volumina secundum Cosmi differentiam divisa* (Oppenheim: Johan-Theodori de Bry, 1619); Thomas Taylor, *The Theoretic Arithmetic of the Pythagoreans* [1816], with an introductory essay by Manly Hall (Los Angeles: Phoenix Press, 1934), ix–xii.

86. Fludd, "De Numero et Numeratione," in *Utriusque*.

87. The context of this statement makes it clear that Augustine was referring to astrologers. Augustine, *De Genesi ad Litteram* [ca. 408 CE], in Iosephi Zycha, *Corpus Scriptorum Ecclesisticorum Latinorum*, vol. 28 (Prague: F. Tempsky, 1894), bk. 2, p. 62, trans. Martínez.

88. Bartholomaeus Agricola, *Symbolum Pythagoricum; sive De Justitia in Forum Reducenda*, 2 lib. (Neapoli: Nemetum, 1619).

89. Galileo seems unaware that by mistakenly attributing the heliocentric theory to Pythagoras, he connected it to many old heresies. To what extent were the clergymen in the proceedings against Bruno and Galilei concerned with such connotations? That remains a direction for future research.

90. Galilei, depositions of 12 and 30 April 1633, in Finocchiaro, *Galileo Affair*, 260, 262, 277.

91. Galilei, "Abjuration," 22 June 1633, in Finocchiaro, *Galileo Affair*, 292.

92. Giuseppe Baretti, *The Italian Library* (London: Millar, 1757), 52.

93. The painting, apparently by Bartolomé Murillo, a Spaniard, is dated 1643 or 1645; see Antonio Favaro, "Eppur si muove," *Il Giornale d'Italia*, 12 July 1911, 3; J. Fahie, *Memorials of Galileo Galilei, 1564–1642* (London: Courier Press, 1929), 72–75, plate 16; Stillman Drake, *Galileo at Work: His Scientific Biography* (Chicago: University of Chicago Press, 1978; reprint: New York: Courier Dover Publications, 2003), 356–57.

94. For the history of this old myth, see Maurice Finocchiaro, "That Galileo Was Imprisoned and Tortured for Advocating Copernicansim," in *Galileo Goes to Jail*, 68–78.

95. Melchior Inchofer, *Tractatus Syllepticus* (*A Summary Treatise Concerning the Motion or Rest of the Earth and the Sun, according to the Teachings of the Sacred Scriptures and the Holy Fathers*) [1633], in Richard J. Blackwell, ed., *Behind the Scenes at Galileo's Trial: Including the First English Translation of Melchior Inchofer's Tractatus Syllepticus* (Notre Dame: University of Notre Dame Press, 2006), 108, 123, 167, 106. Inchofer also ridiculed Pythagoras and the Pythagoreans for believing that Earth has a soul, and that hell fire at its center causes it to move (he cited Hermias as a critical source against the Pythagoreans). See Blackwell, *Behind the Scenes*, 189–91.

96. Following the report of a Catholic Study Commission, in 1992, Pope John Paul II stated that theologians in Galileo's trial did not grasp the distinction between

Scripture and its interpretation: "The error of the theologians of the time, when they maintained the centrality of the Earth, was to think that our understanding of the physical world's structure was, in some way, imposed by the literal sense of the Sacred Scripture." John Paul II, "Allocution," 31 October 1992, in Bernard Pullman, ed., *The Emergence of Complexity in Mathematics, Physics, Chemistry and Biology: Proceedings of the Plenary Session of the Pontifical Academy of Sciences, 27–31 October 1992* (Vatican City: Pontificia Academia Scientiarum/Princeton University Press, 1996), 471.

97. "A Letter from Professor Bessel to Sir J. Herschel, Bart., Dated Königsberg, Oct. 23, 1838," *Monthly Notices of the Royal Astronomical Society* 4, no. 17 (1838), 152–61 and no. 18 (1838), 163. On the basis of thousands of observations over a year, Bessel ascertained hundreds of measurements of the positions of star 61 Cygni against two background stars. In 1837, Wilhelm Struve had reported an estimate of parallax of the star Vega on the basis of just seventeen measurements, but his results were so rough and inconclusive that in 1848 he acknowledged Bessel's priority.

## Chapter 3. Newton's Apple and the Tree of Knowledge

1. Richard G. Olson, *Science & Religion, 1450–1900; From Copernicus to Darwin* (Baltimore: Johns Hopkins University Press, 2006), 18. Many other books still repeat this old mistake, such as the following. Leon Lederman, with Dick Teresi, *The God Particle* (1993; repr., New York: Houghton Mifflin, 2006), 86; James Shipman, Jerry D. Wilson, Aaron Todd, *An Introduction to Physical Science*, 12th ed. (Boston: Houghton Mifflin, 2007), 47; Keith Johnson, *Physics for You*, Revised National Curriculum Edition for GCS, 4th ed. (Cheltenham: Nelson Thornes, 2001), 376; B. R. Hergenhahn, *An Introduction to the History of Psychology*, 6th ed. (Belmont: Wadsworth/Cengage Learning, 2009), 112; Paul A. Tipler and Gene Mosca, *Physics for Scientists and Engineers*, 6th ed. (New York: W. H. Freeman, 2008), 93.

2. Joseph Warton, ed., *The Works of Alexander Pope, Esq.*, vol. 2 (London: B. Law et al., 1797), see editor's note on "Epitaphs. XII. Intended for Sir Isaac Newton, in Westminster-Abbey," 403. See also "Newton," *Walker's Hibernian Magazine, or, Compendium of Entertaining Knowledge* (Dublin: R. Gibson, 1798), 119; and "Newton," *Scots Magazine* 60 (Edinburgh, 1798): 228.

3. Stephen W. Hawking, *God Created the Integers: The Mathematical Breakthroughs that Changed History* (Philadelphia: Running Press, 2005), 365. Hawking retired from the Lucasian Chair of Mathematics in 2009.

4. Michael A. Seeds and Dana E. Backman, *Horizons: Exploring the Universe*, 11th rev. ed. (Belmont: Books/Cole, Cengage Learning, 2010), 58. Current convention for English dates is to use the Old Style Julian dates until the change of calendars in September 1752.

5. Some early accounts (influential and very useful, but yet significantly incomplete) are: Bolton Corney, "Art. XXI.—The Path of the *Woolsthorpe Apple*—Calculat-

ed on Data Not Known to Sir Isaac Newton!" in Corney, *Curiosities of Literature by I. D'Israeli, Illustrated*, 2nd rev. ed. (London: Richard Bentley, 1838), 152–58; Augustus De Morgan, "Newton's Apple," in De Morgan, *A Budget of Paradoxes* (London: Longmans, Green, and Co., 1872), 81–82; Douglas McKie and G. R. de Beer, "Newton's Apple," *Notes and Records of the Royal Society of London* 9, no. 1 (1951): 46–54; D. McKie and G. R. de Beer, "Newton's Apple: An Addendum," *Notes and Records of the Royal Society of London* 9, no. 2 (1952): 333–35. More recent accounts are available, but those also omit quotations, include modifications and errors in transcriptions, do not provide translations, and overall, use fewer primary sources than the present chapter.

6. Isaac Newton, "Before Whitsunday 1662," manuscript, Fitzwilliam Notebook, Fitzwilliam Museum, Cambridge. This manuscript, written in Thomas Shelton's shorthand notation, was deciphered by Richard S. Westfall in "Short Writing and the State of Newton's Conscience," *Notes and Records of the Royal Society* 18, no. 1 (June 1963): 10–16.

7. Wm. Stukeley, manuscript: "Memoirs of Sr. Isaac Newtons Life" [1752Royal Society archives, GB 117, MS 142, p. 42 (hand numbered 15)]. See also McKie and de Beer, "Newton's Apple."

8. Robert Greene, "Miscellanea Qaedam Philosophica," in Greene, *The Principles of the Philosophy of the Expansive and Contractive Forces, An Inquiry into the Principles of the Modern Philosophy* (Cambridge: C. Crownfield, 1727), 972. The original reads: "Quæ Sententia Celeberrima, Originem ducit, uti omnis, ut fertur, Cognitio nostra, a Pomo; id quod Accepi ab Ingeniosissimo & Doctissimo Viro, pariter ac Optimo, mihi autem Amicissimo, *Martino Folkes* Armigero, Regiæ vero Societatis Socio Meritissimo . . ." trans. Martínez. Note that the awkward initial phrase is difficult to clarify, it seems ungrammatical, and the word "Pomo" can be translated otherwise as "fruit." Aside from my translation above, one might very literally write: "which famous proposition, originates, all used, so found, our knowledge, an apple . . ."

9. John Conduitt, "Memoir of Newton" [1727–1728], Keynes Ms. 129 (A), Newton Project Archive, Kings College Library, Cambridge. Conduitt sent his memoir to Bernard le Bovier de Fontenelle for his eulogy of Newton. Fontenelle, "Elóge de M. Neuton," Histoire de l'Academie Royale des Sciences (Paris, 1728), 151–72. (Note that the Newton Project Archive is located at the University of Sussex, where there are copies of Newton's work; many of the original manuscripts are owned by King's College, Cambridge, and located there.)

10. John Conduitt, draft of "Memoir of Newton" [1727–1728], Keynes Ms. 129 (B), Newton Project Archive, King's College Library, Cambridge.

11. John Conduitt, Keynes Ms. 130.4, Newton Project Archive, King's College Library, Cambridge.

12. Mr. de Voltaire, *An Essay upon the Civil Wars of France, extracted from Curious*

*Manuscripts. And also upon the Epick Poetry of the European Nations, from Homer to Milton* [1727], 2nd ed., corrected by Voltaire (London: N. Prevost, 1728), 103.

13. Ibid. Soon, the essay was published in French: M. de Voltaire, *Essay sur la Poësie Epique. Traduit de l'Anglois* (Paris: Chaubert, 1728), see p. 123: "C'est ainsi que Pythagore dût l'invention de la Musique au bruit des marteaux d'une forge, & que de nos jours M. Isaac Newton, en se promenant dans son jardin, conçut la premiere idée de son systême de la gravitation, en voyant tomber une pomme du haut d'un arbre." In the preface to that French translation, Voltaire acknowledged his daring for having tried to write in English, having spent merely eighteen months in England, having an awful pronunciation, and being barely able to understand the language in conversation.

14. M. D. V. [Voltaire], *Lettres Ecrites de Londres sur les Anglois et Autres Sujets* ("Basle" [actually London: William Bowyer], 1734), 15th letter: "Sur l'Attraction," 121–22, trans. Martínez. The preface notes that such letters were written from 1728 until 1730, and, allegedly, were not originally intended for publication. The editors note that an English translation was in circulation in 1732. An English preface (in an edition of 1733) claims that the letters were written "between the end of 1728, and about 1731."

15. Voltaire acknowledged "Madame Conduit" in a similar account: "One day in the year 1666. Newton retired to the countryside & seeing fruits fall from a tree, according to what his niece told me (Madame Conduit), let himself go into a deep meditation about the cause that pulls all bodies in a line which, if it be prolonged, would nearly pass through the center of Earth. What is, he asked himself, that force which cannot come from all the imaginary vortices thus demonstrated to be false? it acts on all bodies in proportion to their masses, & not their surfaces, it would act upon the fruit that fell from that tree." Mr. De Voltaire, *Elemens de Philosophie tirez de Neuton et de Quelques Autres*, revue, corrigée et considerablement augmentée par l'auteur in *Œuvres de Monsieur de V, Nouvelle Édition* (Dresde: George Conrad Walther, 1749), pt. 3, chap. 3, p. 189, trans. Martínez.

16. Henry Pemberton, *A View of Sir Isaac Newton's Philosophy* (London: S. Palmer, 1728), preface.

17. David Brewster, *The Life of Sir Isaac Newton* (London: John Murray, 1831), 344.

18. Joanne Keplero, *Astronomia nova, Aitiologētos: seu Physica Coelestis, tradita Commentariis de Motibus Stellae Martis, ex Observationibus G. V. Tychonis Brahe . . .* (Pragæ: Gotthard Vögelin, 1609), introduction, trans. Martínez.

19. In view of Kepler's passage, one commentator remarked: "Who, after perusing such passages in the works of an author, whose writings were in the hands of every student of astronomy, can believe that Newton waited for the fall of an apple to set him thinking for the first time on the theory which has immortalized his name? An apple may have fallen, and Newton may have seen it; but such speculations as those

which it is asserted to have been the cause of originating in him had been long familiar to the thoughts of every one in Europe pretending to the name of natural philosopher." John Eliot Drinkwater, "Life of Kepler," in *Lives of Eminent Persons* (London: Baldwin and Cradock, 1833), 24.

20. Hooke suspected the inverse square relation by the late 1670s. For his quotations in table 3.1, see the following sources: Robert Hooke, *Lectiones Cutlerianæ* (London: John Martyn, 1674), reprinted in R. T. Gunther, *Early Science in Oxford*, vol. 8, *The Cutler Lectures of Robert Hooke* (Oxford: Oxford University Press, 1931), 27–28; Robert Hooke to Isaac Newton, 6 January 1680, in *Newton, The Correspondence of Isaac Newton*, ed. H. W. Turnbull, vol. 2, 1676–1687 (Cambridge: University Press for the Royal Society, 1959), 309. Additional sources for table 3.1: Kepler, *Astronomia Nova*, chap. 33; and Kepler, *Epitome Astronomiae Copernicanae* (Lentijs ad Danubium: J. Plancus, 1618), bk. 4, pt. 3, question 5. Ismaelis Bvllialdi, *Astronomia Philolaïca. Opvs novvm, in quo Motus Planetarum per Nouam ac Veram Hypothesim Demonstrantur* (Paris: Simeonis Piget, 1645), bk. 1, chap. 12 ("Whether the Sun Moves the Planets"), reprinted in Alexandre Koyré, *The Astronomical Revolution, Copernicus, Kepler, Borelli*, trans. R. Maddison (Ithaca: Cornell University Press, 1973), app. 3; Isaac Newton, Add MS 3958.5 (ca. 1666–1671), f. 87, in A. Rupert Hall, "Newton on the Calculation of Central Forces," *Annals of Science* 13, no. 1 (1957): 62–71.

21. For discussion, see Richard Westfall, *Never at Rest: A Biography of Isaac Newton* (Cambridge: Cambridge University Press, 1980), 387, 402, 449–52, 511.

22. Newton manuscript, early 1690s, quoted in J. McGuire and P. Rattansi, "Newton and the 'Pipes of Pan,'" *Notes and Records of the Royal Society of London* 21 (1966): 118–19.

23. In a draft of the Scholium to Proposition IX, Newton wrote: "Pythagoras, on account of its immense force of attraction, said that the Sun was the prison of Zeus." See Newton, manuscript (no date, early 1690s?), Library of the Royal Society of London (Gregory MS 247); translation from McGuire and Rattansi, "Newton and the 'Pipes,'" 119; original Latin in Paolo Casini, "Newton: the Classical Scholia," *History of Science* 22 (1984): 33. I suspect that Newton's source was Proclus, who wrote: "the Pythagoreans . . . . the centre they called the prison of Jupiter; because since Jupiter has placed a demiurgical guard in the bosom of the world, he has firmly established it in the midst. For, indeed, the center abiding, the universe possesses its immovable ornament, and unceasing convolution." See Proclus, *The Philosophical and Mathematical Commentaries of Proclus, on the First Book of Euclid's Elements*, trans. Thomas Taylor, vol. 1 (London, 1792), 118.

24. Newton manuscript, early 1690s, in McGuire and Rattansi, "Newton and the 'Pipes,'" 116–17.

25. Macrobius, *Commentary on the Dream of Scipio* [ca. 430? CE], trans. William Harris Stahl (New York: Columbia University Press, 1952), 186–87.

26. "Mr. Newton believes that he has discovered pretty clearly that the Ancients like Pythagoras and Plato &c. possessed all the demonstrations that he gives of the true system of the world, and which are based upon gravity diminishing inversely as the squares of the increasing distances." Fatio de Duiller to Christiaan Huygens, February 1692, in A. Rupert Hall, *Isaac Newton: Adventurer in Thought* (Oxford: Blackwell Publishers, 1992), 346.

27. John Conduitt, Keynes Ms. 130.5 (no date): 'Miscellanea,' no. 2, Newton Project Archive, King's College, Cambridge.

28. David Gregory, *The Elements of Astronomy, Physical and Geometrical*, vol. 1 (London: J. Nicholson, 1715), xi. This work was first published as *Astronomiae Physicae et Geometricae Elementa* [1703].

29. The Bible does not specify the fruit from the Tree of Knowledge, but an early allusion to the apple appears in a poem by Alcimus Avitus, the Bishop of Vienne in Gaul, an active defender of the Orthodox Church who denounced heresies, especially the belief that Jesus is inferior to God (a belief that Newton later held in secrecy). Avitus, *De Spiritalis Historia Gestis* [ca. 510? CE], in *Patrologiæ Cursus Completus sive Bibliotheca Universalis, Integra, Uniformis, Commoda, Oeconomica* . . . Series Prima, J.-P. Migne, ed., vol. 59 (Paris: Venit Apud Editoem, 1847), 323–81; Alcimi Ecdicii Aviti, *Poematum Mosaicæ Historiæ Gestis*, Liber Secundus: "De Originali Peccato," 334. Milton paraphrased sentences from Avitus, without giving him credit, see Philip Gengembre Hubert, "A Precursor of Milton," *Atlantic Monthly* 65, no. 387 (January 1890): 33–52.

30. John Milton, *Paradise Lost: A Poem Written in Ten Books* (London: P. Parker, R. Boulter, M. Walker, 1667). I cite line numbers from his revised edition, which has become the standard: Milton, *Paradise Lost: A Poem in Twelve Books* (London: S. Simmons, 1674), bk. 1, lines 286–91; bk. 3, line 583; bk. 8, lines 124, 130; bk. 9, lines 598–605, 679–93, 776–85.

31. The Reverend Joseph Spence recorded this quotation, noting that [Andrew Michael] Ramsay attributed it to Newton "a little before he died." Ramsay knew friends of Newton such as Fatio de Duillier and Samuel Clarke. Spence died in 1768, and his collection of anecdotes remained as a manuscript sometimes used by writers, until it was finally edited and published in 1820. Joseph Spence, *Observations, Anecdotes, and Characters, of Books and Men*, ed. Edmund Malone (London: John Murray, 1820), 158–59. The advertisement to the book argued that (pp. iv-v): "The great value of the present collection must always rest on its authenticity; every particular is sanctioned by the name of the speaker; and from that simplicity of taste and minute correctness which mark the character of the writer, we may confidently infer, that as he never embellishes, he scrupulously delivers the identical language of the speaker." In his biography of Newton, Biot wrote that once "when his surrounding

friends testified to him the just admiration his discoveries had universally excited, he said, 'I know not what the world will think of my labours, but, to myself, it seems that I have been but as a child playing on the sea-shore; now finding some pebble rather more polished, and now some shell rather more agreeably variegated than another, while the immense *ocean of truth* extended itself *unexplored* before me.'" J. B. Biot, "Life of Sir Isaac Newton," in [various authors], *Lives of Eminent Persons* (London: Baldwin and Cradock, 1833), 37. Biot's account includes a footnote that states: "This anecdote is mentioned in a manuscript of Conduitt. Vid. Turner." I have not managed to find or confirm such a manuscript by Conduitt. Biot's article was first published in French: Biot, "Notice Historique sur Newton," *Biographie Universelle*, vol. 31 (1822).

32. John Milton, *Paradise Regained* (1671; reissued, London: Henry Colburn, 1827), bk. 4. In the original and second edition, "pebbles" is spelled "pibles."

33. Anonymous, "Conversations of Maturin.—No. II," *The New Monthly Magazine and Literary Journal*, Part I: Original Papers, vol. 19 (London: Henry Colburn, 1827), 570–77, quotation on p. 573.

34. Leonhard Euler, 3 September 1760; in Euler, *Lettres a une Princesse d'Allemagne sur Divers Sujets de Physique & de Philosophie*, vol. 1 (St. Petersburg: Academie Impériale des Sciences, 1768), 208, 212, trans. Martínez. The letters were translated into German within a few years: Euler, *Briefe an eine Deutsche Prinzessinn*, pt. 1 (Leipzig: Johann Friedrich Junius, 1769), 179, 182.

35. "Poets, Philosophers, and Artists, Made by Accident," in *Curiosities of Literature* (London: J. Murray, 1791); reprinted in various publications, such as *New England Quarterly Magazine* 1, no. 1 (Boston: Hosea Sprague, 1802): 246–48.

36. "Poets, Philosophers," in *New England Quarterly Magazine* 1, no. 1 (1802): 247–48.

37. Baron George Gordon Byron, *Don Juan, Cantos IX. X. XI.* (London: John Hunt, 1823), canto 10, st. 1 and 2, p. 25.

38. David Drummond, *Objections to Phrenology, Being the Substance of a Series of Papers Communicated to the Calcutta Phrenological Society* (Calcutta: Drummond, 1829), 165.

39. Bolton Corney, *Curiosities of Literature by I. D'Israeli, Illustrated*, 2nd rev. ed. (London: Richard Bentley, 1838), v, 64; writing about Isaac D'Israeli, *Curiosities of Literature*, 9th ed. (London: Edward Moxon, 1834).

40. Isaac D'Israeli, *The Illustrator Illustrated* (London: Edward Moxon, 1838).

41. Bolton Corney, "Mr. Corney on D'Israeli's *Illustrator Illustrated* [March 1838]," in Sylvanus Urban, *Gentleman's Magazine*, vol. 9, New Series (London: William Pickering; John Bowyer Nichols and Son, 1838), 371.

42. Meanwhile, the story continued to evolve, as other writers added or modified minor details. The astronomer Rev. Thomas Chalmers, for example, wrote that the

apple fell "at his feet," a version that likewise spread, but was too minor to generate complaints. Thomas Chalmers, "Popular Astronomy. Part I," *Saturday Magazine* 12, no. 369 (Supplement for March 1838), 125.

43. Augustus De Morgan, *A Budget of Paradoxes*, first published in *Assurance Magazine and Journal of the Institute of Actuaries*, vol. 11 (London: Charles & Edwin Layton, 1864), 194.

44. "Art. IX.—*Travels through the Alps of Savoy, and other parts of the Pennine Chain; with Observations of the Phenomena of Glaciers*. By James D. Forbes, 1843," *North British Review*, vol. 3, no. 2 (Edinburgh: W. P. Kennedy, 1844), 527–45; see p. 545.

45. Frederick Bridges, *Phrenology Made Practical and Popularly Explained*, 2nd ed. (Liverpool: George Philip and Son, 1861), 49. A similar claim was made in James Stanley Grimes, *A New System of Phrenology* (Buffalo, N. Y.: Oliver Steele/Wiley & Putnam, 1839), 86.

46. S. R. Wells, Editorial reply to: "Beating Round the Bush. Phrenology Criticised," *Phrenological Journal and Life Illustrated*, vol. 50 old series (April 1870), vol. 1 new series (New York: S. R. Wells, 1870), 261.

47. *Phrenological Journal and Science of Health*, vol. 103 old series (no. 1, June 1897), vol. 55 new series (New York/London: Fowler & Wells/L. Fowler & Co., 1897), 25.

48. George McC. Robson, "A Great Discovery," *Science and Industry*, vol. 4, no. 9 (October 1899) (Scranton: Colliery Engineer Company, 1899), 409. Robson's words are actually a paraphrase of Augustus De Morgan, *Budget of Paradoxes*, 81.

49. Carl Gauss, quoted in W. Sartorius v. Waltershausen, *Gauss, Zum Gedächtniss* (Leipzig: S. Hirzel, 1856), 84, trans. Martínez.

50. Michael White, *Isaac Newton: The Last Sorcerer* (New York: Basic Books, 1997), 214 and 87, respectively.

51. A. Rupert Hall, *Isaac Newton: Eighteenth Century Perspectives* (Oxford: Oxford University Press, 1999), 18.

52. David Brewster, *The Life of Sir Isaac Newton* (London: John Murray, 1831), 344.

53. David Brewster, *Memoirs of the Life, Writings, and Discoveries of Sir Isaac Newton*, vol. 1 (Edinburgh: Thomas Constable and Co., 1855), 27. In the second volume of this work, Brewster repeated his prior claim that the tree "was long ago destroyed by the wind." See Brewster, *Memoirs of the Life*, vol. 2 (Edinburgh: Thomas Constable and Co., 1855), 416.

54. DeMorgan, *A Budget of Paradoxes*, first published in *Assurance Magazine and Journal of the Institute of Actuaries*, vol. 11 (January 1864), 194.

55. R. G. Keesing, "The History of Newton's Apple Tree," *Contemporary Physics* 39, no. 5 (1998): 377–91.

56. Edmund Turnor, *Collections for the History of the Town and Soke of Grantham, Containing Authentic Memoirs of Sir Isaac Newton* (London: W. Bulmer and W. Miller, 1806), 160.

57. Mr. Walker to the Royal Astronomical Society, 12 January 1912, as quoted in McKie and de Beer, "Newton's Apple: An Addendum," 334–35.

58. George Forbes, *History of Astronomy* (New York: G. P. Putnam's Sons, 1909), 65.

59. Keesing, "History," 378.

60. Richard Keesing, "A Brief History of Isaac Newton's Apple Tree," University of York, Department of Physics, last updated on 26 January 2010, http://www.york.ac.uk/ physics/about/newtonsappletree/.

61. "Newton's Famous Apple Tree to Experience Zero Gravity," Royal Society, *Science News*, 10 May 2010, http://royalsociety.org/Newtons-famous-apple-tree -to-experience-zero-gravity.

62. Astronaut Ken Ham, quoted in "NASA's Atlantis Space Shuttle Ready for Final Voyage," BBC News, 14 May 2010, http://historynewsnetwork.org/roundup/ entries/126698.html.

## Chapter 4. The Stone of the Ancients

1. Oswald Crollie, *Philosophy Reformed & Improved in Four Profound Tractates. The I. Discovering the Great and Deep Mysteries of Nature*, trans. Henry Pinnell (London: Lodowick Lloyd, 1657), 31.

2. Although it has become common to write "Philosopher's Stone," it is more accurate to write "Philosophers' Stone," in direct translation of early expressions.

3. Newton, *Commentarium* [1680s], in Betty Jo Teeter Dobbs, *The Janus Faces of Genius: The Role of Alchemy in Newton's Thought* (Cambridge: Cambridge University Press, 1991), 276.

4. Ovid, *Metamorphoses* [ca. 8 CE], ed. Brookes More (Boston: Cornhill Publishing Co., 1922), bk. 11, lines 85–145.

5. Ovid, *Metamorphoses*, bk. 15.

6. Ovid, *Metamorphoses*, bk. 15. See also Giambattista della Porta, *Natural Magick, in Twenty Books . . . wherein Are Set Forth All the Riches and Delights of the Natural Sciences* (London: T. Young and S. Speed, 1658), bk. 2, chap. 2.

7. *Lucian's Science Fiction Novel* True Histories: *Interpretation and Commentary*, ed. Aristoula Georgiadou and David H. Laramour (Leiden: Brill, 1998), 203.

8. Pliny the Elder, *Historia Naturalis* (*Natural History*) [ca. 77 CE], trans. H. Rackham (Cambridge, Mass.: Harvard University Press, 1949–54), bk. 19, sec. 30, also bk. 24, secs. 99 and 101, and bk. 25, sec. 5. Pliny acknowledged that some attributed the book on plants to the physician Cleemporus, but Pliny insisted that "an ancient and unbroken tradition assigns it to Pythagoras," and that an author should be glad to assign his labor to the great Pythagoras, to enhance the book's authority.

9. Ibid., bk. 20, secs. 33, 73, 87, respectively.

10. *Placita Philosophorum* [falsely attributed to Plutarch, actually by another writer, based on a work by Aetius, ca. 50 BCE, as noted by Theodoret], *Peri tōn*

*areskontōn philosophois physikōn dogmatōn* [and falsely attributed to Qustā ibn Lūqā by Ibn al-Nadīm], in Hans Daiber, ed., *Aetius Arabus: Die Vorsokratiker in Arabischer Überlieferung* (Wiesbaden: Franz Steiner Verlag, 1980), 133, trans. Martínez.

11. Heraclides Ponticus (ca. 387 to 312 BCE) as paraphrased by Diogenes Laertius, *The Lives and Opinions of Eminent Philosophers* [ca. 225 CE], trans. C. D. Yonge, bk. 8, *Life of Pythagoras* (London: Henry G. Bohn, 1853), sec. 4.

12. *Auriferae Artis, quam Chemiam Vocant, Antiquissimi Authores, sive Turba Philosophorum* (Convention of Philosophers) (Basel, 1572), dictums 13, 49, 32, trans. Martínez. The Arabic manuscript seems to date from about 900 CE, as shown by Martin Plessner, and it partly derives from Greek sources. See E. J. Holmyard, *Alchemy* (Baltimore: Penguin Books, 1957), 82–83.

13. In addition, some alchemical texts that noted that Pythagoras knew the secret of transmutation are the following. Johann Siebmacher noted that after Hermes Trismegistus, the second sage to know the secret art "aside from those in Holy Scripture" was Pythagoras. See Johann Ambrosius Siebmacher, *Wasserstein der Weysen das ist, ein chymisch Tractätlein, darin der Weg gezeiget, die Materia genennet, und der Process beschrieben wird, zu dem hohen geheymnuss der Universal Tinctur zukommen* (Waterstone of the Wise) (Frankfurt: Lucas Jennis, 1619), pt. 1. *Wasserstein der Weysen* was reprinted for decades. For the Latin translation, see *Musaeum Hermeticum* (Frankfurt: L. Jennisii, 1625). See also Jean Jacques Manget, *Bibliotheca Chemica Curiosa* (Geneva: Sumpt. Choet, G. De Tournes, Cramer, Perachon, Ritter, & S. De Tournes, 1702); *Hermetisches A. B. C. derer ächten Weisen alter und neuer Zeiten vom Stein der Weisen* (Berlin: Christian Ulrich Ringmacher, 1778). Like Siebermacher, Johann Grasshof claimed that against the envious, Pythagoras described the one substance as "the one and true Matter": Hermannus Condeesyanus [Grashhof], *Dyas Chymica Tripartita, Das ist: Sechs Herrliche Teutsche* (Frankfurt am Main: Luca Jennis, 1625). Also, a medieval sonnet claimed, "This is the stone blessed and great, of which spoke Hermes and Gratiano, Elit, Rosir, Pandolfo and Ortolano, Pythagoras with all of his sect." *Codex Riccardiano* N. 946 (Biblioteca Medicea Laurenziana, Florence), in Mario Mazzoni, ed., *Sonetti Alchemici-Ermetici di Frate Elia e Cecco d'Ascoli* (San Gimignano, Tuscany: Casa Editrice Toscana, 1930).

14. Johannes Kepler, *Harmonices Mundi, Libri V* (Lincii Austriae: Godofredi Tambachii, 1619), bk. 3, in Joannis Kepler, *Astonomi Opera Omnia*, vol. 5, ed. C. Frisch (Frankfurt: Heyder & Zimmer, 1864), 132: "quin aut Pythagoras hermetiset, aut Hermes pythagoriset."

15. For example, historian David Lindberg notes, "Herodotus (fifth century B.C.) reported that Pythagoras traveled to Egypt, where he was introduced by priests to the mysteries of Egyptian mathematics." See David C. Lindberg, *The Beginnings of Western Science*, 2nd ed. (Chicago: University of Chicago Press, 2007), 12.

16. Isocrates, *Busiris* [ca. 375 BCE], secs. 28–29; George Norlin, *Isocrates*, 3 vols. (Cambridge, Mass.: Harvard University Press, 1980).

17. Herodotus, *The Histories* [ca. 430 BCE], trans. A. D. Godley (Cambridge, Mass.: Harvard University Press, 1920), e.g., bk. 2, chaps. 49, 50.

18. Ibid., bk. 2, chap. 81.

19. Niall Livingstone, *A Commentary on Isocrates' Busiris* (Leiden: Brill, 2001), 159.

20. Thomas Taylor, "Dissertation on the Platonic Doctrine of Ideas," in Proclus, *The Philosophical and Mathematical Commentaries of Proclus, on the First Book of Euclid's Elements*, trans. Thomas Taylor, vol. 1 (London, 1792), cvi.

21. *Gloria Mundi sonsten Paradeiss Taffel* (Frankfurt, 1620), reprinted in *Musaeum Hermeticum* (Francofurti: Sumptibus Lucae Jennissii, 1625); translated as: *The Glory Of The World; or, Table Of Paradise; A True Account of The Ancient Science which Adam Learned From God Himself; Which Noah, Abraham, And Solomon Held as One of the Greatest Gifts of God; which also All Sages, at All Times, Preferred to the Wealth of the Whole World, Regarded as the Chief Treasure of the Whole World, and Bequeathed Only to Good Men; namely, The Science of the Philosopher's Stone*, pt. 2, in Arthur White, ed., *The Hermetic Museum*, vol. 1 (London: James Elliot, 1893).

22. Will H. L. Ogrinc, "Western Society and Alchemy from 1200 to 1500," *Journal of Medieval History* 6 (1980): 119.

23. Bernardi Trevisanvs, *De Chymico Miracvlo, qvod Lapidem Philosophiæ appellant*, ed. Gerardvum Dornevm (first published in 1567; Basileæ: Hæredum Petri Pernæ, 1583), 3–15.

24. William R. Newman and Lawrence M. Principe, *Alchemy Tried in the Fire: Starkey, Boyle, and the Fate of Helmontian Chymistry* (Chicago: University of Chicago Press, 2002), 229.

25. Basile Valentin, *Les Douze Clefs de la Philosophie* [1599] (Paris: Editions de Minuit, 1956), 118, trans. Martínez.

26. On weaponry, the eagle, and sal ammoniac, see Lyndy Abraham, *A Dictionary of Alchemical Imagery* (Cambridge: Cambridge University Press, 1998), 214, 64, 176, 181, respectively.

27. See Lawrence Principe, "The Gold Process and Boyle's Alchemy," in *Alchemy Revisited*, ed. Z. R. W. M. von Martels (Leiden: E. J. Brill, 1990), 200–205.

28. In 1661, weatherman Richard Towneley and physician Henry Power used a barometer to measure air pressure at different altitudes on Pendle Hill in Lancashire, and they found that as the pressure upon air increases, its volume decreases. Power informed William Croone, in London, who in turn relayed a paper to Boyle. In 1662, Boyle described their conclusion as "Towneley's hypothesis," omitting Power, and then systematically confirmed it with the aid of Robert Hooke. Power's work (dated 1 August 1661) was later published as *Experimental Philosophy, in Three-Books: containing New Experiments, Microscopical, Mercurial, Magnetical. with Some Deductions, and Probable Hypotheses, Raised from Them, in Avouchment and Illustration of the Now Famous Atomical Hypothesis* (London: T. Roycroft, 1664 [actually published in 1663]).

29. Daniel Lysons, *History of the Origin and Progress of the Meeting of the Three Choirs of Gloucester, Worcester, and Hereford, and of the Charity Connected with It* (London: D. Walker, 1812), 55; J. Rutherford Russell, *The History and Heroes of the Art of Medicine* (London: John Murray, 1861), 217; Richard Lodge, *The History of England from the Restoration to the Death of William III. 1660–1702* (London: Longman, Green, and Co., 1910), 476.

30. Robert Boyle, *The Origine of Formes and Qualities*, 2nd ed. (Oxford: H. Hall/ Oxford University, 1667); Principe, "The Gold Process," 204; Basilius Valentinus, *Chymische Schriften*, vol. 1 (1677; repr., Hildesheim: H. A. Gerstenberg, 1976), 31. In Valentine's time, other alchemists also knew how to produce *aqua regia*. Principe's reading of Valentine's keys is supported by the confirmation that other alchemists such as Boyle read them also in such terms.

31. Lawrence M. Principe, *The Aspiring Adept: Robert Boyle and His Alchemical Quest. Including Boyle's "Lost" Dialogue on the Transmutation of Metals* (Princeton: Princeton University Press, 1998), 98–100.

32. Boyle, *Origine*, experiment 7, pp. 14, 233, 244.

33. Some of the Rosicrucians claimed that Pythagoras knew the art of how to communicate with the gods, visible and invisible. See Anonymous [attributed to Johann Valentin Andreä], *The Fame and Confession of the Fraternity of R: C: commonly, of the Rosie Cross*, with a Præface by Eugenius Philalethes [pseudonym for Thomas Vaughan] (London: J.M. for Giles Calvert, 1652), preface. Claims that Pythagoras could speak with the gods were ancient, for example: Philostratus, *The Life of Apollonius of Tyana* (ca. 225 CE) trans. Frederick Cornwallis Conybare, vol. 1 (London: W. Heinemann, 1912), 3, 91.

34. Principe, *Aspiring Adept*, 11.

35. Ibid., 100.

36. H. Carrington Bolton, "Chemical Literature" (Part 2) (address, American Association for the Advancement of Science, Montreal, 23 August 1882), reprinted in *Chemical News* 46, no. 1190 (29 September 1882): 146.

37. Edward Gibbon, *The History of the Decline and Fall of the Roman Empire*, 1st. ed., vol. 1 (London: W. Strahan, 1776), 418.

38. Theophrastus Paracelsus, *The Aurora of the Philosophers* [1575], in *Paracelsus His Aurora, & Treasure of the Philosophers. As also The Water-Stone of the Wise Men; Describing the Matter of, and Manner How to Attain the Universal Tincture*, J. H. Oxon, ed. (London: Giles Calvert, 1659), chap. 5.

39. Ovid, *Metamorphoses*, bk. 15.

40. An initiation rite for the Rosicrucian Society included a procedure for multiplying the red medicine toward transmutation: "The multiplication is performed according to the table of Pythagoras; the ratio of one side of the equilateral triangle to the whole figure, thus. That is, 4 parts of the Medicine to 10 parts of the metallic

water." See Sigismund Bacstrom, "Copy of the Admission of Sigismund Bacstrom into the Fraternity of Rosicrucians by the Comte de Chazal" [1794], transcribed by Frederick Hockley [1839], First Multiplication, 17; Andover Harvard Theological Library, Cambridge, Mass., item bMS 677.

41. Marie Curie, *Pierre Curie,* with an introduction by Mrs. W. Brown Meloney and with autobiographical notes by Marie Curie (New York: Macmillan, 1923), 186.

42. Barbara Goldsmith, *Obsessive Genius: The Inner World of Marie Curie* (New York: W. W. Norton, 2005), 96.

43. Frederick Soddy, "Some Recent Advances in Radioactivity," *Contemporary Review* 83 (May 1903): 720.

44. Ernest Rutherford, quoted in William Cecil Dampier Whetham to Ernest Rutherford, 26 July 1903, Rutherford Papers, Cambridge University, England; microfilm at the Niels Bohr Library, American Institute of Physics, Maryland.

45. For discussion see Goldsmith, *Obsessive Genius,* 191–204.

46. Goldsmith, *Obsessive Genius,* 85. See also Marie Curie, *Revue Scientifique* (July 1900), quoted in Susan Quinn, *Marie Curie: A Life* (New York: Addison-Wesley, 1995), 171.

47. Frederick Soddy, interview by Muriel Howorth, early 1950s, in Muriel Howorth, *Pioneer Research on the Atom; Rutherford and Soddy in a Glorious Chapter of Science: The Life Story of Frederick Soddy* (London: New World, 1958), 83–84. Soddy was mistaken, the product was actually radon, not argon.

48. Ibid.

49. Soddy, letter of 23 December 1950, published in Muriel Howorth, *Atomic Transmutation: The Greatest Discovery Ever Made; from Memoirs of Professor Frederick Soddy* (London: New World, 1953), 74.

50. Marie Sklodowska Curie, "Radium and Radioactivity," *Century Magazine* (January 1904), 461–66.

51. E. Rutherford and F. Soddy, "Radioactive Change," *Philosophical Magazine* 5 (1903): 576–91.

52. Pierre Curie, "Radioactive Substances, Especially Radium" (lecture to the Swedish Academy, 6 June, 1905, for the Nobel Prize in physics of 1903), in *Nobel Lectures, Physics 1901–1921* (Amsterdam: Elsevier, 1967), 77.

53. Mrs. William Brown Meloney [Marie Mattingly Meloney, known as "Missy"], "The Greatest Woman in the World," *Delineator* 98, no. 3 (April 1921): 15–16.

54. H. G. Wells, *The World Set Free* (New York: E. P. Dutton and Company, 1914), 50–51.

55. Mark S. Morrison, *Modern Alchemy: Occultism and the Emergence of Atomic Theory* (Oxford: Oxford University Press, 2007), 143.

56. Edwin McMillan, Martin Kamen, Samuel Rubin, "Neutron-Induced Radioactivity of the Noble Metals," *Physical Review* 52, no. 4 (August 1937): 375–77.

57. J. Cork and J. Halpern, "The Radioactive Isotopes of Gold," *Physical Review* 58, no. 3 (August 1940): 201. J. Lawson and J. Cork, "Internally Converted Gamma-Rays from Radioactive Gold," *Physical Review* 58, no. 6 (September 1940): 580.

58. R. Sherr, K. Bainbridge, H. Anderson, "Transmutation of Mercury by Fast Neutrons," *Physical Review* 60, no. 7 (October 1941): 473–79.

59. K. Aleklett, D. Morrissey, W. Loveland, P. McGaughey, and G. Seaborg, "Energy Dependence of 209Bi Fragmentation in Relativistic Nuclear Collisions," *Physical Review* C 23, no. 3 (March 1981): 1044–46.

60. Frederick Soddy, "The Evolution of Matter" [1917], in Soddy, *Science and Life: Aberdeen Addresses* (London: John Murray, 1920), 107.

61. Marie Curie, quoted in Eve Curie, *Madame Curie, A Biography by Eve Curie*, trans. Vincent Sheean (Garden City, N.J.: Doubleday/Doran, 1937), 341.

62. Joseph Campbell, *The Power of Myth*, with Bill Moyers, and Betty Sue Flowers, ed. (New York: Doubleday, 1988), 143.

### Chapter 5. Darwin's Missing Frogs

1. Frank J. Sulloway, "Darwin and His Finches: The Evolution of a Legend," *Journal of the History of Biology* 15, no. 1 (Spring 1982): 1–53.

2. For example, many students preparing for college admission exams have used study guides that included the statement: "It all began in the Galapagos, with these finches"—accompanied by a question asking to what "It" refers. The "correct answer" is: "(C) Darwin's theory of evolution." Sharon Weiner Green, Ira K. Wolf, eds., *Barron's How to Prepare for the SAT 2007*, 23rd ed. (New York: Barron's Educational Series, 2006), 582–83. Other examples: "What Darwin saw were thirteen distinct finch species, each closely resembling each other in most ways, yet each had a characteristic beak structure well suited to a particular (specialized) food source.... It was the finches that clinched it for Darwin," in Barry Boyce, *A Traveler's Guide to the Galapagos Islands* (Aptos, Calif., and Edison, N.J.: Galapagos Travel/Hunter Publishing, 2004), 15. "The finches were about the same size and all very similar in color. The only differences in the finches Darwin saw were their beaks and what kind of food they ate. There were finches that ate insects, seeds, plant matter, egg yolks and blood," in Liz Thompson, Michelle Gunter, Emily Powell, *Passing the Nevada 8th Grade CRT in Science* (Woodstock, Ga.: American Book Company, 2008), 194; also in Michelle Gunter, *Passing the ILEAP Science Test in Grade 7* (American Book Company, 2006), 132. "Later he saw in these finches the key to understanding the evolutionary process," in Michael Roberts, Michael Reiss, Grace Monger, *Advanced Biology* (Nelson: Delta Place, U.K., 2000), 724. Another biology textbook that remained unaware of Sulloway's findings is Peter H. Raven and George B. Johnson, *Biology*, 5th ed. (Boston: WCB/McGraw-Hill, 1999).

3. Charles Darwin, *Journal of Researches into the Natural History and Geology of the*

*Countries Visited during the Voyage of H.M.S. Beagle*, 2nd ed. (London: John Murray, 1845), 380.

4. Nora Barlow, ed., "Darwin's Ornithological Notes," *Bulletin of the British Museum (Natural History)*, Historical Series, vol. 2, no. 7 (February 1863): 201–78; for table 5.1, Darwin's quotations from 1835 are from pp. 261–62, and are taken from Darwin, "M.S. Notes Made on Board H.M.S. *Beagle*, 1832–36," no. 29: "Birds," pp. 72–74. University Library, Cambridge. See also Gavin de Beer, ed., "Darwin's Notebooks on Transmutation of Species, Part 1, Four Notebooks (B–D: July 1837 to July 1839)," *Bulletin of the British Museum (Natural History)*, Historical Series, vol. 2, nos. 2–5 (January–September 1960): 23–183; and Gavin de Beer, M. Rowlands, and B. Skramovsky, eds., "Darwin's Notebooks on Transmutation of Species, Part VI: Pages Excised by Darwin," *Bulletin of the British Museum (Natural History)*, Historical Series, vol. 3, no. 5 (March 1967): 129–76. Finally, the quotation of 1857 from Darwin's big manuscript on "Natural Selection" is from: R. C. Stauffer, ed., *Charles Darwin's Natural Selection; Being the Second Part of his Big Species Book Written from 1836 to 1858* (Cambridge: Cambridge University Press, 1975), 257, and are taken from Darwin, "Natural Selection" (manuscript), chap. 6, folio 43, Cambridge University Library.

5. David Lack, *Darwin's Finches* (Cambridge: Cambridge University Press, 1947).

6. Robert I. Bowman, *Morphological Differentiation and Adaptation in the Galapagos Finches*, University of California Publications in Zoology, vol. 58 (Berkeley: University of California Press, 1961). Afterward, Peter and Rosemary Grant actually measured changes in the beak sizes of Galápagos finches over time and draught. See, for example, Peter Grant, *Ecology and Evolution of Darwin's Finches* (Princeton: Princeton University Press, 1999).

7. Additional sources for table 5.2: Charles Darwin, Oct. 1835, in *Narrative of the Surveying Voyages of His Majesty's Ships Adventure and Beagle between the years 1826 and 1836*, vol. 3 (London: Henry Colburn, 1839), 462; Darwin, *Journal of Researches*, 380; Darwin, *On the Origin of Species by means of Natural Selection, or the Preservation of Favoured Races in the Struggle for Life* (London: John Murray, 1859), 28.

8. Stephen Jay Gould, "Darwin's Sea Change, or Five Years at the Captain's Table," in *Ever Since Darwin: Reflections in Natural History* (New York: W. W. Norton, 1977), 33.

9. Georges Cuvier, *An Essay on the Theory of the Earth* [1813], trans. Robert Kerr, with notes by Robert Jameson, 3rd ed. (New York: Arno Press, 1977), 17.

10. William Paley, *Natural Theology: or Evidences of the Existence and Attributes of the Deity, Collected from the Appearances of Nature* (London: R. Faulder, 1802), 451, 453, 464–65.

11. Charles Darwin, *The Autobiography of Charles Darwin, 1809–1882*, with original omissions restored [manuscript 1876–1882], ed. Nora Barlow (London: Collins, 1958), 72.

12. Ovid, *Metamorphoses* (ca. 8 CE), bk. 15, ed. Brookes More (Boston: Cornhill Pub., 1922).

13. Charles Lyell, *Principles of Geology, Being an Attempt to Explain the Former Changes of the Earth's Surface, by Reference to Causes Now in Operation*, vol. 1 (London: John Murray, 1830), 12. Lyell speculated that "Pythagoras might have found in the East not only the system of universal and violent catastrophes and periods of repose in endless succession, but also that of periodical revolutions, effected by the continued agency of ordinary causes" (14).

14. John W. Judd, *The Coming of Evolution: The Story of a Great Revolution in Science* (Cambridge: Cambridge University Press, 1910), 16.

15. Charles Darwin, *Charles Darwin's Beagle Diary*, ed. Richard Keynes (Cambridge: Cambridge University Press, 1988), 292.

16. Claudius Ælianus, *Varia Historia* [ca. 220 CE], in *Claudius Ælianus, His Various History*, trans. Thomas Stanley (London: Thomas Dring, 1665), bk. 4, chap. 17.

17. *Narrative of the Surveying Voyages of His Majesty's Ships Adventure and Beagle between the Years 1826 and 1836, vol. 2: Proceedings of the Second Expedition, 1831–1836, under the Command of Captain Robert Fitz-Roy* (London: Henry Colburn, 1839), 486–87.

18. Darwin, *Darwin's Beagle Diary*, 16 September 1835, 351–52.

19. Darwin, *Darwin's Beagle Diary*, 354, 353, 359; Darwin, *Journal of Researches*, 388

20. Darwin, *Narrative*, vol.3; Darwin, *Journal and Remarks, 1832–1835*, 468. See also Stauffer, *Charles Darwin's Natural Selection; being the Second Part of his Big Species Book Written from 1836 to 1858* (Cambridge: Cambridge University Press, 1975), 496: "What a contrast with all amphibious animals in Europe, which, when disturbed by the more dangerous animal, man, instinctively & instantly take to the water."

21. Darwin, *Origin*, 398.

22. Frank J. Sulloway, "Darwin's Conversion: The *Beagle* Voyage and Its Aftermath," *Journal of the History of Biology* 15, no. 3 (1982): 338–45.

23. Charles Darwin, "Ornithological Notes" [June/July 1836], quoted in Sulloway, "Darwin's Conversion," 327–28.

24. Charles Darwin to Otto Zacharias, 24 Febrary 1877, "When I was on board the 'Beagle,' I believed in the permanence of species, but, as far as I can remember, vague doubts occasionally flitted across my mind." Reprinted in Francis Darwin, ed., *Charles Darwin: His Life Told in an Autobiographical Chapter, and in Selected Series of His Published Letters* (London: John Murray, 1892), 166.

25. J. Herschel to Charles Lyell, 20 February 1836, in Charles Babbage, *The Ninth Bridgewater Treatise. A Fragment*, 2nd ed. (London: John Murray, 1838), 226.

26. John Gould, "Observations on the Raptorial Birds in Mr. Darwin's Collection,

with Characters of the New Species," *Proceedings of the Zoological Society of London* 5 (1837): 9

27. Charles Darwin to Otto Zacharias, 1877: "On my return home in the autumn of 1836 I immediately began to prepare my journal for publication, and then saw how many facts indicated the common descent of species, so that in July, 1837, I opened a note-book to record any facts which might bear on the question. But I did not become convinced that species were mutable until I think two or three years had elapsed." Darwin's notebook of 1837, however, seems to give direct evidence that he was pretty convinced of evolution by that year; yet his letter of 1877 claims that it took a while longer.

28. Darwin, *Origin*, 398.

29. Darwin, *Journal and Remarks*, 472. J.B.G.M. [Jean Baptiste Geneviève Marcellin] Bory de St.-Vincent, *Voyage dans les Quatre Principales îles des Mers d'Afrique: fait par ordre du Gouvernement, pendant les années neuf et dix de la République (1801 et 1802), avec l'Histoire de la Traversée du Capitaine Baudin jusqu'au Port-Louis de l'île Maurice*, vol. 3 (Paris: F. Buisson, 1804). See also Darwin, *Origin*, 393.

30. Darwin, *Journal and Remarks* (1839), 472; see also Darwin, *Journal of Researches* (1845), 381.

31. Darwin found no frogs at St. Jago, Cape Verde, St. Helena, and so forth. As evidence for the Sandwich Islands, Darwin referred to "Tyerman and Bennett's *Journal*, Vol. 1, p. 434." See also Daniel Tyerman and George Bennet, *Journal of Voyages and Travels, to Visit Their Various Stations in the South Sea Islands, China, India, &c. between the years 1821 and 1829*, compiled by James Montgomery, vol. 2 (Boston: Crocker and Brewster, 1832), 57, where they briefly note the absence of frogs and toads. Regarding the island of Mauritius, Darwin alluded to Jacques-Henri Bernardin de Saint-Pierre, *Voyage à l'Isle de France, à l'Isle de Bourbon, au Cap de Bonne-Espérance, &c, Avec des Observations nouvelles sur la nature & sur les Hommes, par un Officier du Roi* (Amsterdam, 1773), pt. 1, p. 170, where the author noted that they tried to import frogs to Mauritius (l'Isle de France) but they died. By Darwin's time, frogs imported by settlers into Mauritius, Madeira, and the Azores had proliferated so abundantly that Darwin described them as a nuisance (*Origin*, 393). As for the Canary Islands, Darwin cited Philip Barker Webb and Sabin Berthollet, *Histoire Naturelle des iles Canaries* (Paris: Béthune et Plon, 1840).

32. Darwin, *Journal of Researches* (1845), 382. These words were slightly edited from his *Journal and Remarks* (1839), 472.

33. Darwin, *Journal of Researches*, 378. Darwin referred to Herschel as the source of the phrase "mystery of mysteries," in Darwin, "Notebook E" (Transmutation of species [1838–1839]), 2 December 1832, p. 59, Cambridge University Library, also quoted in DeBeer, "Darwin's Notebooks," 165.

34. Darwin, *Origin*, 392.

35. Likewise, only mammal quadrupeds small enough to be carried by rafts of vegetation could possibly reach the Galápagos, and consequently evolve there—and indeed a few species of mice and rats have been found that are endemic to these islands.

36. Anonymous [Benjamin Franklin], "Observations Concerning the Increase of Mankind" [1751], in *Observations On the Late and Present Conduct of the French, with Regard to their Encroachments upon the British Colonies in North America . . . To which is Added, Wrote by another Hand; Observations Concerning the Increase of Mankind, Peopling of Countries, &c.* [ed. William Clarke] (Boston: S. Kneeland, 1755). Reprinted in Franklin, *Experiments and Observations on Electricity*, 4th ed. (London: David Henry, 1769), 205. At the time, Franklin estimated "One Million English Souls in North-America (tho' 'tis thought scarce 80,000 have been brought over Sea)." He argued that in Europe there was about one marriage per year for every one hundred people, whereas in America there were two per year, and since American marriages occurred at earlier ages, most couples gave birth to an average of eight children, of whom half lived to adulthood and married at around the age of twenty. Such estimates entail that by 1800, there would be 4 million Americans and by 1900 there would be 64 million, and in fact the population then became 5.2 million and 76.6 million respectively. Afterward, world wars and other factors have kept the actual population from doubling every twenty-five years.

37. Benjamin Franklin, *The Interest of Great Britain Considered, with Regard to her Colonies, and the Acquisitions of Canada and Guadaloupe. To which are Added, Observations Concerning the Increase of Mankind, Peopling of Countries, &c.* (London: T. Becket, 1760).

38. Anonymous [Thomas Robert Malthus], *An Essay on the Principle of Population* (London: Johnson, 1798). Malthus's various editions of his book vary greatly, e.g., the second (1803) was greatly enlarged, so it is significant to note which edition Darwin read in 1838. Darwin's notebook on "Books to Read" (dated 1 June 1838) lists the "Last Edit[ion] of Malthus 1826," marks it as "read," and a copy of that edition is in the Darwin Library at Cambridge University Library. But since Darwin listed that (sixth) edition of 1826 on a page *following* various publications of 1839, it might seem that the edition he read in October of 1838 was a prior edition. However, in his Notebook E (Transmutation of Species), page 3 (which is undated, but the following page is dated 4 October 1838), Darwin quotes a passage by Malthus, the wording of which matches only the fifth (1817) and sixth (1826) editions of Malthus's book, and since Darwin also noted the page number (which varied among editions), it shows that in late 1838 he did read the edition of 1826.

39. In his original work of 1798, Malthus's "harshest" conclusion was that poverty and misery among the lower classes could not be remedied, but by 1803 he noted that he had "softened" his views by realizing that knowledge of the principle of population

could help to discourage the natural inclinations that lead to expansion, and thus prevent some vice and misery. See, e.g., Malthus, *An Essay on the Principle of Population* (London: John Murray, 1826), viii, 12–17. Note also that Malthus's work of 1798 was independent of Franklin's, but in his preface of 1803, Malthus acknowledged the contribution of "Dr. Franklin."

40. Malthus, *Essay* (1826), 95.

41. Darwin, 28 September 1938, in de Beer et al. eds, "Darwin's Notebooks on Transmutation of Species, Part VI: Pages excised by Darwin," *Bulletin of the British Museum* (1967): 162.

42. Darwin, *Origin*, 64; see also Stauffer, *Charles Darwin's Natural Selection*, 177.

43. Darwin, *Origin*, 151.

44. Anonymous [Robert Chambers], *Vestiges of the Natural History of Creation* (London: John Churchill, 1844).

## Chapter 6. Ben Franklin's Electric Kite

1. Abbott L. Rotch, "Did Franklin Fly His Kite before He Invented the Lightning Rod?" *American Antiquarian Society, Proceedings* 18 (1907): 115–23; Alexander McAdie, "The Date of Franklin's Kite Experiment," *American Antiquarian Society, Proceedings* 34 (1925): 374–76; Tom Tucker, *Bolt of Fate: Benjamin Franklin and His Electric Kite Hoax* (New York: Public Affairs, 2003). An article that argues that Franklin *did* fly the kite is: I. Bernard Cohen, "The Two Hundredth Anniversary of Benjamin Franklin's Two Lightning Experiments and the Introduction of the Lightning Rod," *American Philosophical Society, Proceedings* 96, no. 3 (June 1952): 331–66.

2. *Pennsylvania Gazette*, 17 June 1731; 10 July 1732.

3. *Pennsylvania Gazette*, 12 August 1736.

4. In 1786, a member of the French Academy of Sciences seriously reported that Franklin several times told him the anecdote: [Jean-Baptiste] Leroy, "Extrait des Registres de l'Académie Royale des Sciences: Du 5 Août 1786," *Observations sur la Physique, sur l'Histoire Naturelle et sur les Arts* 29, pt. 2 (October 1786): 294.

5. *Pennsylvania Gazette*, 29 April 1742.

6. Ovid, *Metamorphoses* [ca. 8 CE], ed. Brookes More (Boston: Cornhill Pub. Co., 1922), bk. 15.

7. Hesiod, *Theogony* [8th century BCE], ed. and trans. Glenn W. Hart (Cambridge, Mass.: Harvard University Press, 2006), p. 49, lines 558–65.

8. M. Dalibard, report to the Académie Royale des Sciences, 13 May 1752; see also, "Analogie de l'Electricité avec le Tonnerre. Découverte nouvelle," *Journal Œeconomique* (Paris: Boudet, June 1752): 71–87. The iron bar was insulated at the bottom so that the electricity would not ground.

9. Franklin to M. Dalibard, 31 January 1768, in *The Works of Benjamin Franklin*, vol. 6 (Boston: Hilliard, Gray, and Company, 1838), 277; Jacques de Romas to the Académie de Bordeaux, letter of 13 July 1752, in Romas, *Mémoire sur les Moyens de se*

*Garantir de la Foudre dans les Maisons* (Bordeaux: Bergeret, 1776), 105–6. This letter was apparently read at the meeting of the Academy of 17 July 1752. Later, de Romas specified that he referred to a kite. In a work published in 1755, de Romas stated that upon communicating his letter of July 1752, he only described his projected experiment to Mr. le "Chevalier de Vivens [a member of the Academy of Sciences of Bordeaux], and other persons who did me the honor of wishing me well." de Romas, "Mémoire, Où après avoir donné un moyen aisé pour élever fort haut, & à peu frais, un corps Électrisable isolé, on rapporte des observations frappantes, qui prouvent que plus le corps isolé est élevé au dessus de la terre, plus le feu de l'Électricité est abondant," *Mémoires de Mathématique et de Physique, présentés à l'Académie Royale des Sciences, par divers Savans, & lûs dans ses Assemblées*, vol. 2 (Paris: Imprimerie Royale, 1755), 394.

10. "Extract of a Letter from Paris," *Pennsylvania Gazette*, 27 August 1752; from a letter printed previously in *Gentleman's Magazine* (May 1752), and in *London Magazine* (May 1752).

11. The length of the string is an important factor; if the string is too short, no effect will be detected; as noted in, for example, Dimitri Prince de Gallitzin to Benjamin Franklin, 28 January 1777, in Benjamin Franklin, *The Papers of Benjamin Franklin*, vol. 23, ed. William B. Willcox (New Haven: Yale University Press, 1983), 250.

12. Benjamin Franklin, "The Kite Experiment," *Pennsylvania Gazette*, 19 October 1752; published also as "A Letter of Benjamin Franklin, Esq; to Mr. Peter Collinson, F. R. S., concerning an electrical Kite," *Philosophical Transactions*, Royal Society 47 (1752): 565–67.

13. Franklin, "Request for Information on Lightning," *Pennsylvania Gazette*, 21 June 1753.

14. Cadwallader Colden to Franklin, 24 October 1752, draft: New York Historical Society. Franklin did not publish a more detailed account.

15. M. R. P., "Lettre au P. R. J. sur une Expérience Electrique," 18 October 1753, in *Memoires pour l'Histoire des Sciences et des Beaux Arts* (Paris: Briasson, 1753), 2969–76; M. de Romas, "Mémoire," 393–407. De Romas noted that the larger the kite, the more it can rise because it can thus support the weight of more string. The higher it rises, the more electricity it can collect.

16. Abbé Nollet, *Lettres sur l'Electricité*, Second Part (Paris: H. Guerin, 1760), 17th letter: 228–32; summarized in *Suite de la Clef, ou Journal Historique sur les Matieres du Tems* 88 (Paris: Ganeau, Dec. 1760), 417. Nollet stated his knowledge that in August of 1752 de Romas had made efforts with Mr. Duthil to obtain a suitable kite, and had told Mr. le Chevalier de Vivens a good acquaintance of Nollet, that de Romas planned to use a kite to collect electricity from the sky. Initially, Nollet had appended an account of Franklin's experiment onto de Romas's account when it was published, but Nollet vowed in 1760 to impartially clarify the question of priority.

17. Pierre-Louis Moreau de Maupertuis, "Lettre sur le Progrès des Sciences," *Œuvres de Mr. de Maupertuis*, vol. 2, new ed. (Lyon: Jean-Marie Bruyset, 1756), 392, trans. Martínez.

18. De Romas cordially wrote to Franklin on 19 October 1753, sending him two memoirs on electricity; Franklin replied months later, briefly expressing gratitude and only adding that "A more particular answer I must defer till the next Opportunity," but de Romas later complained that Franklin never sent any fuller reply. Franklin to Jacques de Romas, 29 July 1754, in Abbot Lawrence Rotch, "Did Benjamin Franklin Fly his Electrical Kite before He Invented the Lightning Rod?" *Proceedings of the American Antiquarian Society*, New Series, vol. 18 (Oct. 1906), 118–23; see pp. 119–20.

19. Joseph Priestley, *The History and Present State of Electricity, with Original Experiments* (London: J. Dodsley, 1767), 180: "dreading the ridicule which too commonly attends unsuccessful attempts in science, he communicated his intended experiment to no body but his son, who assisted him in raising the kite." There seems to be no evidence that his son, William, confirmed or denied the account in Priestley's book.

20. Anonymous [Benjamin Franklin], "The Speech of Miss Polly Baker," *General Advertiser*, 15 April 1747 (London); it was also issued in *Gentleman's Magazine*, the *Boston Weekly Post-Boy*, the *New York Gazette*, and *Maryland Gazette*.

21. "But his Scholar, the Greater PYTHAGORAS, prov'd the Author of the 47th Proposition of Euclid's first Book, which if duly observ'd, is the Foundation of all Masonry, sacred, civil, and Military." James Anderson, *The Constitutions of the Free-Masons. Containing the History, Charges, Regulations, &c. of that Most Ancient and Worshipful Fraternity. For the Use of the Lodges* (London, 1723; reprinted: Philadelphia: Benjamin Franklin, 1734), 22.

22. Tucker, *Bolt of Fate*, 253.

23. I. Bernard Cohen, "The Two Hundredth Anniversary of Benjamin Franklin's Two Lightning Experiments and the Introduction of the Lightning Rod," *American Philosophical Society, Proceedings* 96, no. 3 (June 1952): 366.

24. "Franklin's Kite," *MythBusters*, episode 48, Beyond Television Productions for the Discovery Channel, aired 8 March 2006.

25. "Kite flier electrocuted; used wire instead of string," *News 5* (Belize), 20 March 2006, http://www.channel5belize.com/archive_news_cast.php?news_date=2006-03-20#a2.

### Chapter 7. Coulomb's Impossible Experiment?

1. Jean-Noël Hallé and Jean-Baptiste Biot, "Rapport appouvé par la Classe des Sciences Physique et Mathématiques de l'Institut National," 21 vendémiaire, an 11 (Gregorian calendar date: 13 October 1802); quoted in Jean Aldini, *Essai Théorique et Expérimental sur le Galvanisme, avec une série d'expériences faites en présence des*

*Commissaires de l'Institut National de France, et en divers Amphithéatres Anatomiques de Londres*, vol. 1 (Paris: Fournier Fils, 1804), 115, trans. Martínez.

2. Andrew Knapp and William Baldwin, "George Foster, Executed for the Murder of His Wife and Child," *The Newgate Calendar* [also known as *The Malefactor's Bloody Register*], vol. 3 (London: J. Robins and Co., 1825), 314–18. Abridged versions were later published, including: "George Foster: Executed at Newgate, 18th of January, 1803, for the Murder of his Wife and Child, by drowning them in the Paddington Canal; with a Curious Account of Galvanic Experiments on his Body," *The Complete Newgate Calendar*, vol. 4 (London: Navarre Society, 1926), 257–59.

3. Giovanni Aldini, *General Views on the Application of Galvanism to Medical Purposes: principally in Cases of Suspended Animation* (London, 1819), 80.

4. Knapp and Baldwin, "George Foster" (1825), 318; ibid. (1926), 259.

5. Diogenes claimed that Aristotle and Hippias reported that Thales of Miletus "attributed souls also to lifeless things, forming his conjecture from the nature of the magnet, and of amber." Diogenes Laertius, *Lives of Eminent Philosophers*, trans. C. D. Yonge (London: Henry G. Bohn, 1853), bk. 1, sec. 3.

6. Prior to Coulomb, a few other physicists surmised an inverse square law for electricity, including Joseph Priestley, *The History and Present State of Electricity, with Original Experiments*, vol. 2 [1676]; 3rd edition (London: C. Bathurst et. al, 1775), e.g., 374. Priestley inferred the inverse square relation on the basis of the claim that a body inside a spherical shell would not be attracted in one or another direction, but his actual experiment did not involve a sphere but a metal cup, and it did not involve movable particles to test variations of force at various distances, etc.

7. Samuel Devons, "The Art of Experiment: Coulomb, Volta, Faraday," presentation, 29 June 1984, videotape, Bakken Library Museum for Electricity in Life, Minneapolis, Minnesota.

8. Lacking a lightweight synthetic material such as foam, past experimenters used dehydrated pith. Pith is the lightweight spongy tissue inside the stems of vascular plants; it was often extracted from elderberry shrubs of the *Sambucus* genus, or from chèvrefeuilles.

9. Charles Augustin Coulomb, "Premier Mémoire sur l'Électricité et le Magnétisme. Construction & usage d'une Balance électrique, fondée sur la propriété qu'ont les Fils de métal, d'avoir une force de réaction de Torsion proportionnelle à l'angle de Torsion (1785)," *Mémoires de l'Académie Royale des Sciences*, Paris (1788), 572, trans. Martínez.

10. Peter Heering, "On Coulomb's Inverse Square Law," *American Journal of Physics* 60 (1992): 990.

11. Ibid.

12. Ibid., 991. The *American Journal of Physics* selected Heering's paper as one of its most memorable articles: Robert H. Romer, "Editorial: Sixty Years of the Ameri-

can Journal of Physics—More Memorable Papers," *American Journal of Physics* 61, no. 2 (1993): 103–6.

13. John L. Heilbron, "On Coulomb's Electrostatic Balance," in Christine Blondel and Matthias Dörries, eds., *Restaging Coulomb: Usages, Controverses et Réplications autour de la Balance de Torsion*; Biblioteca di Nuncius, vol. 15 (Firenze: Leo S. Olschki, 1994), 151–61, see p. 151.

14. Christian Licoppe, "Coulomb et la 'Physique Experimentale': Pratique Instrumentale et Organisation Narrative de la Preuve," in *Restaging Coulomb* (1994), 67–83.

15. Heilbron, "On Coulomb's Electrostatic Balance," 156.

16. Christine Blondel and Bertrand Wolff, *Coulomb invente une balance pour l'électricité,* film, narrated by Stéphane Pouyllau (experiments at the Lycée Emile-Zola, Rennes), with documentation by Marie-Hélène Wronecki, www.ampere.cnrs.fr, accessed 1 June 2008, trans. Martínez.

17. For a complete account, see Alberto A. Martínez, "Replication of Coulomb's Torsion Balance Experiment," *Archive for History of Exact Sciences* 60 (2006): 517–63.

18. I do not know why Heering did not obtain results similar to Coulomb's; I haven't inspected his torsion balance. A video of the experiment by Wolff, at least, shows that parts of Wolff's device were defective; e.g., one of the carriers of charge was neither spherical nor smooth; see Blondel and Wolff, "Coulomb invente."

## Chapter 8. Thomson, Plum-Pudding, and Electrons

1. Rubén Martínez, "Plum Pudding and the Folklore of Physics" (paper presented at the annual meeting of the History of Science Society, Cambridge, Mass., 2003); ibid. (unpublished manuscript, University of Texas at Austin, 2007).

2. James Arnold Crowther, *Molecular Physics*, 2nd ed. (Philadelphia: P. Blakiston's Son and Co., 1919), 94: "The problem becomes much simpler if we assume that the positive electrification occupies the whole volume of a sphere co-extensive with the atom, and that the electrons are embedded in it like raisins in a pudding. An atom of this kind was suggested by Lord Kelvin, and has been worked out in detail by Sir J. J. Thomson."

3. Peter Guthrie Tait, *Properties of Matter*, 4th ed. (London: Adam and Charles Black, 1899), 21: "A much more plausible theory is that matter is continuous (*i.e.* not made up of particles situated at a distance from one another) and compressible, but intensely heterogenous; like a plum-pudding, for instance, or a mass of brick-work." In 1900, George FitzGerald rejected Joseph Larmor's claims that mechanical physics should be replaced by purely mathematical expressions, and FitzGerald then voiced "his preference for 'plum-pudding' physics—for brass wheels and bands . . . rather than an equation with integrals in it." Meeting of the British Association, discussed in: *Observatory, Monthly Review of Astronomy* 23, no. 297 (October 1900): 391.

4. Among the historians who have discussed this topic are Isobel Falconer, Stuart

M. Feffer, Nadia Robotti, Theodore Arabatzis, and Graeme Gooday; this chapter owes greatly to their works.

5. J. J. Thomson, "Presidential Address to the British Association," *British Association for the Advancement of Science, Report* (1909): 29.

6. Max Planck to Carl Runge, 9 December 1878, and 4 March 1879, Carl Runge Papers, Staatsbibliothek Preussischerkulturbesitz; see also John Heilbron, *The Dilemmas of an Upright Man: Max Planck and the Fortunes of German Science* (Berkeley: University of California Press, 1986), 10.

7. Robert A. Millikan, *Autobiography* (New York: Prentice Hall, 1950), 269–70.

8. J. J. Thomson, "Cathode Rays," *Electrician* 21 (May 1897), 104–11; J. J. Thomson, "Cathode Rays," *Philosophical Magazine* 44 (October 1897): 293–316.

9. Jean Perrin, "Nouvelles proprietés des Rayons Cathodiques," *Comptes Rendus Hebdomadaires des Séances de l'Académie des Sciences, Paris* 121, no. 27 (30 December 1895): 1130–34; translation: "New Experiments on Kathode Rays," *Nature* 53 (30 January 1896): 298–99.

10. Some of Thomson's cathode ray tubes are on display, see "The Discovery of the Electron: Electrical Discharges in Gases," Museum of the Cavendish Laboratory, www-outreach.phy.cam.ac.uk/camphy/museum/area2/cabinet3.htm, accessed 1 June 2008.

11. William Crookes, *On Radiant Matter. A Lecture Delivered to the British Association for the Advancement of Science* (London: Davey, 1879), 15.

12. Ibid., 30.

13. Arthur Schuster, "The Bakerian Lecture: Experiments on the Discharge of Electricity through Gases. Sketch of a Theory," *Proceedings of the Royal Society of London* 37 (1884): 317–39; see pp. 318, 331–33.

14. Arthur Schuster, "The Bakerian Lecture: The Discharge of Electricity through Gases," *Proceedings of the Royal Society of London* 47 (1889–1890), 526–61; see pp. 545–47. See also Schuster, *The Progress of Physics During 33 Years (1875–1908): Four Lectures Delivered to the University of Calcutta, March 1908* (Cambridge: Cambridge University Press, 1911), 64–67.

15. Schuster, *Progress*, 59.

16. Stoney based his conclusions on the phenomena of electrolysis and the spectra of gases. G. J. Stoney, "On the Cause of Double Lines and of Equidistant Satellites in the Spectra of Gases," *Scientific Transactions of the Royal Dublin Society*, 2nd ser. 4 (1891): 583.

17. Heinrich Hertz, "Über den Durchgang der Kathodenstrahlen durch dünne Metallschichten [November 1891]," *Annalen der Physik und Chemie* 45 (1892): 28–32.

18. Philipp Lenard, "Über Kathodenstrahlen in Gasen von atmosphaersichem Druck und im äussersten Vacuum," *Sitzungsberichte der Königlich Preussischen Akad-*

*emie der Wissenschaften zu Berlin* (January 1893): 3–7; expanded version in *Annalen der Physik und Chemie* 51, no. 2 (1894): 225–67.

19. Philipp Lenard, "Über die magnetische Ablenkung der Kathodenstrahlen," *Annalen der Physik und Chemie* 52 (1894): 23–33.

20. Jean Perrin, "Nouvelles proprietés des Rayons Cathodiques," *Comptes Rendus* 121 (1895): 1130–34.

21. Working at the Institute and the Society for the Promotion of German Science, in Bohemia, Jaumann immersed a pear-shaped cathode ray tube in a vat of oil, running a weak electric current between the cathode and the anode (the latter being outside of the tube, but in the oil). By holding a rubbed glass rod near the tube, he found that the cathode rays were strongly deflected. He also succeeded in stretching the cathode rays and altering their intensity, to brighten or dim. By contrast to magnetic deflection, he found that electrostatic deflection was a rapidly transient effect. G. Jaumann, "Elektrostatische Ablenkung der Kathodenstrahlen," *Sitzungsberichte der Kaiserliche Akademie der Wissenschaften, Wien, Mathematische und Naturwissenschaftliche Klasse* 105 Abt. IIa. (April 1896): 291–306; also in: *Wiener Anzeiger* 111–14 (1896), 121–22; and in *(Wiedemann's) Annalen der Physik und Chemie* 295 no. 10 [alternate numbering 59, no.1] (1896): 252–66. Gustav Jaumann, "Über die Interferenz und die elektrostatische Ablenkung der Kathodenstrahlen," Sitzungsberichte der Kaiserliche Akademie der Wissenschaffen, Wien, Mathematische und Naturwissenschafliche Klasse, 106 Abt. IIa (March and April, 1897), 533–50. Jaumann had developed the view that cathode rays consist of "longitudinal light," that is, longitudinal vibrations of the ether. See Jaumann, "Longitudinales Licht," *Sitzungsberichte der Kaiserliche Akademie der Wissenschaften*, Wien, Abt. IIa, 104, 7–10 (1895): 747–92; also in *Ann. d. Phys.* 293 [alternate numbering 57], no. 1 (1896): 147–84.

22. For example, the History Center of the American Institute of Physics states that before Thomson, "All attempts had failed when physicists tried to bend cathode rays with an electric field." See American Institute of Physics, "The Discovery of the Electron / 3 Experiments, 1 Big Idea," text by Kent Staley, ed. Spencer Weart, http://www.aip.org/history/electron/jj1897.htm, accessed January 2010.

23. Theodore Arabatzis, "Rethinking the 'Discovery' of the Electron," *Studies in the History and Philosophy of Modern Physics* 27, no. 4 (1996): 405–35, see p. 423.

24. Pieter Zeeman, "The Effect of Magnetisation on the Nature of Light Emitted by a Substance," *Nature* 55 (11 February 1897): 347; Zeeman, "On the Influence of Magnetism on the Nature of the Light Emitted by a Substance," *Philosophical Magazine* 43 (March 1897): 226–39; Zeeman, "Doubles and Triplets in the Spectrum Produced by External Magnetic Forces," *Philosophical Magazine* 44 (July 1897): 55–60.

25. Zeeman to Oliver Lodge, 24 January 1897, quoted in Arabatzis, "Rethinking," 424. In his paper of February 1897, Zeeman announced this "direct evidence of the existence of ions" ("Effect," 347).

26. Emil Wiechert, "Ergebniss einer Messung der Geschwindigkeit der Kathodenstrahlen [7 January 1897]" *Schriften der Physikalisch-ökonomisch Gesellschaft zu Königsberg* 38 (1897): 3, trans. Martínez.

27. Walter Kaufmann, "Die magnetische Ablenkbarkeit der Kathodenstrahlen und ihre Abhängigkeit vom Entladungspotential [April 1897]," *Annalen der Physik und Chemie*, series 3, 61 (June 1897): 544–52; W. Kaufmann and E. Aschkinass, "Ueber die Deflexion der Kathodenstrahen," *Ann. der Phys. u. Chem.* 62 (November 1897): 588–95. Kaufmann, "Nachtrag zu der Abhandlung: 'Die magnetische Ablenkbarkeit der Kathodenstrahlen,'" *Ann. der Phys. u. Chem.* 62 (1897): 596–98. Kaufmann willfully abstained from endorsing the claim that cathode rays consist of charged particles.

28. For discussion, see George E. Smith, "J. J. Thomson and the Electron, 1897–1899," in *Histories of the Electron*, ed. Jed Z. Buchwald and Andrew Warwick (Cambridge, Mass.: MIT Press, 2001), 42–43; and Graeme Gooday, "The Questionable Matter of Electricity: The Reception of J. J. Thomson's 'Corpuscle' among Electrical Theorists and Technologists," in Buchwald and Warwick, *Histories of the Electron*, 112.

29. Lord Rayleigh IV, *The Life of Sir J. J. Thomson* (Cambridge: Cambridge University Press, 1942), 91.

30. Philipp Lenard, *Wissenschaftliche Abhandlungen*, vol. 3 (Leipzig: S. Hirzel, 1944), 1.

31. J. J. Thomson, "On the Existence of Masses Smaller than the Atoms" (report of the Sixty-Ninth Meeting of the British Association for the Advancement of Science, Dover, September 1899), published as "On the Masses of the Ions in Gases at Low Pressures," *Philosophical Magazine* 48 (1899): 547–67.

32. Moreover, in the 1899 paper, Thomson rightly characterized the negative electron as the fundamental factor in the ionization of gases and their electrical discharges (denying an imagined positive electron), a major contribution to research on electrical conduction, as emphasized by Smith, "J. J. Thomson," 21–76.

33. Nadia Robotti and Francesca Pastorino, "Zeeman's Discovery and the Mass of the Electron," *Annals of Science* 55 (1998): 161–83.

34. Schuster, *Progress* (1908), 71.

35. P. Curie and M. Curie, "Les nouvelles substances radioactives et les rayons qu'elles émettent," in *Rapports présentés au Congrès international de physique réuni à Paris en 1900*, Tome III: *Électro-optique et ionization*, ed. Lucien Poincaré and Charles-Édouard Guillaume (Paris: Gauthier-Villars), 79–114.

36. Patrick Matthew, "Stewart's Planter's Guide, and Sir Walter Scott's Critique," in *On Naval Timber and Arboriculture* (London: Longman, Orme, Brown, and Green, 1831), 308: "Man's interference, by preventing this natural process of selection among plants, independent of the wider range of circumstances to which he introduces them, has increased the difference in varieties, particularly in the more domesticated kinds;

and even in man himself, the greater uniformity, and more general vigour among savage tribes, is referrible to nearly similar selecting law—the weaker individual sinking under the ill treatment of the stronger, or under the common hardship." See also his appendix, pp. 364–67, 387.

37. Charles Darwin, *The Autobiography of Charles Darwin, 1809–1882,* with original omissions restored [manuscript 1876–1882], ed. Nora Barlow (London: Collins, 1958), 125.

38. George Francis FitzGerald, "Dissociation of Atoms," *Electrician* 39 (1897): 104. Arabatzis also points out that William Sutherland made a similar proposal in 1899 ("Rethinking," 429). See also Gooday, "Questionable Matter," 111.

39. John Zeleny, quoted in George Jaffé, "Recollections of Three Great Laboratories," *Journal of Chemical Education* 29 (1952): 236.

40. Owen W. Richardson, *The Electron Theory of Matter* (Cambridge: Cambridge University Press, 1914), 3.

41. See Gooday, "Questionable Matter," 114; on Armstrong, see Lord Rayleigh, *The Life of Sir J. J. Thomson* (Cambridge: Cambridge University Press, 1942), 113–14. In 1901, Ernest Rutherford noted that chemists at McGill University opposed Thomson's theory; see Rutherford to Thomson, 26 March 1901, in A. S. Eve, *Rutherford* (Cambridge: Cambridge University Press, 1939), 77; "The British Association at Dover," *Electrician* 43 (1899): 772–73.

42. Arabatzis, "Rethinking," 433.

43. Edmund Edward Fournier D'Albe, *The Electron Theory: A Popular Introduction to the New Theory of Electricity and Magnetism,* with a preface by G. Johnstone Stoney (New York: Longmans, Green, and Co., 1906), 4.

44. William Crookes, *Researches in the Phenomena of Spiritualism* (London: J. Burns, 1874). About the medium, the fifteen-year-old girl Florence Cook, Crookes noted, "Every test that I have proposed she has at once agreed to submit to with the utmost willingness."

45. Peter Achinstein, "Who Really Discovered the Electron?" in Buchwald and Warwick, *Histories of the Electron,* 403–24.

46. Bruce J. Hunt, "Review of *Histories of the Electron,*" in *British Journal for the History of Science* 38 (2005): 117–18.

47. E. A. Davis and I. Falconer, *J. J. Thomson and the Discovery of the Electron* (London: Taylor & Francis, 1997), 134.

48. Richard T. Glazebrook, "How Research Has Helped Electrical Engineering," in *Practical Electrical Engineering,* ed. E. Molloy, rev. ed., vol. 1 (1931), 3–7; quoted in Gooday, "The Questionable Matter," 125.

49. Thomas S. Kuhn, *The Structure of Scientific Revolutions,* 2nd ed. (Chicago: University of Chicago Press, 1970), 55.

50. Kaufmann, "Die magnetische" (1897), 544.

### Chapter 9. Did Einstein Believe in God?

1. Albert Einstein, "Autobiographical Notes," in *Albert Einstein: Philosopher-Scientist*, ed. and trans. Paul Arthur Schilpp (Evanston, Ill.: The Library of Living Philosophers/George Banta Publishing Company, 1949), 3.

2. Maja Winteler-Einstein, "Albert Einstein—Beitrag für sein Lebensbild" [1924], in John Stachel, ed., *The Collected Papers of Albert Einstein*, vol. 1 (Princeton: Princeton University Press, 1987), xlvii–xlvi.

3. Abraham Pais, *Subtle Is the Lord . . . The Science and the Life of Albert Einstein* (Oxford: Oxford University Press, 1982), 38.

4. Einstein, "Autobiographical Notes," 9.

5. Einstein, "Autobiographical Notes," 17, 5.

6. John Stachel, "Albert Einstein: The Man Beyond the Myth," *Einstein from 'B' to 'Z'* (Boston: Birkhäuser, 2002), 3–11.

7. Albert Einstein to Ilse Rosenthal-Schneider, 15 September 1919, Albert Einstein Archives, item 22–261, The Hebrew University of Jerusalem and the Einstein Papers Project at the California Institute of Technology, Pasadena, Calif. (hereafter Einstein Archives); Ilse Rosenthal-Schneider, *Reality and Scientific Truth* (Detroit: Wayne State University Press, 1980), 74.

8. Einstein, May 1921, in Princeton, New Jersey (in response to experimental work by D. C. Miller): "Raffiniert ist der Herr Gott, aber boshaft ist er nicht." The phrase is often translated as "Subtle is the Lord, but not malicious." Oscar Veblen, professor of mathematics at Princeton, heard Einstein's remark, and he later asked Einstein whether the words could be chiseled onto the frame of a fireplace in the newly constructed building of mathematics, Fine Hall, room 202. And so they were, in 1930; the mathematics department has since moved. Banesh Hoffmann, Helen Dukas, *Albert Einstein: Creator and Rebel* (New York: Viking Press, 1972), 146.

9. Albert Einstein to Max Born, 4 December 1926, Einstein Archives, item 8–180: "Die Theorie liefert viel, aber dem Geheimnis des Alten bringt sie uns kaum näher. Jedenfalls bin ich überzeugt, dass der nicht würfelt." In a later letter, Einstein wrote, "Es scheint hart, dem Herrgott in seine Karten zu gucken. Aber dass er würfelt und sich 'telepatischer' Mittel bedient (wie es ihm von der gegenwärtigen Quanten-Theorie zugemutet wird) kann ich keinen Augenblick glauben," translation: "It seems hard, to look at the Lord God's cards. But that he rolls dice and uses 'telepathic' means (as is expected of him by the present quantum-theory) I can not believe for a moment." Einstein to Cornelius Lanczos, 21 March 1942, Einstein Archives, item 15–298, trans. Martínez.

10. Albert Einstein, "Science and Religion" (address at the Conference on Science, Philosophy, and Religion, New York, 1940); reissued as Einstein, *Ideas and Opinions* (New York: Crown, 1954), 46.

11. Albert Einstein, "What I Believe," *Forum and Century* 84 (1930): 193–94; reprinted in Einstein, *Ideas and Opinions*, 8.

12. Albert Einstein, *Gelegentliches* (Berlin: Soncino Gesellschaft, 1929), 9; reissued as "On Scientific Truth," in *Ideas and Opinions*, 262.

13. Einstein, interview by George Sylvester Viereck, 1929; in Viereck, *Glimpses of the Great* (New York: Duckworth, 1930), 447. See also Viereck, "What Life Means to Einstein," *Saturday Evening Post*, 26 October 1929, 17.

14. Einstein, handwritten sentence on a letter from A. M. Nickerson to Einstein, 17 July 1953, Einstein Archives, item 36–552. See also Max Jammer, *Einstein and Religion: Physics and Theology* (Princeton: Princeton University Press, 1999), 220.

15. Albert Einstein to Murray W. Gross, 26 April 1947, Einstein Archives, item 58–243.

16. Albert Einstein to Rabbi Herbert S. Goldstein, 25 April 1929, Einstein Archives, item 33–272.

17. Elisabeth Bergner, *Bewundert und Viel Gescholten: Elisabeth Bergners unordentl. Erinnerungen* (Munich: Bertelsmann, 1978), 212.

18. Albert Einstein to P. Wright, 24 January 1936, Einstein Archives, item 52–336 (see also 41–746, 42–599, 42–601, 52–335); *Albert Einstein: The Human Side*, ed. Helen Dukas and Banesh Hoffman (Princeton: Princeton University Press, 1979), 32–33.

19. W. Hermanns, *Einstein and the Poet—In Search of the Cosmic Man* (Brookline, Mass.: BrandenPress, 1983), 132.

20. Esther Salaman, "A Talk with Einstein," *Listener* 54 (1955): 370–71.

21. Einstein, as quoted by Ernst Gabor Strauss (his assistant from 1944 to 1948), in Strauss, "Assistant bei Albert Einstein," in Carl Seelig, *Helle Zeit—Dunkle Zeit* (Zurich: Europa, 1956), 72.

22. Einstein to Oswald Veblen, 30 April 1930, Einstein Archives, item 17–284 (see also 23–152, 23–153); "Die Natur verbirgt ihr Geheimnis durch die Erhabenheit ihres Wesens, aber nicht durch List."

23. Walter Isaacson, *Einstein: His Life and Universe* (New York: Simon & Schuster, 2007), 389.

24. Other writers also portray Einstein as believing in God. For example, Yehuda Elkana claims that one of Einstein's major quests was to understand how God thinks, that Einstein thought himself into God's mind, and that divine thoughts involve great intuitive leaps that help to find universal laws. Yehuda Elkana, "Einstein and God," in *Einstein for the Twenty-First Century*, ed. Peter Galison, Gerald Holton, S. Schweber (Princeton: Princeton University Press, 2008), 35–47.

25. Rabbi Jacob Singer (address, Temple Isaiah Israel, Chicago, 4 January 1931), quoted in Jammer, *Einstein and Religion*, 84.

26. Editorial, *Osservatore Romano* (ca. 1929–1931), in support of Boston's Cardinal O'Connell's critique of Einstein (1929); quoted in Peter Michelmore, *Einstein: Profile of the Man* (New York: Dodd, Mead & Company, 1962), 139.

27. Jammer, *Einstein and Religion*, 96.

28. Albert Einstein to Guy H. Raner Jr. (U.S. Navy ensign), 28 September 1949, Einstein Archives, item 58–702 (see also 58–701, 58–703, 57–288), translation from Guy H. Raner and Lawrence S. Lerner, "Einstein's Belief's," *Nature* 358 (9 July 1992): 102. Einstein claimed that he differed from a typical "freethinker," who is mainly nourished by opposition to naïve superstition, whereas his own outlook was moved by a kind of humility, a consciousness of the insuffiency of the human mind to deeply understand the harmony of the universe; see Einstein to Beatrice F., 12 December 1952; Einstein Archives, item 59–794, quoted in Jammer, *Einstein and Religion*, 121.

29. Albert Einstein to M. Berkowitz, 25 October 1950, Einstein Archives, item 59–215, trans. Martínez.

30. Albert Einstein to Eric B. Gutkind, 3 January 1954, Einstein Archives, item 33–337 (and also 33–338 and 59–897), trans. Martínez. A scanned image of the letter is available at Bloomsbury Auctions, "303. Einstein (Albert, *theoretical physicist*, 1879–1955) Autograph Letter signed to Eric B. Gutkind," http://www.bloomsbury auctions.com/detail/649/303.0, accessed 30 April 2009. I provide an original and very literal word for word translation. In 2009, online news media released small, blurry images of Einstein's letter, along with translations that are not very accurate; mistakes include omitted words, and the insertion of words such as "childish." German versions of the letter, online, include defective re-translations from the English renditions. Moreover, some German transcriptions also include mistakes.

31. Dennis Overbye, "Einstein Letter on God Sells for $404,000," *New York Times*, 17 May 2008.

## Chapter 10. A Myth about the Speed of Light

1. Albert Einstein, "Autobiographical Notes," in *Albert Einstein: Philosopher-Scientist*, ed. and trans. Paul Arthur Schilpp (Evanston, Ill.: The Library of Living Philosophers/George Banta Publishing Company, 1949), 33.

2. Einstein, quoted in John Stachel, "Albert Einstein: The Man Beyond the Myth," in *Einstein from 'B' to 'Z'* (Boston: Birkhäuser, 2002), 11.

3. *Placita Philosophorum* [falsely attributed to Plutarch, actually by another writer, based on a work by Aetius, ca. 50 BCE, as noted by Theodoret], *Peri tōn areskontōn philosophois physikōn dogmatōn* [and falsely attributed to Qustā ibn Lūqā by Ibn al-Nadīm], in Hans Daiber, ed., *Aetius Arabus: Die Vorsokratiker in Arabischer Überlieferung* (Wiesbaden: Franz Steiner Verlag, 1980), 131, trans. Martínez.

4. Immanuel Kant, *De Mundi Sensibilis atque Intelligibilis Forma et Principiis, Dissertatio Pro Loco* [1770], in *Kant's Inaugural Dissertation and Early Writings on Space*, trans. John Handyside (Chicago: Open Court Publishing Company, 1929), 56–57. See also Kant, *Kritik der Reinen Vernuft* [1781], in *Critique of Pure Reason*, trans. and ed. Paul Guyer and Allen Wood, The Cambridge Edition of the Works of Immanuel Kant" (Cambridge: Cambridge University Press, 1998), 164–65.

5. Henri Poincaré, *La Science et l'Hypothèse* (Paris: Flammarion, 1902), 111. See also Karl Pearson, *The Grammar of Science* [1892], 2nd ed. (London: Adam and Charles Black, 1900), sec. 13, p. 186: "there is no such thing as *absolute* time." Both of these works were read by Einstein before 1905. Earlier, Johann Bernhard Stallo had also argued that there exists no absolute time, in J. B. Stallo, *The Concepts and Theories of Modern Physics* (New York: D. Appleton and Co., 1881/1882), 184–85.

6. Einstein, interview by R. S. Shankland, 4 February 1950, in Shankland, "Conversations with Albert Einstein," *American Journal of Physics* 31 (1963): 48.

7. Albert Einstein, interview Max Wertheimer, 1916, in Wertheimer, *Productive Thinking* (New York: Harper & Brothers, 1945), 169.

8. Albert Einstein, "Erinnerungen-Souvenirs," *Schweizerische Hochschulzeitung* 28 *Sonderheft* (1955): 145–53; reprinted as "Autobiographische Skizze," in *Helle Zeit–Dunkle Zeit. In Memoriam Albert Einstein*, ed. Carl Seelig (Zurich: Europa Verlag, 1956), 10.

9. Einstein to Mileva Marić, 10? August 1899, in *The Collected Papers of Albert Einstein*, vol. 1, *The Early Years, 1879–1902*, ed. John Stachel (Princeton: Princeton University Press, 1987), 225.

10. Einstein, "Autobiographical Notes," 8.

11. Hans Byland, "Aus Einsteins Jugendtagen," *Neue Bündner Zeitung*, 7 February 1928.

12. Einstein [1916], quoted in Alexander Moszkowski, *Einstein: Einblicke in seine Gedankenwelt. Gemeinverständliche Betrachtungen über die Relativitätstheorie und ein neues Weltsystem/Entwickelt in Gesprächen mit Einstein* (Hamburg: Hoffmann & Campe, 1921), 18.

13. Einstein, interview by David Reichinstein, in Reichinstein, *Albert Einstein, sein Lebensbild und seine Weltanschaunng* (Prague: Ernst Ganz, 1935), 23. See also Peter Michelmore, *Einstein: Profile of the Man* (New York: Dodd, Mead and Company, 1962), 44.

14. Albert Einstein, "Wie ich die Relativitätstheorie entdeckte" (lecture, University of Kyoto, Japan, 1922), transcribed into Japanese by Jun Ishiwara, "Einstein Kyôzyu-Kôen-roku," *Kaizo* 4, no. 22 (1923): 1–8; also as *Einstein Kyôzyu-Kôen-roku* (Tokyo: Kabushiki Kaisha, 1971), 82.

15. Einstein, interview by R. S. Shankland, 4 February 1950, in Shankland, "Conversations," 48.

16. Einstein to Maurice Solovine, 24 April 1920, in Albert Einstein, *Lettres à Maurice Solovine* (Paris: Gauthier-Villars, 1956), 21.

17. In Newton's *Principia* of 1687, he had claimed: "Absolute, true and mathematical time, of itself, and from its own nature, flows equably without relation to anything external, and by another name is called duration; relative, apparent, and common time, is some sensible and external (whether accurate or unequable) measure of dura-

tion by means of motion, which is commonly used instead of true time; such as an hour, a day, a month, a year." Isaac Newton, *Philosophia Naturalis Principia Mathematica* [1687], in *Mathematical Principles of Natural Philosophy*, trans. Andrew Motte in 1729, rev. Florian Cajori (Berkeley: University of California Press, 1946).

18. Ernst Mach, *Die Mechanik in ihrer Entwickelung historisch-kritisch dargestellt* (Leipzig: F. A. Brockhaus, 1883); 2nd ed. (1889), trans. T. J. McCormack, *The Science of Mechanics: A Critical and Historical Account of Its Development* (1893; rev. ed. 1942; repr., La Salle, Ill.: Open Court Publishing Co., 1960), 127.

19. Einstein to Carl Seelig, 8 April 1952, Albert Einstein Archives, item 39–018, The Hebrew University of Jerusalem and the Einstein Papers Project at the California Institute of Technology, Pasadena, Calif. (hereafter Einstein Archives).

20. Wertheimer, *Productive Thinking*, 174.

21. Anton Reiser [Rudolf Kayser], *Albert Einstein: A Biographical Portrait*, with a preface by Albert Einstein (New York: A. & C. Boni, 1930), 68. Rudolf Kayser, Einstein's stepson-in-law, interviewed him for this biography, which he published under a pseudonym; Einstein described this book's details as accurate.

22. Rømer presented his discovery to the Académie Royale des Sciences on 7 December 1676 and it was described in: "Démonstration touchant le mouvement de la lumière trouvé par M. Römer de l'Académie Royale des Sciences," *Journal des Sçavans, de l'An M.DC.LXXVI.* (Amsterdam: Pierre Le Grand, 1683), 267–70. Rømer noted that Io takes 42.5 hours to complete an orbit around Jupiter (the present value is 42.46 hours); and he stated "that for the distance of about 3000 leagues, such as is very nearly the size of the diameter of the Earth, light needs not one second of time" (268). In August 1676, when Earth was relatively close to Jupiter, observations were carried out from the Paris observatory to measure the time of reappearance of Jupiter's moon when eclipsed, and in early September, Rømer predicted to the Académie that in November Io would take an additional ten minutes to reappear. On November 9, astronomers confirmed his prediction.

23. Henri Poincaré, "La mesure du temps," *Revue de Métaphysique et de Morale* 6 (January 1898): 1–13; reissued in Poincaré, *Foundations of Science*, trans. G. B. Halsted (New York: Science, 1913), 232; words omitted, ellipses added.

24. [Armand] H. Fizeau, "Sur une Expérience Relative à la Vitesse de Propagation de la Lumière," *Comptes Rendus Hebdomadaires des Séances de l'Académie des Sciences, Paris* 29 (1849): 90–92. The length of the double path was $1.7266 \times 10^6$ centimeters. The average of twenty-eight observations gave Fizeau a transit time of $5.5 \times 10^{-5}$ second, so the speed seemed to be $3.14 \times 10^{10}$ cm per second (195,111 miles per second).

25. For a history of the efforts to ascertain any dependence of the speed of light on its source, see Alberto A. Martínez, "Ritz, Einstein, and the Emission Hypothesis," *Physics in Perspective* 6, no. 1 (2004): 4–28.

26. Einstein, "Wie ich die Relativitätstheorie entdeckte," 80, trans. Fumihide Kanaya and A. Martínez.

27. Poincaré, *La Science*, 111. Carl Seelig, *Albert Einstein und die Schweiz* (Zurich: Europa Verlag, 1952), 63.

28. A. Einstein to André Metz, 27 November 1924, Einstein Archives, item 18–255.

29. A. Einstein, *Über die Spezielle und die Allgemeine Relativitätstheorie* (Braunschweig: Vieweg, 1917); Einstein, *Relativity*, trans. R. W. Lawson (New York: P. Smith/H. Holt and Co., 1931), 23, italics in the original.

30. A. Einstein, "Zur Elektrodynamik bewegter Körper," *Annalen der Physik* 17 (1905): 891–921; italics in the original.

31. I've dated Einstein's reading of Hume to March 1905 because in that month his friend Maurice Solovine skipped an appointment with Einstein to read Hume together in order to attend a Bohemian string quartet: "Freitag, 17 März, abends punkt 8 Uhr: Konzert gegeben von berühmten Böhmischen Streichquartett," *Der Bund, Eidgenössisches Zentralblatt* 56 Jahrgang, Nr. 119 (Saturday, 11 March 1905), 4. For Einstein on Hume, see Einstein to Moritz Schlick, 14 December 1915, *The Collected Papers of Albert Einstein*, vol. 8 *The Berlin Years: Correspondence, 1914–1918*, pt. A, ed. Robert Schulmann, A. J. Kox, Michel Janssen, and József Illy (Princeton: Princeton University Press, 1998), 220; Einstein, *Lettres à Maurice Solovine*, x.

32. Solovine, *Lettres à Maurice Solovine*, viii; David Hume, *A Treatise of Human Nature* (London: John Noon, 1739), bk. 1, secs. 2–6, pp. 73–94.

33. Einstein, "Autobiographical Notes," 13. Einstein concluded that human perceptions involve expectations and habits and therefore cannot alone lead to laws of nature; some additional component is necessary. Albert Einstein, "Remarks on Bertrand Russell's Theory of Knowledge," in *The Philosophy of Bertrand Russell*, ed. Paul Arthur Schilpp (Evanston, Ill.: Northwestern University Press, 1944); reprinted in Einstein, *Ideas and Opinions* (New York: Crown Publishers, 1954), 22.

34. For example, N. David Mermin, *Space and Time in Special Relativity* (Prospect Heights, Ill.: Haveland Press, 1968), 1, 4, 19.

35. For example, see Henri Arzeliès, *Relativistic Kinematics* (Oxford: Pergamon Press, 1966); Edwin F. Taylor and John Archibald Wheeler, *Spacetime Physics* (New York: W. H. Freeman and Co., 1963).

36. Jakob Ehrat to Carl Seelig, 20 April 1952, Einstein Archives, item 71–212.

37. Einstein, *Über die Spezielle*, 17–18. Einstein there noted that he formulated this account of the theory of relativity in the sequence and connections in which it actually originated—a claim that was confirmed by the psychologist Wertheimer when he interviewed Einstein to understand his creative process in *Productive Thinking*, 176.

38. Example adapted from Albert Einstein and Leopold Infeld, *Evolution of Physics* (New York: Simon and Schuster, 1938), 178–79.

## Chapter 11. The Cult of the Quiet Wife

1. Melsa Films Pty., Ltd., *Einstein's Wife*, produced in association with the Australian Broadcasting Corporation and Oregon Public Broadcasting in the United States, aired 2003. The present chapter is an expanded version of the article: A. Martínez, "Handling Evidence in History: The Case of Einstein's Wife," *School Science Review* 86, no. 316 (March 2005): 49–56.

2. OPB Interactive for PBS Programming, "Einstein's Wife," http://www.pbs.org/opb/einsteinswife/index.htm, last modified 4 March 2008. Thanks to the dedicated critical efforts of Allen Esterson to point out and correct inaccuracies in the PBS documentary, its website has been revised (see Andrea Gabor, "Editor's Note," http://www.pbs.org/opb/einsteinswife/editor_note.htm, 24 September 2007). Still, certain problems remain; see Allen Esterson, "Articles on Mileva Marić and Sigmund Freud," www.esterson.org, accessed 10 December 2007.

3. Einstein to Marić, letters of 4 April 1901 and 27 March 1901, respectively, in Albert Einstein, Mileva Marić, *The Love Letters*, ed. Jürgen Renn and Robert Schulmann, trans. Shawn Smith (Princeton: Princeton University Press, 1992), 41, 39.

4. Einstein to Mileva Marić, 27 March 1901, in *The Collected Papers of Albert Einstein*, vol. 1, *The Early Years, 1879–1902*, ed. John Stachel (Princeton: Princeton University Press, 1987), 282, trans. Martínez.

5. Einstein to Marić, 28? September 1899, *Collected Papers*, vol. 1, 233.

6. Einstein to Paul Ehrenfest, 25 April 1912, in *The Collected Papers of Albert Einstein*, vol. 5, *The Swiss Years: Correspondence, 1902–1914*, ed. Martin Klein, Anne Kox, Robert Schulmann (Princeton: Princeton University Press, 1993), 450; Einstein to C. O. Hines, February 1952, Einstein Archives, item 12–251; Einstein, interview by R. S. Shankland, 4 February 1950, in Shankland, "Conversations with Albert Einstein," *American Journal of Physics* 31 (1963): 49.

7. Einstein to Mario Viscardini, April 1922, Einstein Archives, item 25–301; Einstein to Ehrenfest, June 1912, in *Collected Papers*, vol. 5, doc. 409, p. 485; Einstein to Albert P. Rippenbeim (draft), 1952, Einstein Archives, item 20–046.

8. Albert Einstein, "Wie ich die Relativitätstheorie entdeckte" (lecture, University of Kyoto, Japan, 1922), transcribed into Japanese by Jun Ishiwara, "Einstein Kyôzyu-Kôen-roku," *Kaizo* 4, no. 22 (1923): 1–8; also as *Einstein Kyôzyu-Kôen-roku* (Tokyo: Kabushiki Kaisha, 1971), 82

9. On the "ten years" of reflection: Einstein, interview, 4 February 1950, in Shankland, "Conversations," 48; and Einstein, Albert Einstein, "Autobiographical Notes," in *Albert Einstein: Philosopher-Scientist*, ed. and trans. Paul Arthur Schilpp (Evanston, Ill.: The Library of Living Philosophers/George Banta Publishing Company, 1949), 53. On "my life for over seven years," see Albert Einstein, interview by R. S. Shankland, 24 October 1952, in Shankland, "Conversations," 56; for "after seven years," see

Einstein to Erika Oppenheimer, 13 September 1932, quoted in *Collected Papers*, vol. 2, *The Swiss Years: Writings 1900–1909*, ed. John Stachel (Princeton: Princeton University Press, 1989), 261–62.

10. Albert Einstein, "Erinnerungen-Souvenirs," *Schweizerische Hochschulzeitung* 28 *Sonderheft* (1955), 145–53, 146; reprinted as "Autobiographische Skizze," *Helle Zeit–Dunkle Zeit. In memoriam Albert Einstein*, ed. Carl Seelig (Zurich: Europa Verlag, 1956), 10; see also Anton Reiser [Rudolf Kayser], *Albert Einstein: A Biographical Portrait*, preface by Albert Einstein (New York: A. & C. Boni, 1930), 49.

11. See, for example, Einstein to Marić, 19 December 1901; in *Collected Papers*, vol. 1, 328.

12. Louis Kollros to Carl Seelig, 26 February 1952, 1, Archives and Private Collections, ETH-Bibliothek, Zurich, Hs 304:740.

13. Helene Savić to her mother, Ida, 14 July 1900, in Milan Popović, ed., *In Albert's Shadow: The Life and Letters of Mileva Marić, Einstein's First Wife* (Baltimore: Johns Hopkins University Press, 2003), 60.

14. Marić to Savić, Spring 1901, in Popović, *In Albert's Shadow*, 76.

15. Marić to Savić, Fall 1901, in Popović, *In Albert's Shadow*, 76–78. For more on Marić, see John Stachel, "Albert Einstein and Mileva Marić: A Collaboration that Failed to Develop, in Stachel, *Einstein from 'B' to 'Z'* (Boston: Birkhäuser, 2002), 39–55.

16. Marić to Savić [Nov.–Dec. 1901] in Popović, *In Albert's Shadow*, 79.

17. Dord Krstić, "Mileva Einstein-Marić," in Elizabeth Roboz Einstein, *Hans Albert Einstein: Reminiscences of His Life and Our Life Together* (Iowa City: Iowa Institute of Hydraulic Research, 1991), 98.

18. Ibid., 85.

19. Marić to Savić, 20 December 1900, in *The Collected Papers of Albert Einstein: English Translation*, vol. 1, trans. Anna Beck (Princeton: Princeton University Press, 1987), 156.

20. Marić to Savić [Nov.-Dec. 1901], *Collected Papers, English*, 183–84.

21. Dennis Overbye, *Einstein in Love: A Scientific Romance* (New York: Penguin, 2000), 110.

22. Maurice Solovine and Albert Einstein, *Lettres à Maurice Solovine* (Paris: Gauthier-Villars, 1956), xii, trans. Martínez.

23. Philipp Frank, Einstein, *Sein Leben und seine Zeit* (Munich: P. List, 1949; Brauschweig/Wiesbaden: F. Vieweg & Sohn, 1979 [with a 1942 foreword by Einstein]), 39, 44, trans. Martínez. A rough English translation appeared before the original German text was published: Frank, *Einstein: His Life and Times* (New York: Knopf, 1947; London: Jonathan Cape, 1948), 32, 34–35.

24. Desanka Trbuhović-Gjurić, *U senci Alberta Ajnstajna* (Krusevac: Bagdala,

1969), trans. *Im Schatten Albert Einsteins, Das tragische Leben der Mileva Einstein-Marić*, ed. Werner Zimmermann (Bern: Paul Haupt, 1993), 97.

25. Evan Harris Walker, "Mileva Marić's Relativistic Role," *Physics Today* 44, no. 2 (February 1991): 123

26. Michele Zackheim, *Einstein's Daughter* (New York: Riverhead/Penguin Putnam, 1999), 19.

27. OPB Interactive for PBS Programming, "The Mileva Question," www.pbs.org/opb/einsteinswife/science/mquest.htm, accessed 4 April 2004; this webpage has since been revised and the quoted text removed.

28. Abram F. Joffe, *Vstrechi s fizikami moi vospominaniia o zarubezhnykh fizikah* (Moscow: Gosudarstvenoye Idatelstvo Fiziko-Matematitsheskoi Literatury, 1962); German trans., A. Joffe, *Begegnungen mit Physikern* (Leipzig: B. G. Teubner, 1967), 88.

29. Abram F. Joffe, "Pamiati Alberta Einsteina," *Uspekhi fizicheskikh nauk* 57, no. 2 (1955): 187, trans. Martínez.

30. Walker, "Mileva Marić's," 123.

31. Carl Seelig, *Albert Einstein, Eine Dokumentarische Biographie* (Zurich: Europa Ver., 1954), 29.

32. I thank Christian Wüthrich and Allen Esterson for confirming this fact.

33. Daniil Semenovich Danin, *Neizbezhnost Strannogo Mira* (Moscow: Molodaia Gvardia, Gosudarstvenaaja Biblioteka SSSR, 1962), 57.

34. Peter Michelmore, *Einstein: Profile of the Man* (New York: Dodd, Mead, and Company, 1962), 36, 45, vii.

35. Ibid., 36.

36. Krstić, "Mileva Einstein-Marić," 94. Krstić dated this letter as being from "the very beginning of 1906."

37. Letter from Marić to Savić, in Popović, *In Albert's Shadow*, 88. Popović dated this letter from December 1906, apparently following notes by Julka Savić, see Popović, *In Albert's Shadow*, xi. Martin Klein, A. Kox, and Robert Schulmann, who edited the *Collected Papers* also dated this letter from December 1906, owing to its contents, see *Collected Papers*, vol. 5, 45.

38. Marić to Savić, 3 September 1909, in Popović, *In Albert's Shadow*, 98.

39. Marić to Savić [Winter 1909/10], Einstein Archives, item 70–726, trans. Martínez.

40. Heinrich A. Medicus, "The Friendship among Three Singular Men: Einstein and His Swiss Friends Besso and Zangger," *Isis* 85 (1994): 456–78, see p. 469.

41. For example, Medicus, "Friendship," 470.

42. Gerald Holton, *Einstein, History and Other Passions* (Cambridge, Mass.: Harvard University Press, 2000), 191.

43. Maurice Solovine to Carl Seelig, 29 April 1952, Archives and Private Collections, ETH-Bibliothek, Zurich, Hs 304:1007, p. 3, trans. Martínez.

## Chapter 12. Einstein and the Clock Towers of Bern

1. For several examples, see Alberto Martínez, *Kinematics: The Lost Origins of Einstein's Relativity* (Baltimore: Johns Hopkins University Press, 2009), 298.

2. Alan Lightman, *Einstein's Dreams* (New York: Pantheon Books, 1993), 3, 19, 33–34, 49, 94, 129, 149, 177.

3. Eric W. Tatham, "'I'll Know What I Want When I See It'—Towards a Creative Assistant," in *People and Computers X: Proceedings of the HCI'95 Conference*, ed. M. A. R. Kirby, A. J. Dix, and J. E. Finlay (Cambridge: Press Syndicate of the University of Cambridge, 1995), 270.

4. Steven Pinker, "His Brain Measured Up," *New York Times*, 24 June 1999, A27.

5. Peter L. Galison, "Einstein's Clocks: The Place of Time," *Critical Inquiry* 26, no. 2 (Winter 2000): 360, 375. See also William R. Everdell, *The First Moderns: Profiles in the Origins of Twentieth-Century Thought* (Chicago: University of Chicago Press, 1997), 237: "During that intense talk with Besso, Einstein has seen the light. . . . The clock tower in Bern could tell Einstein what time it was, but only in Bern. . . . This meant that the time of every clock was a function of the distance between clock and clock-watcher, their relative motion, and the speed of light."

6. Dennis Overbye, *Einstein in Love: A Scientific Romance* (New York: Penguin, 2000), 132.

7. Arthur I. Miller, *Einstein, Picasso* (New York: Basic Books, 2001), 247, 5.

8. Peter L. Galison, *Einstein's Clocks, Poincaré's Maps: Empires of Time* (New York: W. W. Norton, 2003), 101, 104, 105, 122, 125–26, 128, 136, 140.

9. Albrecht Fölsing, *Albert Einstein: A Biography*, trans. Ewald Osers (New York: Viking/Penguin, 1997), 179; Fölsing, *Albert Einstein: Eine Biographie* (Frankfurt: Suhrkamp Verlag, 1993).

10. Galison, *Einstein's Clocks*, 254.

11. Josef Sauter, statement delivered at the Conference 50 Jahre Relativitätstheorie, Bern, 1955; reprinted in Max Flückiger, *Albert Einstein in Bern* (Bern: Paul Haupt, 1974), 156.

12. William R. Everdell, "It's About Time. It's About Space," review of *Einstein's Clocks, Poincaré's Maps: Empires of Time*, by Peter Galison, *New York Times*, 17 August 2003, 10.

13. Alberto A. Martínez, "Material History and Imaginary Clocks: Poincaré, Einstein, and Galison on Simultaneity," *Physics in Perspective* 6, no. 2 (June 2004): 31–48.

14. Alexander Moszkowski, *Einstein, Einblicke in seine Gedankenwelt: gemeinverständliche Betrachtungen über die Relativitätstheorie und ein neues Weltsystem, entwickelt aus Gesprächen mit Einstein* (Hamburg: Hoffmann und Campe, 1921), 227, trans. Martínez.

15. Albert Einstein, "Erinnerungen-Souvenirs," *Schweizerische Hochschulzeitung* 28 *Sonderheft* (1955), 145–53; reprinted as "Autobiographische Skizze," *Helle Zeit–*

*Dunkle Zeit. In memoriam Albert Einstein*, ed. Carl Seelig (Zurich: Europa Verlag, 1956), 12.

16. Franz Paul Habicht to Melania Serbu, 26 October 1943, Albert Einstein Archives, item 39–275, p. 2, The Hebrew University of Jerusalem and the Howard Gottlieb Archival Research Center of Boston University, Boston, Mass., trans. Martínez.

17. Sauter, statement to 50 Jahre Relativitätstheorie; reprinted in Flückiger, *Albert Einstein in Bern*, 154. Anton Reiser [Rudolf Kayser], *Albert Einstein: A Biographical Portrait*, with a preface by Albert Einstein (New York: A. & C. Boni, 1930), 65.

18. Walter Isaacson, *Einstein, His Life and Universe* (New York: Simon and Schuster, 2007), 126, 582; Isaacson misunderstood me as having claimed in my "Material History" that the steeple clock in Muri had not been synchronized with the Bern clocks; but actually I had merely noted that the steeple clock at Muri was not connected electrically to the clocks in Bern, still they were in the same time zone.

19. For example, Patricia Fara, *Science: A Four Thousand Year History* (Oxford: Oxford University Press, 2009), 248; Richard Staley, *Einstein's Generation: The Origins of the Relativity Revolution* (Chicago: University of Chicago Press, 2008), 68. Fara and Staley are historians. See also Richard Panek, *The Invisible Century: Einstein, Freud, and the Search for Hidden Universes* (New York: Penguin Books, 2005), 72.

20. Thibault Damour, *Si Einstein M'Était Conté* (Paris: le Cherche Midi, 2005), 15, 16, trans. Martínez. See also Damour, *Once upon Einstein*, trans. Eric Novak (Wellesley, Mass.: A. K. Peters, Ltd., 2006), 5.

21. Ann Banfield, "Remembrance and Tense Past," in *The Cambridge Companion to the Modernist Novel*, ed. Morag Shiac (Cambridge: Cambridge University Press, 2007), 55.

22. Hans C. Ohanian, *Einstein's Mistakes: The Human Failings of Genius* (New York: W. W. Norton, 2008), 89, 87.

23. Max Jammer, *Concepts of Simultaneity: From Antiquity to Einstein and Beyond* (Baltimore: Johns Hopkins University Press, 2005), 122.

24. Walter C. Mih, *The Fascinating Life and Theory of Albert Einstein*, with a foreword by Bernard Einstein (Commack, N.Y.: Kroshka Books/Nova Science Publishers, 2000), 69.

25. Steven L. Winter, *A Clearing in the Forest: Law, Life, and Mind* (Chicago: University of Chicago Press, 2001), 36. See also Melody Graulich and Paul Crumbley, *The Search for a Common Language: Environmental Writing and Education* (Logan: Utah State University Press, 2005), 1; Brian K. Pinaire, *The Constitution of Electoral Speech Law: The Supreme Court and Freedom of Expression in Campaigns and Elections* (Stanford Law Books, 2008), 287.

26. Michio Kaku, *Einstein's Cosmos: How Albert Einstein's Vision Transformed Our Understanding of Space and Time* (New York: W. W. Norton, 2005), 62.

27. George Will, *One Man's America: The Pleasures and Provocations of Our Singular Nation* (New York: Random House/Three Rivers Press, 2009), 358.

28. G. M. P. Swann, *Putting Econometrics in Its Place: A New Direction in Applied Economics* (Northampton, Mass.: Edward Elgar Publishing, 2006), 195; Len Kurzawa, *The Fundamental Force: How the Universe Works* (Victoria, B.C.: Trafford Publishing, 2009), 7; Robert M. Hazen and James Trefil, *Science Matters: Achieving Scientific Literacy*, 2nd ed. (New York: Random House, 2009), 199; Annette Moser-Wellman, *The Five Faces of Genius: Creative Thinking Styles to Succeed at Work* (New York: Penguin, 2002), 23. For a somewhat similar, brief tale, see also Gregory Mone, "What If Einstein Had Been a Better Violinist," *Popular Science* 266, no. 6 (June 2005): 74.

29. Daniel Simonis, Sarah Johnstone, Nicola Williams, *Lonely Planet: Switzerland*, 5th ed. (Oakland, Calif.: Lonely Planet, 2006), 182. See also *The Cities Book: A Journey Through the Best Cities in the World* (Lonely Planet, 2009), 87.

30. James Trefil and Robert Hazen, *The Sciences: An Integrated Approach*, 5th ed. (Wiley, 2006), 143; Stanislaw D. Głazek and Seymour B. Sarason, *Productive Learning: Science, Art, and Einstein's Relativity in Educational Reform* (Thousand Oaks, Calif.: Corwin Press, 2007), 152.

31. Lucjan Piela, *Ideas of Quantum Chemistry* (Amsterdam: Elsevier, 2007), 94.

Chapter 13: The Secret of Einstein's Creativity?

1. Jeremy Gray, "Finding the Time. The Scientific Struggle to Bring the World's Clocks into Line," *Nature* 424 (2003): 880.

2. Arthur I. Miller, *Einstein, Picasso* (New York: Basic Books, 2001).

3. Besso to Einstein, October to 8 December 1947, in Albert Einstein and Michele Besso, *Correspondance 1903–1955*, German transcriptions with French translations, notes, and introduction by Pierre Speziali (Paris: Hermann, 1972), 386.

4. Einstein to Besso, 1952, in *Correspondance*, 391.

5. John Stachel, "'What Song the Syrens Sang': How Did Einstein Discover Special Relativity?" *Einstein from 'B' to 'Z'* (Boston: Birkhäuser, 2002), 157–69; see p. 166.

6. For example, John Rigden has claimed that the young Einstein wanted to read God's thoughts, and that to him the problem of relative motion was "clearly part of God's thoughts," and that "in 1905, Einstein had a direct line to God's thoughts." John Rigden, *Einstein 1905: The Standard of Genius* (Cambridge, Mass.: Harvard University Press, 2005), 7–8, 150.

7. For example, W. Gordin, "The Philosophy of Relativity," *Journal of Philosophy* 23, no. 19 (September 1926): 517–24; see p. 520.

8. Stachel was alluding to these lines: "What song the Syrens sang, or what name Achilles assumed when he rid himself among women, though puzzling questions, are

not beyond all conjecture." Sir Thomas Browne, *Hydriotaphia: Urne-Buriall* (London: Hen Brome, 1658), reprinted in *Miscellaneous Works of Sir Thomas Browne*, ed. Alexander Young (Cambridge: Hilliard and Brown, 1831), 221.

9. Einstein, quoted in James Franck to Carl Seelig, 16 July 1952, Archives and Private Collections, ETH-Bibliothek, Zurich, Hs 304:637, trans. Martínez.

10. Maja Winteler-Einstein, manuscript, "Albert Einstein-Beitrag für sein Lebensbild" [1924], in *The Collected Papers of Albert Einstein: English Translation*, vol. 1, trans. Anna Beck (Princeton: Princeton University Press, 1987), xviii.

11. The statement about Einstein at age five is according to Antonina Vallentin, who interviewed Einstein: Antonina Vallentin, *Le Drame d'Albert Einstein* (Paris: Libraire Plon, 1954), 15.

12. Maja Winteler-Einstein, in *Collected Papers, English*, vol. 1, p. xviii.

13. "As an infant he had started to talk so late that his parents had been in some alarm about the possibility of an abnormality in their child. At the age of eight or nine years, Einstein presented the picture of a shy, hesitating, unsociable boy, who passed on his way alone, dreaming to himself, and going to and from school without feeling the need of a comrade." Alexander Moszkowski, *Einstein, Einblicke in seine Gedankenwelt* [1921], trans. Henry L. Brose, *Conversations with Einstein* (New York: Horizon Press, 1970), 222.

14. Hans Albert Einstein, interviewed by Bela Kornitzer, in "Einstein Is My Father," *Ladies' Home Journal* 68, no. 4 (April 1951): 47, 134, 136, 139, 141, 255–56, quotation on p. 134.

15. Jean Piaget, *Le Développement de la Notion de Temps Chez l'Enfant* (Paris: Presses Universitaires de France, 1946), trans. Martínez.

16. "Einstein employait avec prédilection la méthode génétique dans l'examen des notions fondamentales. Il se servait pour les éclaircir de ce qu'il a pu observer chez les enfants." Maurice Solovine and Albert Einstein, *Lettres à Maurice Solovine* (Paris: Gauthier-Villars, 1956), viii–ix, trans. Martínez.

17. James Mark Baldwin, *Mental Development in the Child and the Race, Methods and Processes* [1895], 3rd ed. (1906; repr., New York: Augustus Kelley, 1968), 5

18. "Kann es schon bald seine Augen nach etwas hinwenden? Jetzt kannst Beobachtungen machen. Ich möcht auch einmal selber ein Lieserl machen, es muß doch zu interessant sein! Es kann gewiß schon weinen, aber lachen lernt es erst viel später. Darin liegt eine tiefe Wahrheit." Einstein to Marić, Tuesday [4 February 1902], in John Stachel, ed., *The Collected Papers of Albert Einstein*, vol. 1, *The Early Years, 1879–1902* (Princeton: Princeton University Press, 1987), 332, trans. Martínez.

19. Marić to Einstein [after 20 October 1897], *Collected Papers*, vol. 1, 34.

20. Einstein, "Ernst Mach," *Physikalische Zeitschrift* 17 (April 1916): 101–4.

21. Ernst Mach, *Beiträge zur Analyse der Empfindungen* (Jena: G. Fischer, 1886), reprinted as *Contributions to the Analysis of the Sensations*, trans. C. M. Williams (Chicago: Open Court Publishing, 1897), 156.

22. Ibid., 156; see also 161, 170.

23. Hermann von Helmholtz, "Origin and Significance of Geometrical Axioms" (lecture, Docenten Werein, Heidelberg, 1870), trans. and reprinted in David Cahan, ed., *Hermann von Helmholtz, Science and Culture: Popular and Philosophical Essays* (Chicago: University of Chicago Press, 1995), 228–29, 245.

24. Hermann von Helmholtz, "On the Facts in Perception" (speech, Commemoration Celebration of the Frederick Wilhelm University of Berlin, 3 August 1878), also in Cahan, *Hermann von Helmholtz*, 354–58.

25. Charles Darwin, *The Expression of the Emotions in Man and Animals* (London: John Murray, 1872), 211–12.

26. Arthur Schopenhauer, *Parerga und Paralipomena* [1851], selections reissued in Schopenhauer, *The Wisdom of Life and Counsels and Maxims*, trans. T. Bailey Saunders (Amherst, N.Y.: Prometheus Books, 1995), 96 (back pagination). Spinoza used the expression "sub specie aeternitatis" repeatedly in his *Ethics*, a book that Einstein also read before 1905 and greatly admired. See Benedicti de Spinoza, *Ethica Ordine Geometrico Demonstrata* [1677], in Spinoza, *Opera*, vol. 1, ed. Carolus Hermannus Bruder (Lopsiae: Bernh. Tauchnitz, 1843), 185, 403–10. Einstein used the expression "sub specie aeterni" in Einstein, *Geometrie und Erfahrung* (Berlin: Julius Springer, 1921), 8.

27. Anton Reiser [Rudolf Kayser], *Albert Einstein: A Biographical Portrait*, with a preface by Albert Einstein (New York: A. & C. Boni, 1930), 40.

28. Schopenhauer, *Wisdom of Life*, 96–97.

29. Moszkowski, *Conversations with Einstein*, 96. See also Albert Einstein, "H. A. Lorentz, Creator and Personality," *Mein Weltbild* (Zurich: Europa Verlag, 1953); reprinted in Einstein, *Ideas and Opinions* (New York: Crown Pub., 1954), 73–76.

30. Peter Michelmore, *Einstein: Profile of the Man* (New York: Dodd, Mead, and Company, 1962), 44.

31. Albert Einstein, "Autobiographische Skizze," *Helle Zeit–Dunkle Zeit. In memoriam Albert Einstein*, ed. Carl Seelig (Zurich: Europa Verlag, 1956), 10.

32. Einstein to Solovine, 3 April 1953, in *Lettres à Maurice*, 125, trans. Martínez.

33. For an example of Einstein's interest in how children learn, see Max Talmey, *The Relativity Theory Simplified; And the Formative Period of Its Inventor* (New York: Falcon Press/Darwin Press, 1932), 176.

34. Alfred Russel Wallace, "Review," *Quarterly Journal of Science* (January 1873), quoted in Charles Darwin, *The Life and Letters of Charles Darwin*, ed. Francis Darwin, vol. 3 (London: J. Murray, 1887), 172: "the restless curiosity of the child

to know the 'what for?' the 'why?' and the 'how?' of everything" seems "never to have abated its force."

35. Einstein, quoted in Moszkowski, *Conversations with Einstein*, 69.

### Eugenics and the Myth of Equality

1. Iamblichi, *De Vita Pythagorica* [ca. 300 CE], reprinted as Iamblichus, *On the Pythagorean Way of Life*, ed. and trans. John Dillon and Jackson Hershbell (Atlanta: Scholars Press, 1991), chap. 17.

2. E. Cobham Brewer, *Dictionary of Phrase and Fable: Giving the Derivation, Source, or Origin of Common Phrases, Allusions and Words that Have a Tale to Tell*, rev. ed. (London: Cassell and Company, 1900), 831.

3. There are mistaken claims about FitzRoy's death; some say that "he shot himself," e.g., Stephen Jay Gould, *Ever Since Darwin* (New York: W. W. Norton, 1973), 33. His death, however, was described in a contemporary account, "Vice-Admiral FitzRoy," *Gentleman's Magazine and Historical Review* 18, no. 218 (January–June, 1865) (London: John Henry and James Parker, 1865), 789. It read: "The family, finding that he remained longer than usual, knocked several times at the door, but receiving no answer, the door was at length broken down, when the Admiral was found weltering in his blood, having cut his throat. . . . the coroner's jury returned a verdict to the effect that [the] deceased destroyed himself while in an unsound state of mind."

4. Francis Galton, *Hereditary Genius* (London: Macmillan and Co., 1869).

5. Charles Darwin, *The Descent of Man* (London: John Murray, 1871), 111. See also Darwin to Francis Galton, 23 December [1869 or 1870], in Darwin, *More Letters of Charles Darwin*, ed. Francis Darwin, vol. 2 (London: John Murray, 1903), 41.

6. Daniel Kevles, *In the Name of Eugenics*, with a new introduction (1985; Cambridge, Mass.: Harvard University Press, 1995), 4.

7. Ruth Schwartz Cowan, "Francis Galton's Statistical Ideas: The Influence of Eugenics," *Isis* 63, no. 4 (December 1972): 509–28.

8. Francis Galton, *Inquiries into Human Faculty and Its Development* (London: Macmillan and Co., 1883), 24–25.

9. Karl Pearson, "Discussion," *American Journal of Sociology* 10, no. 1 (July 1904): 7.

10. The phrase "We hold these truths to be self-evident, that all men are created equal," was paraphrased by Thomas Jefferson from his friend, Philip Mazzei, an Italian-born patriot and pamphleteer, and edited by Benjamin Franklin.

11. R. C. Olby, "Mendel no Mendelian?" *History of Science* 17 (1979): 53–57. See also Allan Franklin, A. W. F. Edwards, Daniel J. Fairbanks, Daniel L. Hartl, Teddy Seidenfeld, *Ending the Mendel-Fisher Controversy* (Pittsburgh: University of Pittsburgh Press, 2008).

12. Charles B. Davenport, "Crime, Heredity and Environment," *Journal of Heredity* 19, no. 7 (July 1928): 307–13.

13. Garland Allen, "The Biological Basis of Crime: An Historical and Methodological Study," *Historical Studies in the Physical Sciences* 31, pt. 2 (2001): 183–222.

14. John Franklin Bobbitt, "Practical Eugenics," *Pedagogical Seminary* 16 (September 1909): 388.

15. Charles B. Davenport, "Marriage Laws and Customs," in *Problems in Eugenics: Papers Communicated to the First International Eugenics Congress* (London: C. Knight & Co., 1912), 154.

16. Kevles, *In the Name*, 93.

17. Lewis Terman, *The Measurement of Intelligence* (New York: Arno Press, 1916), 91–92.

18. As defined by Lewis Terman, the intelligence quotient (IQ) was the numerical result of taking an individual's test result (the "mental age") divided by the person's chronological age (in years) multiplied by one hundred.

19. Alfred P. Schultz, *Race or Mongrel* (Boston: L. C. Page and Co., 1908), 259.

20. Henry Fairfield Osborn, "Address of Welcome," in *Eugenics, Genetics and the Family: Scientific Papers of the Second International Congress of Eugenics*, vol. 1 (Baltimore: Williams & Wilkins Co., 1923), 2. The congress was held at the American Museum of Natural History, New York, September 22–28, 1921.

21. Calvin Coolidge, "Whose Country Is This?" *Good Housekeeping* 72, no. 2 (February 1921): 13–14, 109. Writers often misquote this passage.

22. The Immigration Restriction Act of 1924 continued in effect until 1965.

23. Dolan DNA Learning Center, "Image Archive on the American Eugenics Movement," Cold Spring Harbor Laboratory, www.eugenicsarchive.org/eugenics/, accessed 1 June 2008.

24. Leta Hollingworth, *Gifted Children: Their Nature and Nurture* (New York: Macmillan, 1926), 69–75, 198, 199.

25. Harry Laughlin, "Family History," in *The Legal Status of Eugenical Sterilization: History and Analysis of Litigation under the Virginia Sterilization Statute, which Led to a Decision of the Supreme Court of the United States upholding the Statute* (Chicago: Fred J. Ringley Co., 1930), 17.

26. Robert J. Cynkar, "Buck v. Bell: 'Felt Necessities v. Fundamental Values?,'" *Columbia Law Review* 81 (November 1981): 1435–53.

27. Hamilton Cravens, *The Triumph of Evolution: American Scientists and the Heredity-Environment Controversy, 1900–1941* (Philadelphia: University of Pennsylvania Press, 1978), 53.

28. It took decades for some of the sterilization laws to be repealed. The Virginia law of 1924, for example, was in effect until 1979, such that more than seven thousand individuals were sterilized.

29. Reginald C. Punnet, "Eliminating Feeblemindedness," *Journal of Heredity* 8 (1917): 464–65.

30. Albert Einstein to Heinrich Zangger, 16 February 1917, *The Collected Papers of Albert Einstein*, vol. 10, *The Berlin Years: Correspondence, May-December 1920, and Supplementary Correspondence, 1909–1920*, ed. Diana Kormos Buchwald, Tilman Sauer, Ze'ev Rosenkranz, József Illy, Virginia Iris Holmes (Princeton: Princeton University Press, 2006), 43.

31. Einstein to Besso, 21 October 1932, in Albert Einstein and Michele Besso, *Correspondence 1903–1955*, German transcriptions with French translations, notes, and introduction by Pierre Speziali (Paris: Hermann, 1972), 290.

32. Einstein to Zangger, 16 February 1917, *Collected Papers*, vol. 10, p. 43.

33. Adolf Hitler, *Mein Kampf* [1925–27], English translation, 3rd ed. (New York: Reynal and Hitchcock, 1941), 649, 640, 660.

34. Ibid., 609, 601, 608.

35. Adolf Hitler, *Mein Kampf: Zwei Bände in einem Band Ungekürtze Ausgabe* (1925; repr., Munich: Franz Eher Nacht, 1943), vol. 1, chap. 10, p. 282, trans. Martínez.

36. Ibid., 636, 656–658, 594, italics original.

37. "Eugenical Sterilization in Germany," *Eugenical News* 18 (1933): 91–93; "Human Sterilization in Germany and the United States," *Journal of the American Medical Association* 102, no. 18 (1934): 1501.

38. Robert J. Lifton, *The Nazi Doctors: Medical Killing and the Psychology of Genocide* (New York: Basic Books, 1986), 31.

39. Joseph DeJarnette, "Delegates Urge Wider Practice of Sterilization," *Richmond (Virginia) Times-Dispatch*, 16 January 1934.

40. Willi Heidinger, in *Denkschrift zur Einweihung der neuen Arbeitsstätte der Deutschen Hollerith Maschinen Gesellschaft m.b.H in Berlin-Lichterfelde*, 8 January 1934, 39–40, quoted in Edwin Black, *War Against the Weak: Eugenics and America's Campaign to Create a Master Race* (New York: Four Walls Eight Windows, 2003), 309.

41. Johannes Stark, "The Pragmatic and the Dogmatic Spirit in Physics," *Nature* 141, no. 3574 (30 April 1938): 770–72.

42. W. A. Oldfather, "Pythagoras on Individual Differences and the Authoritarian Principle," *Classical Journal* 33, no. 9 (June 1938): 537–39.

43. Iamblichus, *Pythagorean Way*, chap. 31.

44. Plato, *The Republic of Plato* [ca. 375 BCE], trans. Benjamin Jowett (London: Oxford University Press, 1881), bk. 3, p. 101, line 415.

45. Anonymous [Benjamin Franklin], "Observations Concerning the Increase of Mankind" [1751], in *Observations On the Late and Present Conduct of the French, with Regard to their Encroachments upon the British Colonies in North America. . . . To which is added, wrote by another Hand; Observations concerning the Increase of Mankind, Peopling of Countries, &c.*, ed. William Clarke (Boston: S. Kneeland, 1755), reprinted in

Franklin, *The Writings of Benjamin Franklin*, vol. 3, ed. Albert Henry Smith (London: Macmillan and Co., 1905), 73.

46. Steven Selden, *Inheriting Shame: The Story of Eugenics in America* (New York: Teachers College Press, 1999), 64.

47. Francis Galton, "Eugenics: Its Definition, Scope, and Aims," *American Journal of Sociology* 10, no.1 (July 1904): 6

48. John Baker to Julian Huxley, 17 December 1960, quoted by Michael G. Kenny, "Racial Science in Social Context: John R. Baker on Eugenics, Race, and the Public Role of Scientist," *Isis* 95 (2004): 409.

49. "Report of the Ad Hoc Committee," *Genetics* 83 (1976): 99–101, quoted in Kevles, *In the Name*, 283.

50. Luigi Luca Cavalli-Sforza, Paolo Menozzi, and Alberto Piazza, *The History and Geography of Human Genes* (Princeton: Princeton University Press, 1996), chap. 2.

51. Albert Einstein to Eduard Einstein, 23 February 1927, Einstein Archives 75–654, translation from Jürgen Neffe, *Einstein, A Biography*, trans. Shelley Frisch (New York: Farrar, Straus and Giroux, 2007), 193.

52. Albert Einstein to Mileva Marić, 15 October 1926, Einstein Archives, item 75–658; Albert Einstein to Hans Albert Einstein, February 1927, and 7 September 1927, Einstein Archives, items 75–738, 75–657.

53. Bernard D. Davis, "Pythagoras, Genetics, and Workers' Rights," *New York Times*, 14 August 1980, A23; C. R. Scriver et al., "Glucose-6-Phosphate Dehydrogenase Deficiency," in *The Metabolic and Molecular Bases of Inherited Disease*, 7th ed. (McGraw-Hill, 1995), 3367–98; A. Mehta, P. Mason, T. Vulliamy, "Glucose-6-Phosphate Dehydrogenase Deficiency," *Baillière's Best Practice & Research. Clinical Haematology* 13, no. 1 (March 2000): 21–38.

54. Lionel Penrose, "Human Chromosomes" [1959], quoted in Kevles, *In the Name*, 248.

55. D. K. Belyaev, "Destabilizing Selection as a Factor in Domestication," *Journal of Heredity* 70 (1979): 301–8.

56. L. N. Trut, "Early Canid Domestication: The Farm Fox Experiment," trans. Anna Fadeeva, *American Scientist* 87 (1999): 160–69; L. N. Trut, "Experimental Studies of Early Canid Domestication," in *The Genetics of the Dog*, ed. Anatoly Ruvinsky, Jeff Sampson (Wallingford, UK: CABI, 2001), 15–43.

57. Tecumseh Fitch, quoted in Nicolas Wade, "Nice Rats, Nasty Rats: Maybe It's All in the Genes," *New York Times*, July 25, 2006.

## Epilogue

1. Marx-Engels Institute, "Letter of Charles Darwin to Karl Marx [sic.]," *Pod znamenem Marksizma* (Moscow), nos. 1–2 (Jan.–Feb., 1931), 203–4; Ernst Kilman,

"About the So-Called 'Agnosticism' of Darwin," *Pod znamenem Marksizma* nos. 1–2 (1931), 205–6; V. Adoratsky, ed., "Biochronik," in *Karl Marks. Datyzhizni I deyatel'nosti* (Moscow: Institut Marksa-Engelsa-Lenina, 1934), 366. For a detailed account of how these claims arose and propagated, see Ralph Colp Jr., "The Myth of the Darwin-Marx Letter," *History of Political Economy* 14, no. 4 (1982): 461–82.

2. For example, Isaiah Berlin, *Karl Marx: His Life and His Environment* (New York, 1959), 252.

3. Erhard Lucas, "Marx' und Engels' Auseinandersetzung mit Darwin: zur Differenz zwischen Marx und Engels," *International Review of Social History* 9 (1964): 468–69; Shlomo Avineri, "From Hoax to Dogma: A Footnote on Marx and Darwin," *Encounter* (March 1967): 32; Ralph Colp Jr., "The Contacts between Karl Marx and Charles Darwin," *Journal of the History of Ideas* 35 (1974): 329–38; David McLellan, *Karl Marx: His Life and Thought* (New York, 1973), 424.

4. Erhard Lucas, "Marx' und Engels'," 464.

5. E. M. Ureña, "Marx and Darwin," *History of Political Economy* 9, no. 4 (Winter 1977): 548–59.

6. Valentino Gerratana, "Marx and Darwin," *New Left Review*, no. 82 (Nov.–Dec. 1973): 79–80.

7. Lewis S. Feuer, "Is the 'Darwin-Marx Correspondence' Authentic?" *Annals of Science* 32 (1975): 1–12. See also Lewis Feuer, P. Thomas Carroll, Ralph Colp Jr., "On the Darwin-Marx Correspondence," *Annals of Science* 33 (1976): 383–94; Margaret A. Fay, "Did Marx Offer to Dedicate *Capital* to Darwin?: A Reassessment of the Evidence," *Journal of the History of Ideas* 39, no. 1 (January–March, 1978): 133–46.

8. Edward Aveling to Charles Darwin, 11 October 1880, quoted in Fay, "Did Marx Offer," 145. This letter was discovered by P. Thomas Carroll and Ralph Colp Jr., in January 1975.

9. Howard E. Gruber, "Marx and *Das Kapital*," *Isis* 52 (1961): 582.

10. Einstein to Max and Hedwig Born, 9 September 1920, in Albert Einstein and Max Born, *Briefwechsel 1916–1955*, ed. Max Born (Munich: Nymphenburger Verlagshandlung, 1969), 59, trans. A. Martínez.

11. Jay Weidner, speaking, in *Nostradamus: 2012*, The History Channel, directed by Andy Pickard, produced by 1080 Entertainment and 2009 A&E Television Networks, aired 8 January 2009.

# *Illustration Sources and Credits*

Figures 2.1, 2.2, 2.3, 2.4, 2.5, 2.8, 2.9, 5.3, 7.3, 7.4, 7.5, 7.6, 7.7, 10.1, 10.2, 10.3, 10.4, 10.5, 10.6, 10.7, 10.8, 10.9, 10.10, 10.11, 10.12, 10.13, 10.14, 10.15, 10.16, 10.17, 10.18, 10.19, 12.1: © Alberto A. Martínez.

Figure 1.1: Frederick Gorton, *A High School Course in Physics* (New York: D. Appleton and Company, 1918), 69.

Figures 1.2 and 3.1: F. J. Rowbotham, *Story-Lives of Great Scientists* (Wells, England: Gardner, Darton, and Company, 1918), 28.

Figure 2.6: Tychonis Brahe, *De Mundi Aetherei Recentioribus Phaenomenis, Liber Secundus* (Prague: Uraniburgi Daniae, 1588), 191.

Figure 2.7: Johannes Kepler, *Mysterivm Cosmographicvm* (Frankfurt: Erasmi Kempferi, Godefridi Tampachii, 1621). Courtesy of the Harry Ransom Humanities Research Center, The University of Texas at Austin.

Figure 3.2: *Biblia Pauperum* (Netherlands: n.p., ca. 1465), 1.

Figure 4.1: Basilius Valentinus, *Les Douze Clefs de Philosophie* (Paris: J. et C. Perier, 1624). Courtesy of the Houghton Library of Harvard University. Call # Houghton 24226.28.13.

Figure 4.2: Newton Ms. 416, undated. Courtesy of the Grace K. Babson Collection, Babson College.

Figure 4.3: *Philadelphia Press*, early 1900s.

Figure 4.4: *Arizona Blade and Florence Tribune*, 12 December 1903.

Figure 4.5: *New York Times*, 19 February 1911.

Figure 4.6: *New York Times*, 8 January 1922.

Figure 5.1: Charles Darwin, *Journal of Researches into the Natural History and Geology of the Countries Visited during the Voyage of H.M.S.* Beagle *Round the World*, 2nd ed. (London: John Murray, 1845), 379.

Figure 5.2: Drawing by Édouard de Montulé, 1816, in Montulé, *Voyage en Amérique*, Atlas Volume (Paris: Dalannay, 1821). Courtesy of the Houghton Library of Harvard University. Call # Typ 815.21.5790.

Figure 6.1: Vignette engraving of Benjamin Franklin, national bank note, second charter type, ten dollars, design 485, signed by W. Rosecrans and E. Nebeker, 1870s–1890s, issued by banks in Hagerstown, Maryland, and Washington, D.C.

Figure 6.2: Woodcut in M. Weems, *The Life of Benjamin Franklin; with Many Choice Anecdotes* (Philadelphia: Lippincott, 1884), 167.

Figure 7.1: Louis Figuier, *Les Merveilles de la Science*, vol. 1 (Paris: Furne, Jouvet et cie., 1867). Courtesy of the Harry Ransom Humanities Research Center, The University of Texas at Austin.

Figure 7.2: Coulomb, "Premier Mémoire sur l'Électricité et le Magnétisme," *Mémoires de l'Académie Royale des Sciences* (1785; Paris: Imprimerie Royale, 1788), 576. Courtesy of the Houghton Library of Harvard University. Call # Houghton Phys 3037.1.

Figures 8.1 and 8.2: William Crookes, *On Radiant Matter. A Lecture Delivered to the British Association for the Advancement of Science* (London: E. J. Davey, 1879), 16. Courtesy of the Harry Ransom Humanities Research Center, The University of Texas at Austin.

Figure 13.1: German Bible (Nuremberg: Anton Koberger, 1483), chap. 6.

# Index